Jianfeng Luo · David Dornfeld

Integrated Modeling of Chemical Mechanical Planarization for Sub-Micron IC Fabrication

Jianfeng Luo · David Dornfeld

Integrated Modeling of Chemical Mechanical Planarization for Sub-Micron IC Fabrication

From Particle Scale to Feature, Die and Wafer Scales

With 175 figures and 14 tables

Springer

Jianfeng Luo
R & D
Cypress Semiconductor
3901 North First St.
San Jose, CA 95134-1599
USA

David A. Dornfeld
University of California at Berkeley
Dept. Mechanical Engineering
5100A Etcheverry Hall
Berkeley, CA 94720-1740
USA

ISBN 3-540-22369-x Springer Berlin Heidelberg New York

Library of Congress Control Number: 2004109174

Bibliographic information published by Die Deutsche Bibliothek
Die Deutsche Bibliothek lists this publication in the Deutsche Nationalbibliografie; detailed bibliographic data is available in the Internet at <http://dnb.ddb.de>.

Springer is a part of Springer Science+Business Media
springeronline.com

Printed in Germany

Coverdesign: deblik, Berlin
Printed on acid-free paper 62/3020/ iw – 5 4 3 2 1 0 –

To Berkeley

Preface

Chemical mechanical planarization, or chemical mechanical polishing as it is simultaneously referred to, has emerged as one of the critical processes in semiconductor manufacturing and in the production of other related products and devices, MEMS for example. Since its introduction some 15+ years ago CMP, as it is commonly called, has moved steadily into new and challenging areas of semiconductor fabrication. Demands on it for consistent, efficient and cost-effective processing have been steady. This has continued in the face of steadily decreasing feature sizes, impressive increases in wafer size and a continuing array of new materials used in devices today.

There are a number of excellent existing references and monographs on CMP in circulation and we defer to them for detailed background information. They are cited in the text. Our focus here is on the important area of process models which have not kept pace with the tremendous expansion of applications of CMP. Preston's equation is a valuable start but represents none of the subtleties of the process. Specifically, we refer to the development of models with sufficient detail to allow the evaluation and tradeoff of process inputs and parameters to assess impact on quality or quantity of production. We call that an "integrated model" and, more specifically, we include the important role of the mechanical elements of the process. This is not meant to downplay any chemical effects, which are fundamental to the successful operation of CMP. Our objective here is to introduce some background on, what seemed to us, the overlooked mechanical aspects of the process - including pad surface topography and abrasive particles.

This book is really the product of a developing research focus on CMP at Berkeley and, as such, owes much to a host of hard working graduate student researchers. Early work led to basic mechanical element measurements during planarization such as friction and the validation of Preston's equation. More recent work is focusing on novel pad concepts for optimizing CMP. This work on modeling represents our first analysis in detail of the process from a mechanical effects point of view and was substantially conducted by the first author.

However, research on CMP is a never ending story. As we write this a host of researchers worldwide are busy advancing the understanding of CMP in various applications. It is our hope that this contribution on modeling and our approach to the integrated model will provide some useful insights to this complex process. This will be specially important as we look towards "optimization" of the process, design of consumables and, importantly, looking to minimize the environmental affects of CMP.

The book is the product of research, suggestions and technical assistance of many individuals. The authors would like to thank Prof. Costas Spanos and Fiona Doyle at UC-Berkeley, Prof. Ronald Gutmann at RPI, Prof. Jan Talbot at UC-San Diego, Dr. Yongsik Moon at Applied Materials, Dr. Serdar Aksu at Suleyman Demiril University and Mr. Kiha Lee at UC-Berkeley for their helpful suggestions and discussions. On a more personal level, Jianfeng Luo would like to thank his wife, Na Xu for her assistance on the type setting of the manuscript. Without her support, understanding and encouragement the book writing would have not been possible.

The research on CMP at UC-Berkeley is supported by National Science Foundation, University of California Industry-University Research Program Discovery Grant under contract 97-01 and semiconductor industry companies including Advanced Energy, AMD, Applied Materials, ASML, Atmel Corporation, Cadence, Canon, Cymer, Dupont, Ebara Technology, EVGroup, Intel, KLA-Tencor, Mentor Graphics, Mykrolis Corp., Nikon Research Corporation of America, Novellus Systems, Panoramic Technology, Photronics, Synopsys and TEL USA. This support is greatly appreciated.

Cypress Semiconductor, San Jose, CA, U. S. A. *Jianfeng Luo*
<lbj@cypress.com>
University of California, Berkeley, CA, U. S. A. *David Dornfeld*
<dornfeld@me.berkeley.edu>
May 2004

Contents

List of Figures

List of Tables

1

INTRODUCTION

1.1 Overview

The last fifteen years have seen the broad application of chemical-mechanical planarization, also known as chemical-mechanical polishing (CMP), in sub-micron integrated circuit (IC) fabrication. During a CMP process, a silicon wafer with size (diameter) ranging from 4 inch (100mm) to 12 inch (300mm), patterned or blanket, is rotated about its axis while being pressed face-down by a carrier against a polishing platen covered with a soft polymer pad/belt. Slurries with nano-scale abrasive particles and special chemicals are distributed into the wafer and pad/belt interface. These slurry particles and chemicals work with the pressure P applied on the wafer top surface and the relative movement between the wafer and the pad/belt to remove the wafer materials and planarize the wafer surface. A typical configuration of a rotational type CMP machine, where the relative movement is obtained from the rotations of the wafer carrier (polishing head) and polishing platen around their own centers, is shown schematically in Fig. 1.1 [1].

Currently CMP is primarily applied in three areas of IC fabrications: i. inter-level dielectric (ILD) and inter-metal dielectric (IMD) planarization [2], ii. the copper damascene process and iii. shallow trench isolation (STI).

[1]A linear type configuration where the place of the rotating pad is taken by a continuous belt which moves below the wafer in a linear motion and a orbital type configuration where the platen rotates around the center of the wafer carrier (polishing head) in an orbital path instead of its own center will be discussed in Chapter 8.

[2]ILD usually refers to the dielectrics deposited over polysilicon after gate-patterning, and IMD usually refers to the dielectrics deposited over interconnect metal layers [1.2].

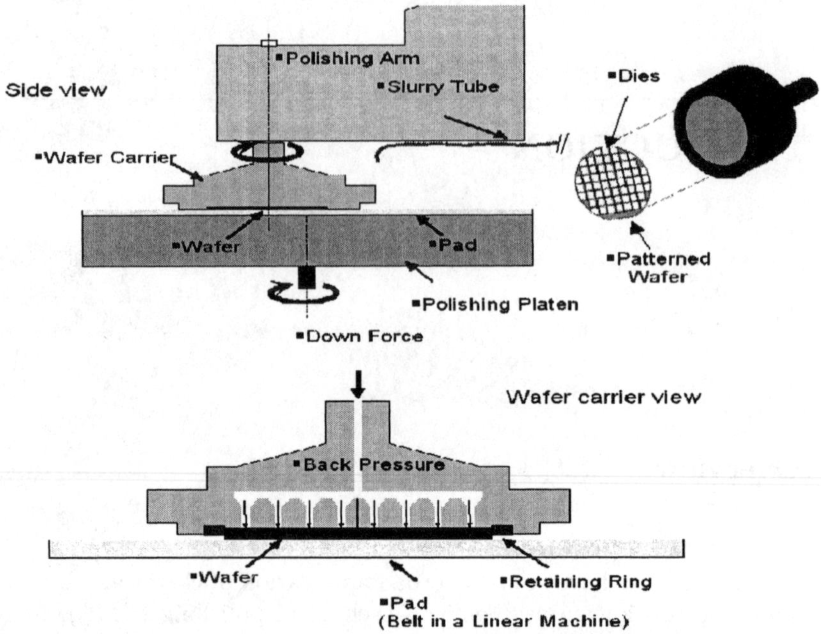

Fig. 1.1. Typical configuration of a rotational type CMP machine

1.1.1 ILD/IMD Planarization

The ILD/IMD planarization is introduced into IC fabrication to address increased device density requirements. With the device density in IC chips increased to almost eight million transistors per cm^2 of chip area and with the corresponding increase in circuit functionality, multi levels of metal interconnect are required to facilitate device communication. To date, up to six levels of metal interconnect have been reported, and it is predicted that a nine-level interconnect will appear by the year 2012 [1.1]. The ability to effectively planarize the dielectric layers which insulate the interconnect levels is indispensable for such multi-level metallization. Conventionally, thermal flow of oxide layers, spin-on-glass and sacrificial-resist-etch back are applied in IC fabrication as planarization techniques. However, they can only provide a local planarization over distances smaller than $10\mu m$. This is insufficient for the fabrication of multi-level integrated circuits with more than three layers of metal, where a global planarization of the dielectric and metal layers over a distance up to several millimeters is needed [1.2]. CMP is the only technology that is able to realize such a global planarization.

Fig. 1.2 (a) shows schematically the usage of CMP in the planarization of ILD/IMD in a multi-level chip. Without CMP, the nonplanarization of the ILD/IMD layers will be accumulated from the lower levels to the higher levels, Fig. 1.2 (b). With the scaling of the IC feature size, the depth of focus (DOF) of the high-resolution lithography is limited to the range of $\pm$ 0.5μm [1.2], a value which the uneveness of the ILD/IMD and metal layer surface should be smaller than to be patterned accurately. It is hard to realize such an eveness at the high levels of multi-level chips without CMP. Besides the DOF problem arising from the lithography, a planarization of the ILD/IMD surfaces by CMP is also needed to address the problems of step-coverage of metal lines as they cross over the steep steps at high levels [1.2].

While chemical vapor deposition (CVD) silicon oxide is conventionally used as the ILD/IMD material due to its ease of processing as well as its good dielectric and mechanical strength, the dielectric constant or k-value of silicon oxide is relatively high (in the range of 3.9-4.5). The International Technology Roadmap for Semiconductors [1.1] predicts that the interconnect RC (resistance-capacitance) delay (or global delay, Fig. 1.3), instead of the gate delay, will dominate the circuit performance in the sub-micron and deep submicron era, Fig. 1.3. To reduce the interconnect delay, one way/requirement is to reduce the interconnect resistivity. This can be realized by replacing the conventional interconnect material, aluminum (Al) alloy, with lower resistance materials such as copper. This leads to the copper damascene process, to be discussed later. The other way/requirement is to replace the silicon oxide with materials with a low-k value ($<$ 3.9) as the ILD/IMD materials. This can reduce the parasitic capacitance between metal lines. Low-k dielectric materials are usually much softer than silicon oxide, and therefore their ability to endure the mechanical force from CMP is relatively low. This generates a lot of unique problems for CMP. Therefore, besides CMP of oxide, CMP of low-k materials has recently become an active research topic in both industry and academia.

1.1.2 Copper Damascene Process

Another important application of CMP in recent years is in the copper damascene process. Replacing the Al alloys (Al-Cu alloys), which have been used for over thirty years as the interconnect metal with copper, has already been attempted in order to reduce the interconnect delay because of the copper's lower resistance. Fig. 1.4 shows a multi-level copper interconnect IC chip fabricated by Intel [1.3]. However, the copper is difficult to be patterned using the

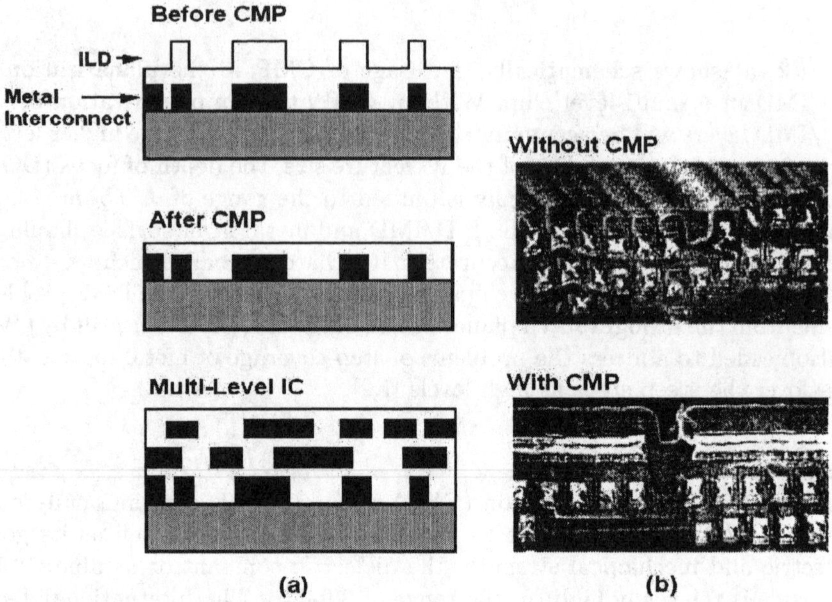

Fig. 1.2. (a) Process flow of CMP of ILD/IMD and (b) Comparison of ILD/IMD surface with and without CMP

conventional dry and wet etching methods. Instead, a method called the damascene process has been used. During the process, the copper is first deposited over a dielectric layer with trenches and/or vias patterned using conventional etching methods. To avoid the metal diffusion, a barrier layer/liner film with materials such as Ti and TiN is used as an interfacial layer between metal interconnects and dielectrics. The coppers deposited in the dielectric trenches and vias are the interconnection lines. Overburden copper left over the barrier surface has to be removed to isolate the lines in different trenches. CMP is currently the only method that is able to remove the overburden copper with good global planarity. Fig. 1.5 shows schematically the process flow of a single copper damascene process [3] and a dual copper damascene process [4]. More details on the process flow can be found in [1.2].

Besides copper, CMP is used in the planarization of aluminum interconnects and tungsten plugs under a similar inverse metallisation scheme as well.

[3]In the single damascene process, metals are deposited to fill trenches and vias in two separated steps.

[4]In the dual damascene process, metals are deposited to fill trenches and vias in one single step.

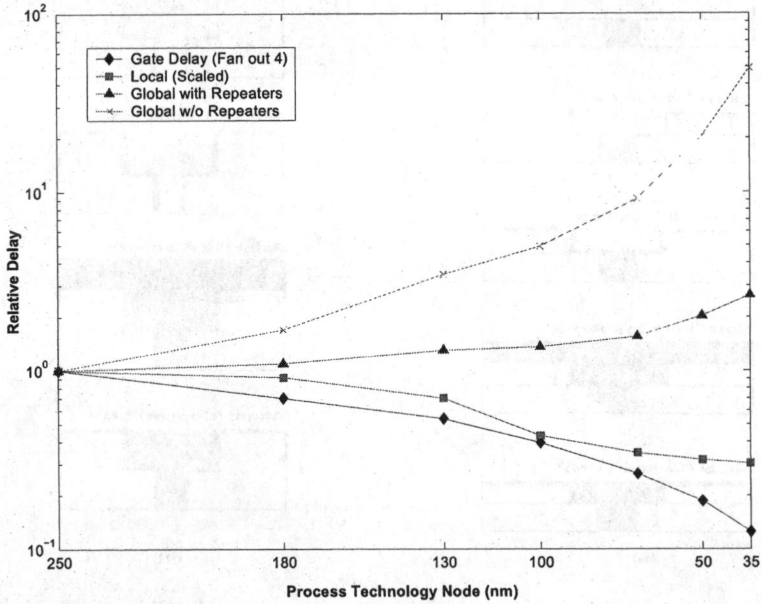

Fig. 1.3. The relative IC delay of different process technology nodes [1.1]

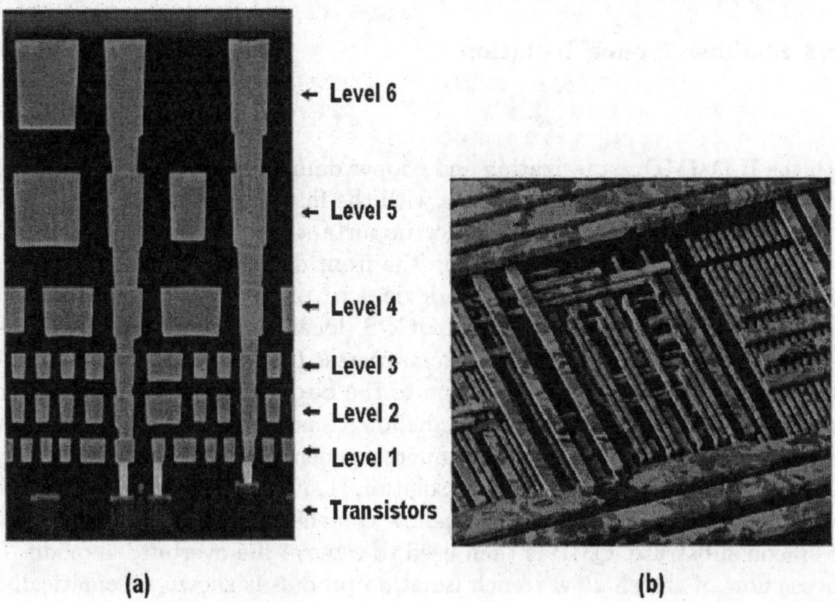

Fig. 1.4. (a) Cross sectional SEM photography of 6-level copper interconnect IC chip (b) SEM photography of orthogonal architectures of copper damascene interconnects [1.3]

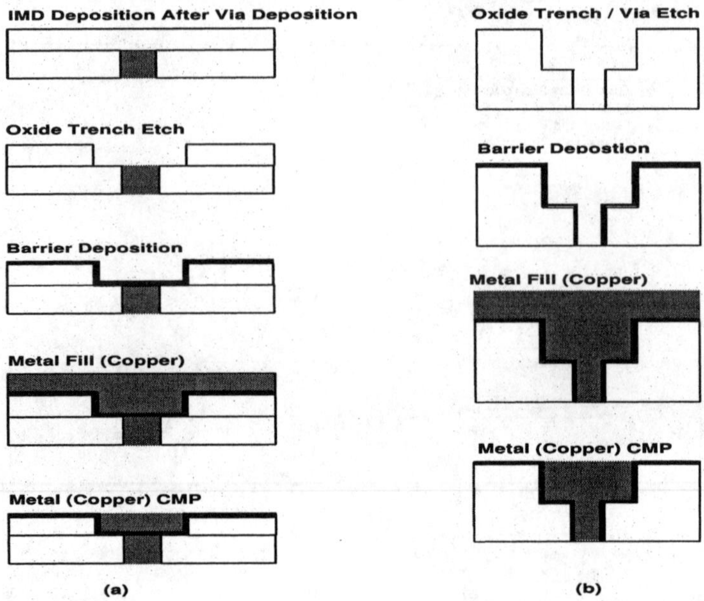

Fig. 1.5. (a) Process flow of single damascene process (b) Process flow of dual damascene process [1.4]

1.1.3 Shallow Trench Isolation

Both the ILD/IMD planarization and copper damascene process belong to the back-end of the processes which deal with the interconnection between different devices and/or modules. Another important application of CMP, called shallow trench isolation, is applied in the front-end of IC processes. The isolation is needed between adjacent devices to prevent the establishment of parasitic channels. Conventionally, LOCOS (local oxidation of silicon) is used which involves the formation of a semi-recessed oxide in the non-active (or field) areas of the substrate [1.4]. Due to the bird's beak encroachment problem, LOCOS has been replaced by shallow trench isolation with the CMOS technology scaled down to the sub-micron generation. Similar to the copper damascene process, shallow trench isolation is also a damascene process. Silicon oxides are deposited into the shallow trenches etched on the field area of the silicon substrate. CMP is then used to remove the overburden oxide. The process flow of the shallow trench isolation process is shown schematically in Fig. 1.6. A silicon nitride layer is deposited on a pad oxide to act as a CMP stop material, Fig. 1.6 (a) and (e). Hence, the CMP of silicon nitride also needs investigation.

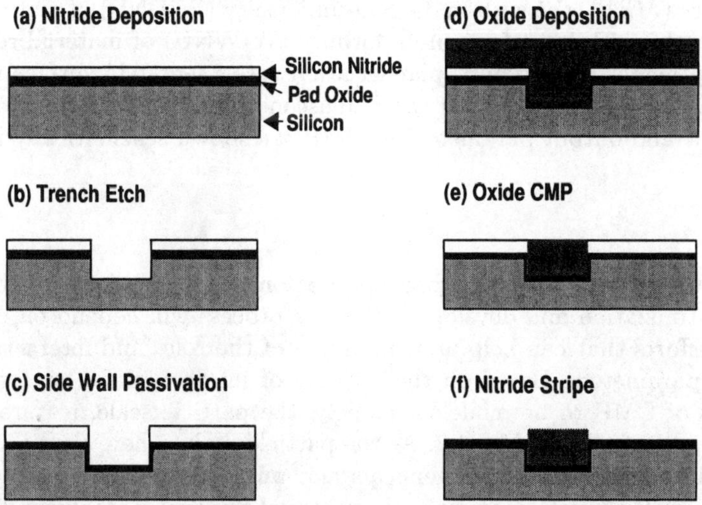

Fig. 1.6. Process flow of shallow trench isolation

Besides ILD/IMD planarization, the copper damascene process and STI, CMP is also applied in the planarization of polysilicon and photoresist. Many of them involve the fabrication of trench-capacitor based DRAMs (Dynamic Random-Access Memories). In addition, CMP is obtaining pervasive applications outside of IC fabrication, as in hard disk drive and MEMS (micro-electrical-mechanical systems) fabrications. These include CMP of aluminum film as a mirror in the micro-mirror array of MEMS, three-dimensional multi-level MEMS fabrication and so on.

1.1.4 CMP Input and Output Parameters

The CMP process is quite complicated. Part of the complication comes from the large number of input parameters. These include consumable related parameters such as slurry pH values, slurry oxidizer concentrations, slurry abrasive size and size distribution, slurry abrasive geometry and materials, slurry temperature, polishing pad topography and polishing pad materials, wafer related parameters such as wafer materials and topography parameters such as feature line width, pattern density [5] and so on, and process and machine

[5] see Chapters 6 and/or 8 for definitions

configuration related parameters such as down pressure, wafer-pad relative velocity, retaining ring size, retaining ring pressure and so on. There are also a number of critical output parameters with CMP. They include the material removal rate (MRR), the with-in die non-uniformity (WIDNU) of material removal rate, the with-in wafer non-uniformity (WIWNU) of material removal rate [5], step height reduction or planarization rate [5], scratch, surface roughness, and dishing & erosion [5] in the damascene process. An incomplete list of the input and output parameters in CMP are shown schematically in Fig. 1.7.

Although CMP has obtained broad applications in IC fabrication, the systematic optimization and development of the process will depend on mature modeling efforts that can help understanding of the roles and interactions of the input parameters. Based on the outputs of interest, there are issues at four scales of CMP to be modeled, namely, the particle scale, feature scale, die scale and wafer scale, Fig. 1.7. At the particle-scale, where the nano-scale abrasives interact with slurry chemicals and micro-scale pad topography to remove the wafer surface materials, the material removal mechanism describing the roles and interactions of slurry abrasives, chemicals, pad and wafer material properties is the primary interest. The outputs include the material removal rate at each single point, surface scratch and roughness, Fig. 1.7. The feature- and die-scale issues are usually related to each other. In this work they are discussed together and referred to as issues at feature- and die-scales. The evolution of topography including step height, dishing and erosion, and with-in die non-uniformity of material removal rate and their dependence on pattern density, polishing time and feature line width are the major concerns at these scales. The final circuit performance, such as the interconnect delay, depends on the topography of metal lines and dielectrics. A successful model at this scale can help circuit design from the perspective of design for manufacturability. At the wafer scale, the major interest is the with-in wafer non-uniformity of material removal rate across the wafer, which is a function of wafer size and CMP equipment configuration and consumable parameters. A small non-uniformity of material removal is preferred to obtain high production yields and uniform circuit performance. The average material removal rate, i.e. the average value of material removal rates at each single wafer point, is a wafer-scale parameter. However, it is the sole function of input parameters at particle-scales and not dependent on wafer-scale inputs such as wafer size and pressure distribution. Therefore it can be studied under the category of particle-scale models.

It is difficult to cover all the above parameters in one single model given their huge number and complicated interactions. This work aims to develop models to identify the most important input variables and explain their roles and interactions on the CMP material removal and topography evolution mechanism. Based on the outputs of the interests, models at three scales, namely, particle-scale, feature- and die-scales and wafer scale, will be developed.

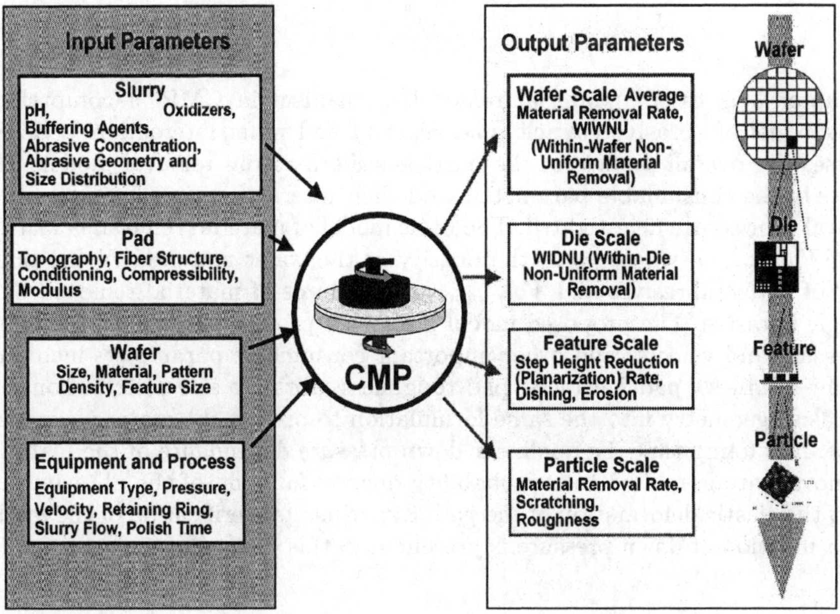

Fig. 1.7. An incomplete list of input and output parameters in CMP

1.2 Structure of the Book

This book is composed of the following four parts:

1.2.1 Review of CMP Modeling

The first part is a review of the previous CMP modeling efforts. A systematic review of the CMP models developed in last decade and, particularly, their connections with this work is presented in Chapter 2.

1.2.2 Particle-Scale Modeling

The second part is on a material removal mechanism model at the particle scale. This includes Chapter 3, Chapter 4 and Chapter 5.

Chapter 3 is titled "Material removal mechanism in CMP: a comprehensive model of abrasive particle, pad asperity and wafer interactions". It discusses the overall picture of the particle-scale material removal model. The roles of the consumable parameters and their interactions in the material removal process are investigated. The basic model of material removal is $MRR = \rho_w N Vol_{removed}$, where ρ_w is the density of the wafer material, N the number of *active* abrasives and $Vol_{removed}$ the volume of material removed by a single abrasive. The proposed model integrates process parameters including pressure and velocity and other important consumable parameters including wafer hardness, pad hardness, pad roughness, abrasive size distribution and abrasive geometry into the same formulation to predict the material removal rate. It is found that the nonlinear down pressure dependence of the material removal rate is related to a probability density function of the abrasive size and the elastic deformation of the pad. Experimental verification of the model as a function of down pressure is presented in this chapter.

Chapter 4 is titled "Effects of abrasive size distribution in CMP". It further addresses the roles of abrasive size distribution in the CMP material removal process. The effects of the abrasive size distribution on the material removal are attributed to two parameters, one, the number of *active* abrasives, and the other, the size of the *active* abrasives. A detailed model of wafer-abrasive-pad contact is developed to explain how and where abrasive size distributions come into the particle-scale material removal model. Experimental evidence supporting the proposed abrasive size effects is presented. A material removal rate formulation as a function of abrasive size distribution is proposed and verified.

Chapter 5 is titled "Material removal regions in CMP: coupling effects of slurry chemicals, abrasive size distribution and wafer-pad contact area". The material removal model as a function of abrasive weight concentration has been proposed in this chapter by extending the material removal model developed in Chapters 3 and 4. With an increase of the weight concentration of abrasives, three regions of material removal are proposed: first, a chemically dominant and rapid increasing region, whose range is determined by the generation/passivation rate and hardness of the surface passivation layer; second, a mechanically dominant linear region, where the material removal is proportional to the weight concentration; and third, a mechanical dominant saturation region, where the material removal saturates because the total contact area is fully occupied by the abrasives. The passive layer of the wafer surface is proposed to be a bi-layer structure. A detailed model is proposed to explain that the transition from the first to the second region is due to a transition from a wafer surface covered with a single soft material to a surface covered with both soft and hard materials. The chemicals contribute to the material removal through the generation rate of the upper softer layer of the passive films. Experimental evidences supporting this model is discussed in this chapter.

1.2.3 Feature- and Die-Scale Modeling

The third part is on topography evolution models at the feature- and die-scales. This is covered in Chapters 6 and 7.

Chapter 6 is titled "One and semi-two dimensional feature- and die-scale modeling for the damascene process". Pad models based on both linear elasticity and linear viscoelasticity assumptions are first used to develop a one-dimensional topography evolution model for the damascene process. The two-dimensional boundary element model and contact mechanics model are applied to evaluate the pad deformation over the features. The deformation is then used to evaluate the *effective* pad stiffness as a function of pattern density and line width, which is input into a one-dimensional topography model to evaluate the dishing and erosion. The vertical directional die and feature scale topography interactions are implemented in the model. In addition, a scheme of lateral die-feature scale integration considering the random distribution of feature density based on a physical understanding of the concept of the *effective* pattern density is proposed. In comparison with the previous linear elastic model, the simulation results in this work show that the pad model based on viscoelasticity can capture the pattern density dependency of dishing more accurately than that based on the linear elasticity assumption. The effects of oxide and copper material removal properties and pad properties are simulated based on the linear viscoelasticity assumption. The topography model proposed in this chapter can be used to model the feature- and die-scale topography evolution in the ILD/IMD planarization and STI process as well.

In Chapter 7, titled "Three-dimensional feature-scale modeling of CMP", a three-dimensional model based on contact mechanics is presented to evaluate the topography evolution at the feature scale. Numerical implementation details of the model and simulation results of the topography evolution for a single material CMP are presented. Topography evolution information in both vertical and lateral directions can be obtained. This model can be extended to cover the damascene process. In addition, linear viscoelasticity can be included in the future.

1.2.4 Wafer-Scale Modeling

The fourth part is on wafer scale modeling and is covered in Chapter 8. The with-in wafer non-uniformity of the material removal rate and its optimization are discussed from the viewpoint of both the machine and consumable effects. On the machine side, the pressure and velocity distribution are identified as the major contributors of the non-uniform material removal rate across the wafer. Therefore, the velocity and pressure non-uniformity in both cases of rotational and linear types of machines and the cases of the solid-solid contact and the solid-fluid-solid contact are modeled. Adding the retaining ring is found to be an effective way to reduce the pressure non-uniformity at the edge of the wafer. In the solid-fluid-solid contact mode, the friction coefficient is identified as the major contributor to the non-uniformity of the pressure distribution. A process window of pressure and velocity can be obtained by combining the model and the Stribeck curve [6] of the friction coefficients to obtain small non-uniformity. On the consumable side, a systematic method is proposed to optimize the WIWNU in solid-solid contact mode based on the particle-scale material removal model developed in previous chapters. The basic idea is to reduce the sensitivity of the material removal rate on pressure and velocity distribution by adjusting consumable parameters.

As a summary, the major contributions of this work are listed in Table 1.1. The major conclusion and future work will be detailed in Chapter 9.

1.3 References

[1.1] *The International Technology Roadmap for Semiconductors*, Semiconductor Industry Association, San Jose, CA, U. S. A., 1999.

[6] see Chapter 8

Table 1.1. List of major contributions of this work

Particle-Scale Model	Identify and verify the roles and interactions of consumable parameters related to polishing pad, abrasives and slurry chemicals (**Chapter 3**) Further identify and verify the roles of abrasive size distribution on the number and size of active abrasives in the material removal process (**Chapter 4**). Connect the passivation rate to the material removal rate (**Chapter 5**)
Feature- and Die-Scale Model	Develop a one and semi-two dimensional topography evolution model to explain the dependency of dishing and erosion on the pattern density and line width based on linear viscoelasticity (**Chapter 6**). Develop a contact mechanics based three-dimensional model to evaluate the topography evolution at feature scale. (**Chapter 7**)
Wafer-Scale Model	Develop a boundary element model to evaluate the pressure distribution over wafer-pad interface in solid-solid contact mode (**Chapter 8**) Develop a lubrication model to predict the pressure distribution in solid-fluid-solid contact mode. The friction coefficient is identified as the major contributor of pressure non-uniformity (**Chapter 8**).

[1.2] S. Wolf, *Silicon Processing for VLSI Era: Deep Sub-Micron Process Technology*, Vol. **4**, Lattice Press, Sunset Beach, CA, U. S. A., 2003.

[1.3] Intel Corporation Website: http://www.intel.com/research/silicon, 2002.

[1.4] S. Wolf, *Silicon Processing for VLSI Era: Process Integration*, Vol. **2**, Lattice Press, Sunset Beach, CA, U. S. A., 1998.

2

REVIEW OF CMP MODELING

2.1 Introduction

Although chemical mechanical planarization (CMP) is widely applied in integrated circuit fabrication, it is still a process of trial and error. Understanding the basic mechanism of this process has initiated research efforts from both industry and academia in the last decade. This chapter reviews previous CMP modeling efforts.

There are issues at three scales of CMP to be modeled, namely, the particle-scale, feature- and die-scale and wafer-scale. The models at the particle scale are needed to address the roles and interactions of slurry particles, slurry chemicals, polishing pads and wafer materials. The material removal process in CMP can be considered as a sliding of the slurry particles over a chemically influenced thin layer over the wafer surface. The mechanical action, including slurry abrasive indentation into the wafer surface under the down force and its sliding due to the relative movement of the wafer and polishing pad, and the chemical action, including the passivation of the wafer materials, the etching of the wafer materials, the hardening/softening of the slurry particles and polishing pad by the slurry chemicals, work together to remove the wafer surface material. Finding out the parameters and how they contribute to the mechanical-chemical actions is critical for understanding the issues related to the material removal mechanisms, e.g. the material removal rate (*MRR*) and surface quality (roughness and scratches).

The models at the feature and die scales are needed to address the topography evolution of IC chips as a function of pattern density, line width, pitch width and polishing time. In the earlier stage, CMP is used in the planarization of inter-layer or inter-metal dielectrics (ILD/IMD) in the multi-layer ICs to satisfy the depth of focus (DOF) requirements of the lithography process. With the copper and low-k dielectrics taking the place of aluminum and silicon oxide as the primary interconnection and inter-layer/metal dielectric materials, CMP is becoming an indispensable process for the patterning of copper in the IC fabrication. As predicted by the Semiconductor Industry Association (SIA)'s Technology Roadmap [2.1], the electrical properties such as the time delay of the chips will depend on the resistivity and coupling capacitances of interconnection lines in the sub-micron era, which are functions of the geometry/topography of the interconnection metal lines and dielectrics [2.49]. A better understanding of the elements of the formation mechanism of the topography such as the pattern density effects, dishing and erosion, can help to optimize IC design, for instance the design of the pattern layout or the dummy filling in the lower density area from the beginning of the design stage, and the process parameters, for instance the selectivity of material removal rates of the metal to the dielectrics/barriers at the fabrication stage. The final goal of modeling efforts at this scale is to facilitate the development of electronics design automation (EDA) or electronics computer aided design (ECAD) software tools in terms of design for manufacturability .

The wafer-scale model is to address the issues related to the material removal non-uniformity across the wafer. To satisfy the manufacturability requirements in the sub-micron IC fabrication, the margin of the profile non-uniformity is usually limited in the range of several hundred nanometers. The material removal non-uniformity across the wafer is rooted in the non-uniform distributions of pressure and velocity, which in a large degree, are determined by the configurations of the CMP equipment. A model at this scale, therefore, needs to address the machine configurations properly.

In the last decade, various kinds of models have been proposed by researchers, concentrating on different aspects of CMP. All of them can be put in one of the above three categories and succeed in a certain respect. They will be reviewed in this chapter, one by one, from the particle scale to the feature, die and wafer scales. A framework of an integrated model to consider the interactions among the models at the three scales will be proposed at the end of this chapter to address the issues including material removal rate, scratch, dishing and erosion, with-in die non-uniformity and with-in wafer non-uniformity, in a comprehensive way.

2.2 Particle-Scale Models

The two most important issues at the particle scale are the material removal rate (*MRR*) and the surface quality (surface roughness and scratches). The material removal rate determines the production rate of the process. Moreover, it is most sensible to process conditions, and therefore an output which is most of interest in process optimization. An understanding of the material removal mechanism first requires understanding how the material removal rate varies with the process conditions. Surfaces qualities determine the yield of the process. The micro-scratches diminish the reliability and reflectivity as well as the endurance of the surface. With the surface scratches, which are more a concern in metal CMP than oxide CMP, the corrosion resistance in the thin film may be lowered and a short circuit caused by electromigration may occur easily. Therefore, an understanding of the formation mechanism of scratches and their reduction is needed. With low-k dielectrics taking the place of oxide dielectrics as the primary dielectric materials, the delamination of the dielectrics and interconnection metals during CMP is becoming a critical surface quality issue as well. A full understanding of these problems requires proper answers to the following three questions:

What are the most important consumable and wafer parameters in the CMP material removal process?

What roles are these parameters playing once they are identified as critical parameters?

How do these parameters interact to affect the material removal?

The four most important elements in the CMP material removal process are the slurry abrasive particles, slurry chemicals, polishing pad and wafer materials. There are six interactions among these four elements, namely, wafer-abrasive interaction, wafer-pad interaction, wafer-chemical interaction, abrasive-pad interaction, abrasive-chemical interaction and pad-chemical interaction. To answer the above three questions, the first step is to understand these interactions independently. In the following sections, we review the previous particle-scale modeling efforts, which are mostly concentrated on one to two of the interactions. Comprehensive modeling efforts and a framework considering all the above interactions will be discussed at the end of this section.

2.2.1 Models Considering Slurry Particle-Wafer Surface Interactions

The most famous material removal model is the experimental Preston's Equation [2.2], which was initially introduced for glass polishing,

$$MRR = K_e PV \tag{2.1}$$

where *MRR* is the material removal rate, K_e an all-purpose coefficient, P, the down pressure and V the relative velocity between the wafer and pad. It demonstrates a linear dependency of material removal rate on the pressure and velocity. An equivalent equation is the Archard's equation [2.3] in the area of *wear*. Not all experimental *MRR* data in CMP, specially, in metal CMP, supports the linear pressure times velocity dependency. Revised Preston's equations were therefore proposed by different researchers. For example, considering that the material removal rate does not extrapolate to zero, Maury et al. [2.4] introduces a fitting parameter MRR_0 into Preston's equation:

$$MRR = K_e PV + MRR_0 \tag{2.2}$$

Later, the nonlinear experimental equation

$$MRR = K_e P^\alpha V^\beta \tag{2.3}$$

where α, β are two fitting parameters, was proposed by Wrschka et al. [2.5] to get a better fit of the experimental data.

The major problem with Preston's equation and its revision is that consumable and wafer parameters are not included explicitly. Therefore, the process window in terms of consumable effects cannot be obtained. By extending Brown's model [2.6] of the metal polishing to the silicon polishing, Cook [2.7] developed a physical model to address this limitation. The interactions between the abrasive particles and the wafer surface is proposed as a Hertzian elastic penetration [2.8] of a spherical particle under uniform pressure P into the wafer surface, sliding along the surface with a velocity V and removing glass volume proportional to the penetration, Fig. 2.1. A *MRR* formulation was proposed as:

$$MRR = (2E_w)^{-1} PV \tag{2.4}$$

where E_w is the Young's modulus of the wafer materials. This model can be taken as a theoretical verification of the Preston's equation since it supports the linear dependency of *MRR* on pressure and velocity. The relationship between the wafer surface roughness R_a and the down pressure P and abrasive size can also be obtained based on this model:

$$R_a = \frac{3}{4}x(P/2C_k E_w)^{2/3} \tag{2.5}$$

where C_k is a particle concentration dependent parameter and equal to 1 for a fully filled close hexagonal packing [2.7] and x the diameter of the slurry particles. A similar model was developed by Liu et al. [2.9] based on the statistical method and Herzian elastic penetration. Besides the wafer material parameter, including wafer hardness H_w and wafer Young's modulus E_w, this model includes pad hardness H_p and abrasive Young's modulus E_a:

$$MRR = C_e \left(\frac{H_w}{H_w + H_p}\right)\left(\frac{E_a + E_w}{E_a E_w}\right) PV \tag{2.6}$$

where C_e is a coefficient to account for the effects of slurry chemicals and other consumable parameters. This model, similar to Cook's model, suggests that the material removal is proportional to the applied pressure and relative speed.

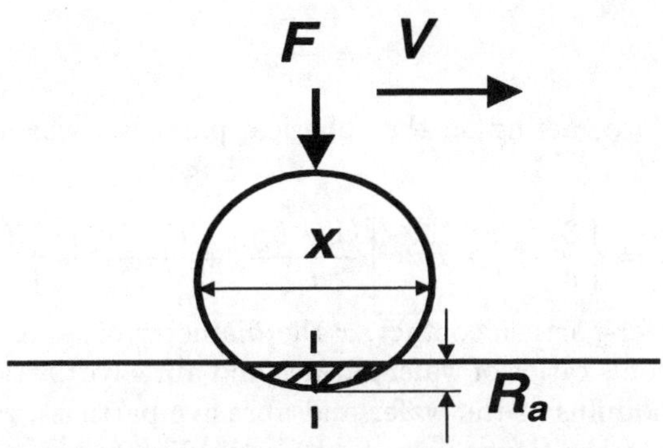

Fig. 2.1. Mechanics of particle-glass contact [2.7]

The advantages of Cook's and Liu's models over Preston's equation are that they provide insights into the roles and interactions of the consumable parameters. The contributions of the slurry abrasives and pad, for example, have been attributed to their size and hardness. An additional benefit is that not only material removal rate, but also surface quality such as roughness (Eq. 2.5), can be addressed using these models.

In Cook's and Liu's models, the mechanical removal by abrasive particles is the dominant mechanism. Some researchers, however, believe that the material removal is due to a mechanical-enhanced erosion. Runnel et al. [2.10] follows this idea and develop an erosion-based model for CMP. They assume that a fluid film exists between the wafer and pad interface, which affects the erosion/material removal rates at each single point through fluid stress tensors:

$$MRR = C_e \sigma_t \sigma_n \tag{2.7}$$

where C_e is an all purpose coefficient, σ_t is the shear stress due to the slurry flow and σ_n the normal stress.

Runnel's model has been integrated into several particle-scale models [2.11-2.12]. Tseng and Wang attributed the normal stress at the particle-wafer contact to the elastic indentation of the particle into the wafer surface, which is similar to the model proposed by Cook [2.7], and calculated the normal stress over the wafer-particle interface as

$$\sigma_n = \frac{F}{\pi r_c^2} \tag{2.8}$$

where F is the force acting on the spherical particles, which is proportional to the pressure P,

$$r_c = \left\{ \frac{3}{4} \times \frac{x}{2} \times F \times \left[\frac{(1-\upsilon_w^2)}{E_w} + \frac{(1-\upsilon_a^2)}{E_a} \right] \right\}^{1/3} \tag{2.9}$$

the radius of wafer-particle contact, x the diameter of abrasive particles, υ_w and υ_a the Poisson's ratios of wafer surface and abrasive particles and E_w and E_a the elastic modulus of the wafer and abrasive particles, respectively. The shear stress due to the slurry flow can be approximated as

$$\sigma_t = C_e \sqrt{\mu V P A_0} \tag{2.10}$$

where μ is the dynamic viscosity of the slurry and A_0 the area of wafer surface. Substitution of Eqs. 2.8 and 2.10 into Eq. 2.7 yields:

$$MRR = K_e P^{5/6} V^{1/2} \tag{2.11}$$

where K_e is the parameter to account for material properties, slurry abrasive concentration and chemical processes. This model demonstrated a non-linear relationship between the material removal and the pressure times velocity. In comparison to Cook's and Liu's models, Tseng's model attempts to connect the elastic indentation to the erosion rate instead of the mechanical abrasion. While the down pressure dependency (an exponent of 5/6) is close to a linear dependency, the velocity dependency (an exponent of 1/2) is quite nonlinear. This is because the contribution of velocity has been attributed to the slurry flow instead of a sliding of abrasive particles.

Cook's, Liu's and Tseng and Wang's models attributed the penetration of the abrasive particles to Hertzian elastic contact. Zhang and Busnaina [2.12] estimated the contact pressure between the particle and the contact surface and found that it is larger than the yielding stress of the polished materials. Therefore, they proposed that plastic deformation is a more likely deformation mechanism of polishing surfaces. The contact pressure over the particle-wafer interfaces is suggested to be equal to the hardness H_w of the wafer materials. Replacing the normal stress (Eq. 2.8) in Tseng and Wang's model with the hardness H_w yields the following material removal rate formulation:

$$MRR = K_e(PV)^{1/2} \tag{2.12}$$

where K_e is the parameter to account for materials properties, slurry abrasive concentration and chemical processes. Both Tseng and Wang and Zhang and Busnaina's models suggest a nonlinear pressure times velocity dependency of material removal rate. However, Zhang and Busnaina attribute all the non-linearity to the fluid flow while part of the non-linearity in Tseng and Wang's model is from the elastic indentation of the abrasives. Moreover, it is noted that besides the external force applied on the particles from the pad, Zhang and Busnaina [2.12] also proposed that an adhesion force, either van der Waals force or electrostatic force depending on the separation distance between the particle and the wafer, contributes to the indentation. This has been integrated into another particle-scale model by Zhang et al. [2.13]. In a series of papers by Ahmadi and Xia [2.14] and Mazaheri and Ahmadi [2.15, 2.16], a thermodynamic work parameter W_a of adhesion is used to account for its effects on the indentation of abrasive particles. Lately, Mazaheri and Ahmadi [2.16] have introduced a double layer (dl) force into the indentation force, whose magnitude is a function of the zeta potential of the abrasives and the wafer. Beside the abrasive wear, Ahmadi and Xia [2.14] also consider the adhesion wear of wafer in their model. Moreover, while most of the models treated the abrasives as spherical shape, Mazaheri and Ahmadi [2.15, 2.16] treated them as spheres with a number of hemispherical bumps around their surface. The penetrations of slurry abrasives are modeled as the penetrations of the bumps. Besides the analytical models, it is noted that recently several particle-wafer interaction models based on molecular dynamics (MD) simulation have been developed for both silicon oxide and copper CMP [2.71, 2.72, 2.73]. They may offer some insights into the material removal from the detailed nanometric cutting perspective.

The above models imply that the abrasives are embedded into the pad and indented into the wafer surface. Beside these kinds of 'two-body' based models, it is noted that there are modeling efforts which assume that the abrasive particles float in the slurry and impact the wafer surface from time to time. It is these impacts that remove the materials. One model of this aspect has been proposed by Su [2.17], assuming a three-body abrasion of materials. Models attributing the material removal to the erosion enhanced by the 'three-body' abrasive impact, instead of a 'two-body' indentation as that by Tseng and Wang [2.11], may be developed.

2.2.2 Models Considering Polishing Pad-Wafer Surface Interactions

The pad is assumed to be smooth in the earlier particle-wafer interaction models. It has been observed that the pad topography and pad material play an important role in material removal process. For example, the material removal rate increases with the pad surface roughness [2.18]. A softer pad yields a larger material removal rate [2.18]. Without conditioning of the polishing pad, the material removal rate decreases exponentially with polishing time [2.19]. In consideration of this, Yu et al. [2.20] developed a pad-based model. They approximated the peaks on the pad surface by hemispherical asperities with constant radius β; see Fig. 2.2. The asperity height is assumed to follow a Gaussian distribution $\Phi_Z(\mu_Z, \sigma_Z)$ where μ_Z is the mean and σ_Z the standard deviation. Based on the model, the real contact area is smaller than the nominal contact area and proportional to the down pressure. Steigerwald et al. [2.21] proposed that the material removal rate is proportional to the number of abrasive particles over the contact area. Combining this argument with Yu's model yields a linear dependency of the material removal rate on the down pressure. This agrees with Preston's Equation.

Zhao and Shi [2.22] also proposed a model based on the wafer-asperity contact. Unlike Yu's model [2.20], the model does not consider the Gaussian distribution of the asperity heights. The contact area between an asperity and the wafer is given by $A \propto P^{2/3}$ based on Hertz elastic contact theory. By combining Steigerwald's argument, the material removal rate formulation can be obtained as

$$MRR = K_e(V)P^{2/3} \tag{2.13}$$

where $K_e(V)$ is a function of the relative velocity V and other CMP parameters. It is further considered by Zhao and Shi [2.22] that when the particles are rolling against the wafer surface, their contribution to material removal will be negligible. They argued that whether the particle is rolling or not is determined by the surface friction between the particles and the wafer, and only when the down pressure P is larger than a threshold down pressure P_{th}, the pure rolling can be avoided. This leads to the following material removal rate formulation:

$$MRR = \begin{cases} K_e(V)(P^{2/3} - P_{th}^{2/3}) & P \geq P_{th} \\ 0 & P < P_{th} \end{cases} \tag{2.14}$$

Besides the above analytical models, it is noted that recently numerical models based on the finite element method have been used to investigate the wafer-pad contact as well [2.35, 2.60].

The fundamental difference between the above pad-based models and the particle-based models by Cook and others is that pad-based models attribute the material removal rate to the number of abrasive particle captured by the polishing pad while the latter attributes the material removal rate to the interaction between a single abrasive and the wafer. Therefore, the nonlinear pressure dependency in Yu et al. [2.20] and Zhao and Shi's models [2.22] is due to the pressure dependency of the contact area while in the particle-based models it is due to the pressure dependency of the indentation of a single abrasive. Neither of these may be sufficient. A complete model should consider that the material removal rate is equal to the number of abrasives times the material removed by a single abrasive. A particle-pad interaction model is critical for this purpose considering that the function of the polishing pad is to hold the abrasive particles, transmit load forces to the particle-wafer surface, as well as conform to the wafer being polished.

2.2.3 Models Considering Slurry Particle-Polishing Pad Interactions

Since the force supported by the 'cutting tools'-slurry abrasives is critical to determine the material removed by a single abrasive and is obtained from the polishing pad or slurry film, a successful particle-pad interaction model should first be able to evaluate the force. Several possible contact modes between the particles and pad exist. The first mode is that a slurry film is formed over the wafer-pad interface and therefore the particles are never embedded into the pad but impact the pad only. In this case, the pad contributes to the force through the slurry film. A detailed fluid mechanics model considering the topography and deformation of the pad is needed to evaluate this force. Su's model [2.17] may be helpful in this respect. The second possible mode is that the abrasives are embedded into the polishing pad. This is the case of the

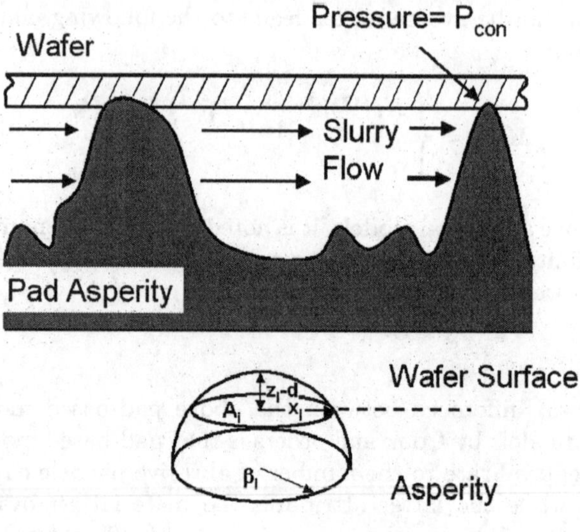

Fig. 2.2. Wafer-pad contact model of Yu et al. [2.20]

'two-body' removal of materials. Cook's model [2.7] suggested a close packing of spherical abrasives into the pad. It is assumed that the wafer and pad are separated completely by the abrasives and no pad-wafer direct contact exists. The force applied on a single abrasive under these assumptions is given by

$$F = \frac{2\sqrt{3}Px^2}{k_1} \tag{2.15}$$

where P is the polishing pressure, x the abrasive size and k_1 the particle fill fraction on the pad. This particle-pad interaction model has been integrated into the material removal model of Cook [2.7]. It is also used by Ahmadi and Xia [2.14] to evaluate the force on a particle in their case of a hard pad and a larger concentration of abrasive particles. Later, Zhao and Shi [2.22] proposed that when the pad is soft enough, the abrasive particles will be embedded into the pad deeply and the force from the wafer is supported by the pad and abrasives together. This idea has been applied by Luo and Dornfeld [2.23, 2.24, 2.25] and Fu et al. [2.26] in their integrated material removal model. Luo and Dornfeld's model suggests that this force is proportional to the contact pressure times the abrasive size by assuming that the abrasives are closely packed to each other and that these closely packed abrasives are enwrapped by the pads so that the effective contact area between wafer and pad is equal to that without abrasives. Moreover, the size of the abrasives that may be captured by the pad is a function of abrasive size distribution and pad properties. Details of this abrasive particle-pad interaction model will be addressed in Chapters 4 and 5 of this book. Fu et al. [2.26] later assume that the abrasives are dispersed evenly over the pad surface and use a beam model to evaluate the wafer-pad direct contact between each two single abrasives. The force supported by a single abrasive can be obtained from the beam model and is a function of abrasive size, down pressure and pad material properties [2.26].

Besides the force, the second purpose of an abrasive-pad interaction model is to evaluate the number of abrasives involved in the material removal process. Is this number simply proportional to the wafer-pad contact area, as that in the pad-wafer contact model by Yu et al. [2.20] and Zhao and Shi [2.22], or is it buried in more complicated scenarios? Fu et al. [2.26] simply took the number as an independent parameter in their model. This may be misleading considering that various parameters, say, the abrasive weight concentration and abrasive size, may have an influence on the number [2.23, 2.24, 2.25]. Luo and Dornfeld [2.23, 2.24, 2.25] considered more complex scenario and suggested that only a portion of abrasives are involved in the material removal process. Similar to the size of the active abrasives, the portion is a function of the abrasive size distribution, pad topography and material properties.

In summary, the pad-abrasive interaction is one of the most important interactions in CMP process. One of the two material removal components, namely, the abrasive number,is a direct output of this interaction. The other, namely, the material removed by a single abrasive, is an output of the interaction through the wafer-abrasive interaction. An accurate model of this interaction will be critical for a successful particle-scale material removal model. There are not many modeling efforts from this aspect before Luo and Dornfeld [2.24, 2.25] and more attention should be paid to it in the future.

2.2.4 Models Considering Slurry Chemical-Wafer Interactions

The contribution of slurry chemicals to the material removal is either neglected in earlier models or represented by an all-purpose coefficient. Cook [2.7] suggested a complete but complicated scenario of the chemical effects. Besides the mechanical removal, it is proposed that the surface removal during polishing should include the following five chemical processes: i. the slurry chemical diffusion into the wafer surface; ii. the subsequent wafer material dissolution under the load imposed by the abrasive particles; iii. the adsorption of the dissolution product onto the surface of the polishing grain; iv. the re-deposition of the polishing materials back onto the wafer surface; and v. the corrosion rate between particle impacts. These steps are not considered in most of the early models for two reasons: first, they are hard to model quantitatively, and second, the contribution of these processes to the total material removal rate is believed to be minimal. From the knowledge of the authors, a recent model by Osseo-Asare [2.27] is the first to treat the adsorption rate of the dissolution product onto the surface of the polishing abrasives.

The major contribution of slurry chemicals to the material removal process has been attributed to the formation of a surface layer. This idea is well demonstrated by Kaufman et al. in their Tungsten CMP model [2.28]. They used the following formulation to describe the formation of a tungsten passivation layer in the presence of ferricyanide:

$$W + 6Fe(CN)_6^3 + 3H_2O \rightarrow WO_3 + 6Fe(CN)_6^{-4} + 8H^+.$$

Kaufman et al. proposed that the passivation layer is removed by slurry abrasives and the fresh tungsten surfaces are exposed, which is subsequently passivated and removed. This mechanism of passivation-removal-repassivation can be used to explain the copper, aluminum and other metal CMP as well. A similar mechanism of surface modification-removal-remodification is supposed to work on silicon, silicon oxide and low-k material CMP. Paul [2.29] and Zhao et al. [2.30] have proposed detailed surface kinetics models to connect the slurry chemical concentrations and fresh metal surface sites available to the formation rate of the surface layer. Their models can explain the material removal rate as a function of chemical and abrasive weight concentration. Recently, Borst et al. [2.31] have proposed a five-step model for CMP of SiLK dielectrics: i. mass transport of the reactant from the slurry to the slurry/wafer interface; ii. adsorption of the reactant to available surface site; iii. reaction between the adsorbed reactant and the specific wafer surface to form an altered wafer surface layer; iv. mechanical removal of the altered wafer surface layer; and v. mass transport of removed material to the bulk slurry. This five-step mechanism is shown in Figure 2.3 schematically. In their work, formulations to cover the steps i, ii and iii, which relate the mass transportation, slurry chemical concentration and reaction rate to the formation of the surface layer were presented. This detailed model is supposed to be able to be extended to the CMP of other materials.

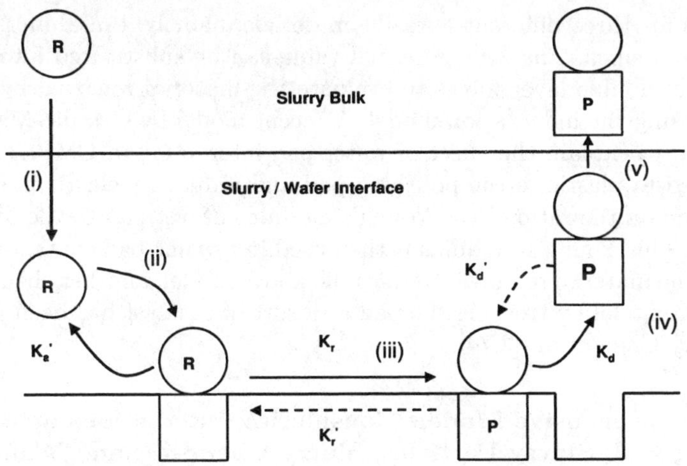

Fig. 2.3. The five-step surface kinetics model [2.31]

The idea of surface modification, removal and re-modification has been applied in Luo and Dornfeld's model [2.25] as well. They extended Kaufman's model by proposing that the surface layer is a bi-layer structure, one, a softer hydrated upper layer, and the other, a harder bottom layer. Detail of this is to be discussed in Chapter 5. They did not cover the details on how the formation rate of the surface layer is affected by the slurry chemicals. However, the model does connect the mechanical removal and chemical passivation rate and propose *MRR* formulations as a coupling function of the surface generation rate, abrasive weight concentrations and wafer-pad contact area.

2.2.5 Models Considering Slurry Chemical-Slurry Particle Interactions and Slurry Chemical-Polishing Pad Interactions

The slurry chemicals affect not only the wafer but also the slurry abrasives and polishing pad. Their contribution to material removal is therefore not only reflected through the surface kinetics, but also through the alteration of the abrasive and pad properties such as the abrasive shape, abrasive size and Young's modulus of the pad. There are not many modeling efforts on these two interactions yet. One of the efforts on the chemical-abrasive interaction is by Mazaheri and Ahmadi [2.16]. Mazaheri and Ahmadi proposed that the indentation of abrasives into the wafer surface is determined not only by the load from the pad but also by the double layer (dl) forces F^{Ψ}, which are a function of abrasive size x and abrasive zeta potential Ψ. They proposed experimental equations of abrasive zeta potential Ψ as a function of slurry pH values for three different abrasive materials, namely, tantalum penoxide, alumina and silica. The zeta potential value can be substituted into the formulation of double layer forces to evaluate the material removed by a single abrasive using the indentation model. A recent model by Castillo-Mejia et al. [2.32] tried to explain the effect of water-pad interaction in CMP. The water is proposed to plasticize the polishing pad and reduce its elastic modulus. A formulation on the ratio of the Young's modulus of wet pad to the dry pad is proposed. This Young's modulus is then used in a wafer-pad contact model to evaluate the material removal. Besides the above model, another chemical-pad interaction model by treating the pad as a sort of catalyst has been proposed recently by Chen et al. [2.74].

2.2.6 Comprehensive Models Considering Interactions among Polishing Pad, Slurry Particles, Slurry Chemicals and Wafer Surface

Most of the earlier models have been concentrated on one or two interactions of the CMP process. They are useful for identifying the input parameters. However, they may not be sufficient for understanding the overall material removal process. The effects of the same input parameters may be contrary to each other when acting in different interactions. For example, in the abrasive-particle interaction, larger abrasive sizes yield larger indentation depth and therefore higher material removal rates, while in the abrasive-pad interaction, a larger abrasive size, however, may yield smaller number of abrasives and therefore smaller material removal rate. Relying on either the abrasive-wafer interaction model or the abrasive-pad interaction model to explain the effects of abrasive size may be misleading. A comprehensive model of CMP integrating the six interactions together is therefore needed. In this section, several integrated modeling efforts at the particle-scale are discussed. Some of them have been mentioned earlier in above sections.

Xie and Bhushan [2.33] were one of the first to consider the comprehensive effects of mechanical elements including the abrasive particles and pad topography in the polishing process. However, the slurry chemicals are not considered. The model cannot explain the effects of abrasive size distribution either. Chapters 2, 3 and 4 of this work propose a model integrating the following four interactions, namely, the interactions of the wafer and pad, the interactions of the abrasive and wafer, the interactions of the abrasives and the polishing pad and interactions of the slurry chemicals and wafer surface, into one single material removal model.

The overall picture of the model can be described briefly here; see Fig. 2.4. The slurry is delivered into the wafer-pad interface, with slurry chemicals reacting with the wafer materials and forming a passive layer over the wafer surface. Since the pad surface is rough, only part of the pad asperities contact the wafer. This part of the pad asperities captures the slurry abrasives and the abrasives captured will be deeply embedded into the pad and share the down force with the pad. These captured abrasives, called *active* abrasives, then remove the chemically influenced surface layer plastically. In the four interactions, the wafer-abrasive interaction is modeled as a plastic indentation. The wafer-pad interaction is modeled as an elastic Hertz contact of sphere with a half space by assuming a uniform distribution of pad asperities. The abrasive-pad interaction is the key part of the model; the pad plays a role as a filter and only abrasives large enough can be captured. The chemical-wafer interaction is modeled as a process of surface modification, removal and surface remodification.

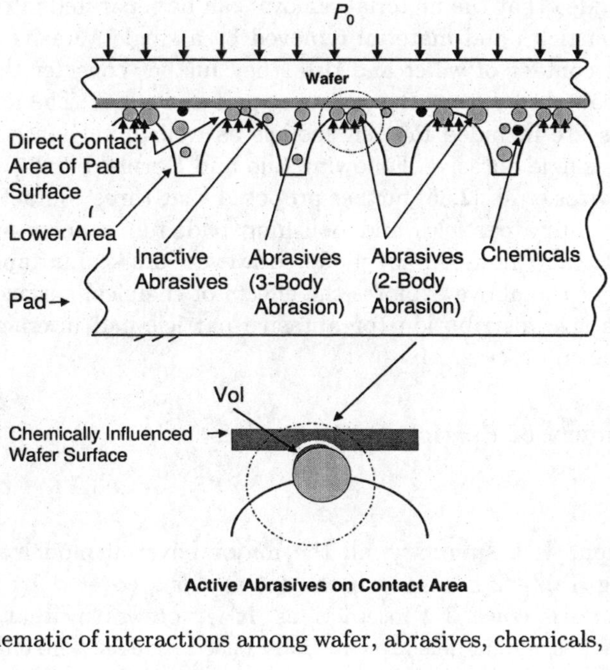

Fig. 2.4. Schematic of interactions among wafer, abrasives, chemicals, and polishing pad

One key idea of the model is that the material removal rate can be decomposed into two parts: one, the number N of the *active* abrasives, and the other, the material removed by a single abrasive per unit time. The advantage of separating the material removal into two parts is that the physical meanings of them are clearer and therefore it is easier to identify and connect the input parameters to them. An additional benefit is that the connections between the four interactions can be easily obtained through their contribution to these two parameters. A framework of the comprehensive particle-scale modeling connecting the interactions has been proposed and shown schematically in Fig. 2.5. The abrasive-chemical and pad-chemical interactions are not included in the model yet. However, they are put in the framework for the purpose of completeness. For more details of the framework, the readers are referred to Fig. 2.A in the appendix of this chapter.

It is noted that following work of Luo and Dornfeld [2.23], similar integrated models have been proposed by Fu et al. [2.26], Zhao and Chang [2.34] and Ahmadi and Xia [2.14]. Works by Mazaheri and Ahmadi [2.15, 2.16] are an extension to the modeling efforts of Ahmadi and Xia [2.14]. They are all based on the idea that the material removal can be separated into the number of abrasive particles and material removed by a single abrasive. Besides the elastic Hertz contact of wafer and pad, they further consider the possibility of a plastic contact between the polishing pad and wafer. The adhesion force and dl forces are included in their models as well to calculate the material removal by a single abrasive. Following Luo and Dornfeld [2.23] and Fu et al. [2.26], Bastawros et al. [2.35] further proposed that three contact modes exist between the slurry particles and polishing pad: full contact mode, partial contact mode, and non-contact mode. However, unlike Luo and Dornfeld's model, none of the above explores the effects of chemical surface passivation and abrasive size distribution (or abrasive particle-pad interaction) on the material removal process.

2.2.7 Summary of Particle-Scale Models

In this section, as a summary, all the major physical models are listed in the following Table 2.1 based on the interactions covered by them. Some integrated models cover 3-4 interactions. It is noteworthy that most of the current particle-scale models are two-body based models. The effects of slurry are to distribute slurry chemicals and particles to the wafer-pad interface. Due to its low selectivity of the material removal on high and low features, the existence of a slurry film that separates the wafer and pad may be undesirable in term of planarization capability. However, with the low-k materials applied as dielectrics, the existence of a slurry film may lead to a less aggressive wafer-pad contact and therefore reduce the delamination of metal and dielectrics

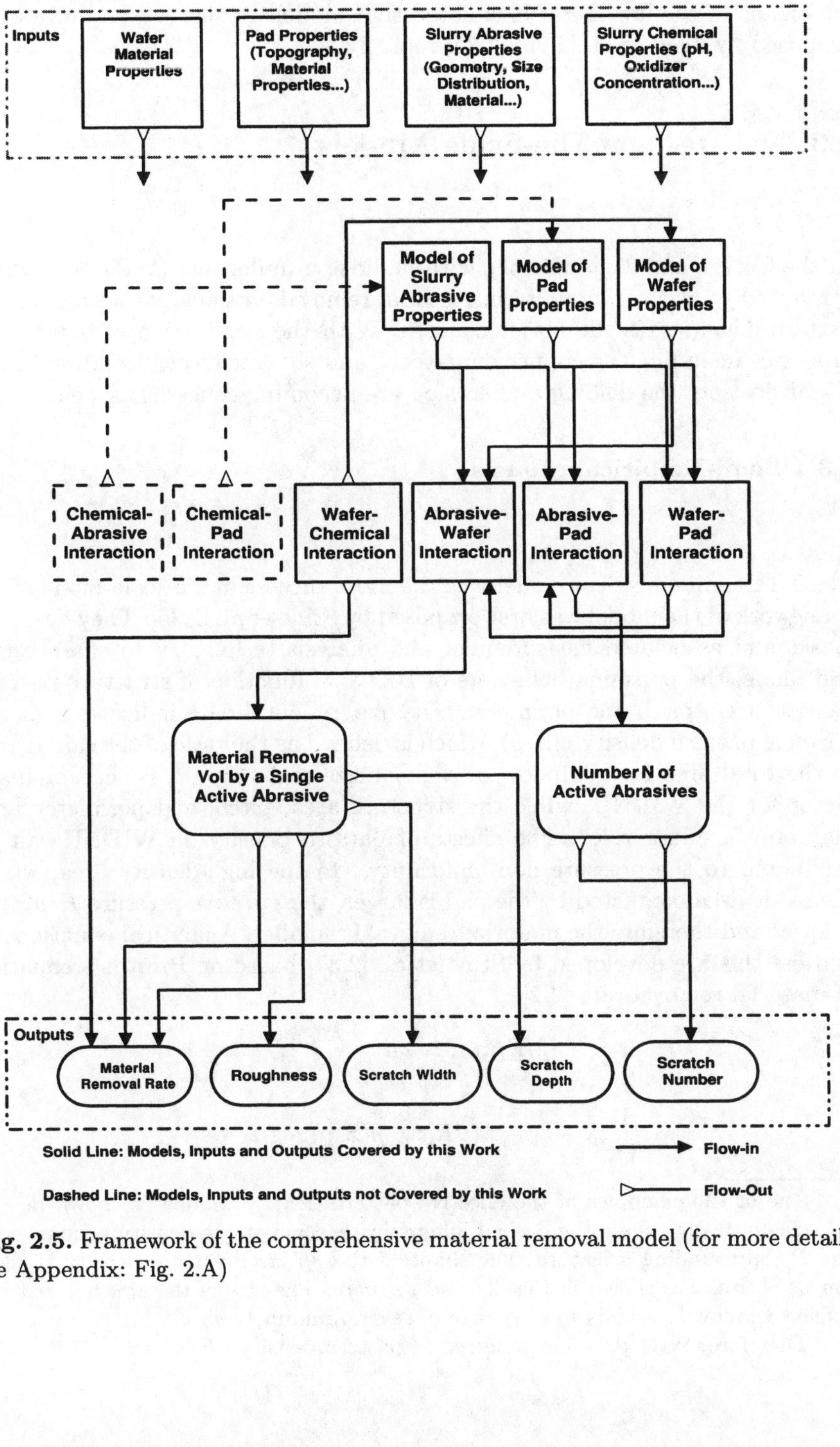

Fig. 2.5. Framework of the comprehensive material removal model (for more details, see Appendix: Fig. 2.A)

during the polishing. Therefore, models considering the slurry film and three body abrasions/removals such as that in Su's work [2.17] may become more important in the low-k era. It is noted that models in this respect have been proposed by Jeng et al. [2.76] [2.77] recently.

2.3 Feature- and Die-Scale Models

At the feature and die scales, the with-in die non-uniformity (WIDNU), which is referred to as the non-uniform material removal on the areas with different pattern densities, is the major concern. With the application of damascene processes including the copper damascene and shallow trench isolation in the IC fabrication, the dishing and erosion are becoming concerns as well.

2.3.1 Semi-Empirical Models

The MIT semi-empirical model was the most successful die-scale model. The framework of the model was first proposed by Stine et al. [2.36]. They designed masks and associated measurement and analysis technology to characterize and model the polishing behaviors of ILD as a function of structure pattern density, area, pitch and perimeter. The experimental data indicates that the *effective* pattern density $\rho(x,y)$, which is defined as the ratio of the raised area to the total area for a window of a given size and shape [8], is the dominant factor for the WIDNU, while the structure area, pitch, and perimeter/area play only a minor role[9]. The effects of pattern density on WIDNU can be attributed to the pressure non-uniformity. On the high-density area, where the oxide area contacted by the pad is larger, the *effective* pressure $P/\rho(x,y)$ is lower and therefore the material removal is smaller. Analytical equations to address this are developed by Stine et al. [2.37] based on Preston's equation of material removal rate [2.2]:

$$z = \begin{cases} z_0 - \frac{MRR_i t}{\rho(x,y)} & t < \frac{\rho(x,y) z_1}{MRR_i} \\ z_0 - z_1 - MRR_i t + \rho(x,y) z_1 & t > \frac{\rho(x,y) z_1}{MRR_i} \end{cases} \tag{2.16}$$

[8]One of the examples of the effective pattern density window is shown in Fig. 2.6 schematically, where the dashed line indicates the pattern density window with size R_p surrounding a feature. The shape of the *effective* pattern density window can be a circle, as shown in Fig. 2.6, or a square. The size of the *effective* pattern density window is usually in the range of several minimeters.

[9]Therefore, WIDNU is also referred as pattern density effects.

Table 2.1. List of particle-scale models

Model Category		AWI[1]	WPI[2]	API[3]	CWI[4]	CAI[5] CPI[6]	CMRM[7]
Identified Input Parameters – Abrasives	Size	[2.6] [2.7] [2.9]		[2.23-2.26] [2.35]		[2.16]	[2.14-2.16] [2.23-2.26] [2.33-2.34]
	Size Distribution	[2.23-2.25]		[2.23-2.25]			[2.23-2.25]
	Shape	[2.6-2.7] [2.9] [2.11-2.17] [2.23-2.26] [2.34-2.35]		[2.23-2.26] [2.35]			[2.14-2.16] [2.23-2.26] [2.33-2.34]
	Weight Concentration	[2.14-2.16] [2.23-2.25] [2.34]		[2.23-2.25]			[2.23-2.25] [2.33-2.34]
	Dilution Ratio	[2.24-2.26]		[2.23-2.25]			[2.23-2.25]
	Number			[2.26] [2.35]			[2.26]
	Young's Modulus	[2.9]					
	Zeta Potential	[2.16]					[2.16]
	Materials					[2.16]	[2.16]
Pad	Pad Topography		[2.23-2.26] [2.14-2.16] [2.34] [2.60]	[2.23-2.26] [2.35]			[2.14-2.16] [2.23-2.26] [2.33-2.34]
	Hardness		[2.14] [2.23-2.35]	[2.23-2.25]			[2.14-2.16] [2.23-2.25]
	Young's Modulus		[2.14-2.16] [2.23-2.26] [2.34]	[2.23-2.26] [2.35]		[2.32]	[2.14-2.16] [2.23-2.26] [2.33-2.34]
	Catalytic Property					[2.74]	
Wafer	Young's Modulus	[2.5-2.7] [2.9] [2.11]					
	Hardness	[2.9] [2.12] [2.14-2.16] [2.23-2.26] [2.34]					[2.14-2.16] [2.23-2.26] [2.33-2.34]
	Zeta Potential	[2.16]					[2.16]
Chemicals	Oxidizer Concentration				[2.28-2.31]		[2.25]
	pH value					[2.16]	[2.16]
	Passivation Rate				[2.25]		[2.25]

[1] Abrasive Particle-Wafer Interaction [2] Pad-Wafer Interaction [3] Abrasive Particle-Pad Interaction [4] Chemical-Wafer Interaction [5] Chemical-Abrasive Particle Interaction [6] Chemical-Pad Interaction [7] Comprehensive Material Removal Model

where z is the height of the oxide pattern feature to the substrate, z_1 the as-deposited step height, $\rho(x,y)$ the *effective* pattern density, $MRR_i = K_e PV$ the blanket wafer polishing rate and t the polishing time. Before $t = \rho z_1 / MRR_i$, the step is not totally removed. The reduction rate/removal rate is $MRR_i/\rho(x,y)$ based on Preston's equation. After that, the removal rate is equal to the blanket removal rate. Note here the usage of Preston's equation provides an interface to integrate the previous particle-scale models into the feature- and die-scale models.

The *effective* pattern density is a concept at the die-scale. Therefore, Eq. 2.16 is a die-scale model. A physical feature-scale step height reduction model under it is

$$z = z_0 - K_e \frac{P_d(x,y)}{PD(x,y)} Vt \tag{2.17}$$

where $PD(x,y)$ is the feature pattern density, or layout pattern density at the feature location (x,y)[10], P_d the die-scale pressure at the feature location, K_e the Preston's coefficient and V the relative velocity. P_d is a function of the *effective* pattern density $\rho(x,y)$ and their relationship will be given in Chapter 6 as $P_d = P \times PD(x,y)/\rho(x,y)$ based on a rough physical understanding of the *effective* pattern density concept. We noted that for the ILD/IMD step height reduction model, replacing the die-scale pressure P_d and feature density $PD(x,y)$ in Eq. 2.17 with the down pressure P and *effective* pattern density $\rho(x,y)$ yields the first die-scale step reduction equation in Eq. 2.16. This is because that there is a simple linear relationship between die-scale pressure P_d and the step height reduction. However, this does not mean that for other step height reduction models, where the step height dependency on P_d may be nonlinear, we can still replace the feature pattern density with the *effective* pattern density to obtain the die-scale topography evolution model. Details of a scheme of die-feature scale integration will be given in Chapter 6 for other non-linear step height evolution models.

The above closed-form analytical equation by Stine et al. [2.37] concerns the topography evolution in the vertical direction. The complexity of the model is buried in the determination of the size of the *effective* pattern density window. The *effective* pattern density ρ and the corresponding pattern density window

[10] The feature pattern density, or layout pattern density is calculated as the ratio of the raised area to total area for a small window with size close to the feature size, which is usually in the range of 20~40μm, as shown in Fig. 2.6. The physical meanings of the *effective* pattern density and the feature layout pattern density are totally different: the first one is an indicator of pressure distribution across the die, while the the second one is an indicator of the feature geometry.

are needed because the pressure supported by a feature at location (x, y) is determined not only by the local wafer topography, but also by the wafer topography in the neighborhood. This is due to the macroscopic deformation of the pad. The polishing time t is usually larger than $\rho z_1/MRR_i$ so that the as-deposit steps, by the second case of Eq. 2.16, are totally removed. If the *effective* pattern density is calculated in the proper size of window, the slope of the oxide thickness as a function of the pattern density will be equal to the as-deposit step height z_1. This window size can, therefore, be extracted from experimental data using an iteration method until the condition of slope $= z_1$ is satisfied [2.36].

An implication in Stine's model [2.36], where the *effective* pattern density is calculated as the ratio of the raised area to the total area of an *effective* pattern density window with a given size, is that all features in the pattern density window make the same contribution to the *effective* pattern density. Ouma et al. [2.38] realized that due to the smaller pad deformation, the features at the edge of the *effective* pattern density window surrounding a feature may have fewer contributions on the pressure. A weight density function $w(x, y)$ based on contact mechanics, which is to be given in Chapter 6, is applied to account for this. The *effective* density $\rho(x, y)$ can be given by the convolution of the layout density $PD\ (x, y)$ and the weight density function in a pattern density window with a circular shape and radius R_p; see Fig. 2.6:

$$\rho(x, y) = PD(x, y) \otimes w(x, y) \qquad (2.18)$$

A rough physical explanation which connects the above definition of the *effective* pattern density with the die-scale pressure distribution will be given in Chapter 6.

If a die is discretized into N feature windows as shown in Fig. 2.6, Eq. 2.18 can be written as

$$\rho_k = \sum_{i=1}^{M} (PD_i w_{ki}) \qquad (2.19)$$

where PD_i is the layout pattern density at the ith feature window, M the number of the feature windows in the pattern density window surrounding the kth feature window and w_{ki} the value of the weight density function at the ith window relative to the kth window [11]. Following the same methods of Stine et al. [2.36], the size R_p of the pattern density window can be extracted

[11] In the case of Stine et al's model [2.36], w_{ki} is equal to constant $1/M$ for all features in the pattern density window.

from the experimental data. More accurate results than those from the Stine et al.'s model can be obtained.

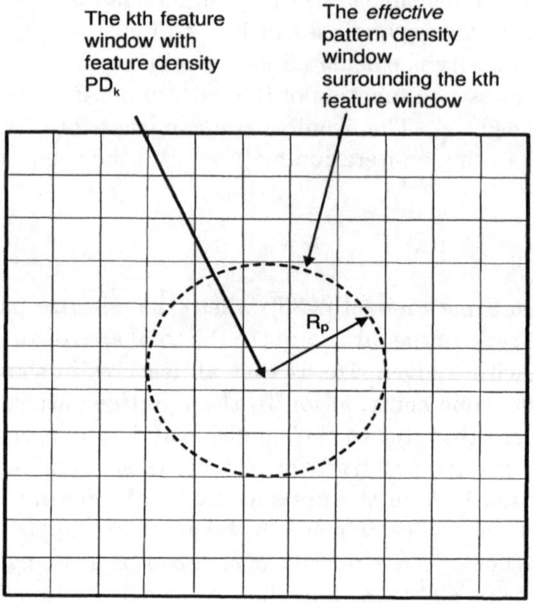

Fig. 2.6. Top view of a die (tens of minimeters × tens of minimeters) discretized into N small feature windows (tens of micrometers × tens of micrometers)

Under the framework of Stine's semi-empirical model, an accurate feature-scale step height reduction model as Eq. 2.17 is of the greatest interest. With the die-scale pressure $P_d(x, y)$, or the *effective* pattern density $\rho(x, y)$, and the layout pattern density $PD(x, y)$ as inputs, the feature-scale model can provide topography information across the whole die, in both vertical and lateral directions. In Eq. 2.17, no local compression of the polishing pad is assumed, so a step is completely removed before the pad touches the down areas. This may not be the case when the pad is soft and the step height is small. It is proposed by Tseng et al. [2.39] that for a soft pad the upper and bottom parts of the steps may be in contact with the pad simultaneously. The pressure difference over the upper and bottom parts of the steps is proportional to the step height times the pad stiffness. Based on this model the step height reduction is obtained as an exponential function of time. This is to be discussed in detail in Chapter 6 of this book. Combining Stine et al. [2.37] and Tseng et al. [2.39]'s models together, Smith et al. [2.40] proposed that the pad may touch the upper part of the step only at the beginning stage of the planarization, followed by a simultaneous touch of upper and bottom parts after the step height is reduced to a certain value (a function of pad materials, down pressure and feature geometry). Step height reduction formulations combining Tseng et al. and Stine et al.'s models were proposed.

By extending the exponential step height reduction model of Tseng et al. [2.39] from a single material over the upper and bottom parts of the steps to two different materials, Grillaert et al. [2.41] and Tseng et al. [2.42] developed a formulation to model the dishing and erosion in shallow trench isolation (STI). Tugbawa et al. [2.43] further developed a step height evolution model for the copper damascene process by combining the above models of Stine [2.36], Tseng [2.39], and Smith [2.40], which covers the bulk/overburden copper removal before the dielectrics are exposed, and the models of Grillaert [2.41] and Tseng [2.42], which cover the overpolishing stage after the dielectrics are exposed.

It is noted, however, that while the model by Tugbawa [2.43] can predict the oxide erosion as a function of pattern density, it cannot capture the pattern density dependency of dishing in the copper damascene process correctly. In Chapter 6 of this book, we propose a novel step height model for the copper damascene process based on linear viscoelasticity, which can capture this feature and therefore may be more accurate for predicting the feature-level topography evolution in the copper damascene and STI process. Moreover, it was realized by Maury et al. [2.4] that revised Preston's equations can capture the pressure dependency of the material removal more accurately. Hence, in the development of the topography evolution model in Chapter 6, the revised Preston's equation is used.

With the feature-scale step height evolution model developed, the next question is how to integrate it into a die-scale model to evaluate the topography across the die. Can we simply replace the pressure $P_d(x, y)$ and feature pattern density $PD(x, y)$ with down pressure P and the die-scale *effective* pattern density $\rho(x, y)$, as in Eq. 2.16 for ILD/IMD CMP? Although this scheme works well for ILD/IMD CMP, as mentioned earlier, it seems not make sense from a physical point of view for other feature-scale step height reduction models. In Chapter 6 of this work, we will propose a new scheme of feature-die integration based on a rough physical understanding of the concept of *effective* pattern density.

A final point worth mention is that all of the above models are based on the assumption of a direct solid-solid contact between the wafer and pad.

2.3.2 Physical Models

While the semi-empirical models of Stine et al. [2.36] and Ouma et al. [2.38] are the most successful at feature and die scales, other researchers are trying to developed physical models to explain the topography evolution. For example, Ouyang et al. [2.44] assumed that a slurry film exists between the die and pad and developed equations based on fluid mechanics to predict the pressure at the location (x, y, z). Preston's equation is then used to evaluate the topography evolution. Another fluid-mechanics based feature-scale topography model was developed by Yao et al. [2.69].

The step height reduction model applied in Stine's semi-empirical die-scale models can be taken as a one-dimensional feature-scale topography evolution model. Two-dimensional models considering the lateral deformation of pad over the features were developed by Chekina et al. [2.45]. Following the same contact mechanics scheme, three-dimensional numerical models using boundary element methods (BEM) are developed by Yoshida [2.67-2.68]. The models assume the pad as a half space. The primary contact mechanics equation applied in the models is based on linear elasticity and can be written as

$$\overline{u}_z(x, y) = \frac{1 - v_h^2}{\pi E_h} \iint\limits_D \frac{P(\xi, \eta)}{r(x, y, \xi, \eta)} d\xi d\eta \tag{2.20}$$

which relates the deformation $\overline{u}_z$ of the pad at an arbitrary location $(x,\ y)$ to the pressure distribution $P(\xi,\ \eta)$ applied in an arbitrary domain D over the half space, where E_h and v_h are the Young's modulus and Poisson's ratio of the half space, respectively, and $r\ (x,\ y,\ \xi,\ \eta)$ the distance between locations $(x,\ y)$ and $(\xi,\ \eta)$. For uniform distribution of the pressure, the deformation at any arbitrary point (x, y) can be obtained by evaluating the integral in Eq. 2.20 analytically. For an arbitrary pressure distribution, the equation has to be solved numerically. Johnson [2.8] has presented the basic idea of the numerical solution of the above equation. In CMP, if assuming the wafer as a rigid body and an intimate contact between wafer and pad during the polishing, the pad deformation $\overline{u}_z$ at location $(x,\ y)$ will follow the feature topography. By substituting the known distribution of $\overline{u}_z$ into the above Eq. 2.20 and solving it numerically, the pressure distribution over the features can be obtained. Then the pressure distribution is applied in the Preston's equation or other particle-scale model to update the wafer topography, which is then used in Eq. 2.20 to update the pressure distribution over the features. This iteration process is continued until the final planarization is finished. In Chapter 7 of this work, we discuss in detail the numerical solution of Eq. 2.20 for 3-D CMP topography evolution under the specific boundary conditions in the CMP process. It is basically an extension of the work of Johnson [2.8] and Yoshida [2.67-2.68].

In comparison with the one-dimensional step height evolution model discussed in earlier sections, the two and three-dimensional physical models are more useful when accurate topography information is needed. The disadvantage is that they are computationally intense and can not be integrated directly into an electronics computer aided design (ECAD) software to address the issue of design for manufacturability.

2.3.3 Electronics Computer Aided Design

One motivation for CMP modeling at the die-scale is to facilitate the computer-aided design of circuits. In interconnect-dominant circuits, the topography, such as the ILD thickness variation, can lead to a more than 20% circuit performance variation, which is significant enough to cause concerns in design of high-performance microprocessors [2.46]. Since the underlying reason for the variation is the pattern density non-uniformity, dummy patterns have been inserted in the low-density area to increase the density there. The dummy filling should not hurt the original circuit function and design rule-correctness [2.47]. To satisfy this requirement, complicated filling algorithms have been developed by different researchers based on the *effective* pattern density models by Stine [2.36] and Ouma [2.38]. These include a series of recent works published by Kahn et al. [2.47], Tian et al. [2.48] [2.70] and Chen et al. [2.49] in *IEEE Transactions on CAD.*

In Kahn et al. [2.47], Tian et al. [2.48] [2.70] and Chen et al. [2.49]'s work, the dummy filling was formulated as an optimization problem. In this case, there were primarily two different kinds of objective functions. One is from the 'process side' and referred to as mini-variation, which is aimed at the smallest with-in die non-uniformity. The other is from the 'design side' and referred to as mini-filling, which is aimed at minimizing the coupling capacitance and uncertainty generated by dummy fillings [2.49]. Dummy filling algorithms aimed at the above two objectives were proposed by Tian et al. [2.48] for ILD/IMD CMP. They can be summarized as follows:

i. Mini-Variance:
Minimize

$$\rho_H - \rho_L \tag{2.21}$$

subject to

$$\begin{aligned} 0 \le \rho_L \le \rho_k \le \rho_H \le 1 \\ 0 \le PD'_k \le PD_k + PD^a_k \\ 1 \le k \le N \end{aligned} \tag{2.22}$$

ii. Mini-Filling:
Minimize

$$\sum_{k=1}^{N} (PD'_k - PD_k) \tag{2.23}$$

subject to

$$\begin{aligned} 0 \le \rho_L \le \rho_k \le \rho_H \le 1 \\ \rho_H - \rho_L \le \epsilon \\ 0 \le PD_k \le PD^a_k \\ 1 \le k \le N \end{aligned} \tag{2.24}$$

where ρ_H is the maximum *effective* pattern density and ρ_L the minimum *effective* pattern density of the N feature windows across the die; see Fig. 2.6, PD_k the original pattern density at the kth window, PD'_k the feature pattern density at the kth window after dummy filling, PD^a_k the maximum incremental change of the pattern density that can be introduced at the kth window based on design rules and ϵ a positive bound for $\rho_H - \rho_L$ for the min-fill objective. For ILD/IMD CMP, there is a monotonic relationship between z and the *effective* pattern density ρ, Eq. 2.16. Therefore, a minimization of $\rho_H - \rho_L$ yields a minimization of the topography variance, Eq. 2.21, while a bound ϵ for $\rho_H - \rho_L$ indicates a constraint on the topography variation, Eq. 2.24. The *effective* pattern density ρ_k can be written as a linear function of the layout densities; see Eq. 2.19. After applying Eq. 2.19, the above minimization problems can be taken as a linear optimization problem for ILD/IMD CMP. For details on the implementation of the algorithm, the readers are referred to [2.48].

Besides the above optimization based algorithm, alternative filling algorithms are proposed by Lee [2.75]. For instance, an algorithm called the 'maximum effective density algorithm' is proposed to assign each density window with a value of the maximum effective density of the original layout. The variance of the *effective* pattern density can be reduced. These algorithms are not aimed at the 'best' solution but a 'better' solution than the original layout. The advantage is that the complicated computer implementation of the linear optimization algorithm may be avoided.

While the algorithms by Tian et al. [2.48] work well for ILD/IMD CMP, the validity of their algorithm for the damascene process [2.70] is in question since they applied the feature-scale topography models at the die-scale by simply replacing the feature pattern density PD with the *effective* pattern density ρ. As mentioned before, this scheme of die-feature integration may be incorrect from a physical point of view. In Chapter 6, a new scheme of feature-die scale integration will be proposed. An integration of this scheme with the filling algorithms proposed in [2.70] may generate more reasonable and accurate dummy filling results for the damascene process.

2.4 Wafer-Scale Models

The with-in wafer non-uniformity (WIWNU) of material removal rate is the major concern at the wafer scale. The primary reasons for the WIWNU is the non-uniform distribution of pressure and velocity across the wafer. To model these two distributions, it is needed to understand the configuration of the CMP equipment. There are two typical CMP configurations, namely, the rotational type and the linear type. For a rotational type of machine (Applied Materials Inc., Novellus Inc. and Ebara Tech.), the polishing head is rotated on a polishing platen, which is rotated around the center of itself. For a linear type of machine, the polishing head is rotated while a belt is moving under it linearly (LAM Research Inc.). The kinetics/velocity distributions associated with them were straightforward. Several researchers have investigated them for a rotational type of machine [2.50]. A conclusion of these analyzes is that the platen rotational speed times the offset of the head and platen centers is the major components of the relative velocity. When the platen speed is close to the head speed, the velocity distribution is more uniform. Similar analysis can be done for a linear type of machine. They are given in Chapter 8 of this work. The pressure distribution modeling is more complicated and can be put in two categories: models based on solid mechanics and models based on fluid mechanics.

2.4.1 Pressure Distribution Models Based on Solid-Solid Contact

The earliest wafer-scale CMP model for pressure distribution is that by Wang et al. [2.51] for the rotational type of machine. They developed a finite element model (FEM) to evaluate the pressure distribution based on the assumption of wafer-pad direct contact. The model results correlate with the material removal non-uniformity profile well qualitatively. The edge effects, i.e., the large material removal rate at the wafer edge, can be well explained by the pressure singularity there. This implies that solid-solid contact between wafer and pad dominates the pressure distribution. Several finite element models were developed after Wang et al., for example, to investigate the improvement of the edge effects by a retaining ring. A model based on the boundary element method, which is advantageous over FEM in thin structure analysis [2.63, 2.64, 2.65], can be used to predict the pressure distribution with and without rings. This is discussed in Chapter 8. Besides the numerical models, analytical models based on Hertz contact theory were proposed by Tseng et al. [2.52]. A unique property of their models is that the effects of wafer curvature/film stress on the pressure distribution can be captured. More recently, Fu and Chandra [2.53-2.54] have developed analytical equations based on both linear elasticity and linear viscoelasticity. It is claimed that their linear viscoelastic model can explain the material removal degeneration with no pad conditioning. These models can be considered as analytical solutions of Eq. 2.20 at wafer-scale under specified boundary conditions, for example, wafer curvature. As that at feature scale [2.67-2.68], numerical models based on Eq. 2.20 can also be applied to predict the wafer-scale pressure distribution.

2.4.2 Pressure Distribution Models Based on Semi-Solid-Solid Contact and Solid-Fluid-Solid Contact

Some researchers believe that a slurry film may be formed over the wafer-pad interface under certain process conditions, say, when the down pressure is small and the relative velocity is large. Due to the small material removal selectivity between higher and lower regions, the existence of such a film may be undesirable in terms of planarization ability. With the soft low-k materials replacing the hard silicon oxide as the dielectric materials in the sub-micron IC fabrication, a formation of slurry film, which makes the mechanical contact between wafer and pad less aggressive may be needed to avoid/reduce the delaminations of metal and dielectrics. Several models have been developed to address the pressure distribution under the condition of a slurry film. Tichy et al. [2.55] assume that the thickness of the film is equal to the height of the pad asperities. Contact mechanics based on linear elasticity is first used to evaluate the pad asperity deformations. The slurry film thickness is supposed to be equal to the asperity height after the deformation, which is then put into the Reynold's equation to predict the fluid pressure. This model can be considered as a semi solid-solid contact or elastohydrodynamic contact model. One conclusion obtained is that the fluid pressure is negative under certain process conditions. Therefore the wafer is sucked against the pad. This implies that the solid-solid contact dominates the process in the process conditions. The fluid pressure influences the contact pressure distribution along the wafer surface. Flowing a similar scheme, an elastohydrodynamic contact model considering the hyperelastic deformation of pad asperities has recently been developed by Kim et al. [2.66]. Besides the semi-solid-solid contact models, Thakurta et al. [2.56-2.58] and Cho et al. [2.59] developed both two-dimensional and three-dimensional fluid models based on Renold's equation. Not only slurry flow in the lateral direction, but also slurry flow in the vertical direction, which is the case in Novellus CMP configuration where slurry is fed from the bottom of the platen, is considered. The pad deformation due to the fluid flow is considered as well and added into the film thickness. One boundary condition in the model is that the gimbal moment of the polishing head is zero under a zero friction force assumption. In Chapter 8 of this work, this model is extended to consider the friction force. The moment due to the friction force has to be balanced by an asymmetric distribution of normal pressure, leading to a change of the average normal pressure profile. The mature theories from bearing lubrication can be borrowed to model the pressure distribution in the case of CMP with a slurry film. It is worth pointing out that most of the previous 'two-body' based particle- scale models may not work any more in the case of solid-fluid-solid contact.

2.5 Discussions and Concluding Remarks

From the viewpoint of industry, CMP modeling should satisfy the following requirements: i. reliability to be used as verification of process; ii. ability to give feedback for 'what-if' scenarios in lieu of time-consuming design of experiment (DOE) tests; iii. performance prediction for realistic, heterogeneous pattern effects; iv. prediction on not only wafer scale phenomena but also feature/chip scale interaction; v. integration of multi-scale (wafer-, die-, feature-scale) interactions for global CMP modeling to be useful; vi. linkage of models to upstream (deposition, etc.) and downstream (lithography, etc.) processes; vii. ability to address defectivity; and viii. ability to address new materials, consumables (pad, slurry, etc.) modeling and characterization [2.61].

Great progress in CMP modeling has been made in the last decade, which can help to address the above requirements from different perspectives. The particle-scale models can give feedback for 'what-if' scenarios to save up the DOE time. A more systematic design of experiments, instead of trial and error, can now be applied in the process development and optimization [2.61]. A successful particle-scale model to identify the critical input parameters of consumables, and to explain the roles and describe the interactions among them, will be particularly helpful for the consumable design. Optimal designs of pad topography and abrasive size distribution have been identified as two promising areas in this respect. This is on-going research at UC-Berkeley [2.61]. The defectivity issues such as scratching and roughness can also be addressed by the particle-scale models. The number of particles and their indentation depths into a wafer are the two dominant factors in this respect based on Luo and Dornfeld's model.

The die-scale model of MIT can now predict the performance of realistic pattern effects. Linkage of the copper models to the upstream copper electrochemical plating process is underway [2.62]. Preston's equation and its revisions provide the interface for particle-scale, feature- and die-scale and wafer-scale model integrations. Combining the particle-scale material removal rate formulation with the pressure distribution prediction at feature, die and wafer-scales can yield the dishing, erosion, with-in die and with-in wafer non-uniform material removal models. A framework of an model integrating the models at the three scales is shown schematically in Fig. 2.7, where the shaded parts represent elements covered in this book.

There are still a number of problems to be solved. First, the particle-scale models, at the current stage, are more qualitative in nature. Calibration of the models under a manufacturing environment has not yet been done. Obtaining many of the input parameters directly, say, the pad Young's modulus, abrasive size distribution and so on by experiment and their tabulation may be needed in the future. In addition, most current particle-scale models are 'two-body' based. A 'three-body' removal model for CMP of low-k materials is needed. Second, the step height reduction model at the die-scale is based on linear elasticity. A viscoelastic model to calculate the weight function of effective pattern density may be needed for more accurate time dependency prediction. With the application of low-k materials, a fluid mechanics based model to understand the step height reduction analytically is needed as well. These kinds of research will lead to a more predictable and reliable CMP process for the sub-micron IC fabrication.

2.6 References

[2.1] *The International Technology Roadmap for Semiconductors*, Semiconductor Industry Association, San Jose, CA, U. S. A., 1999.

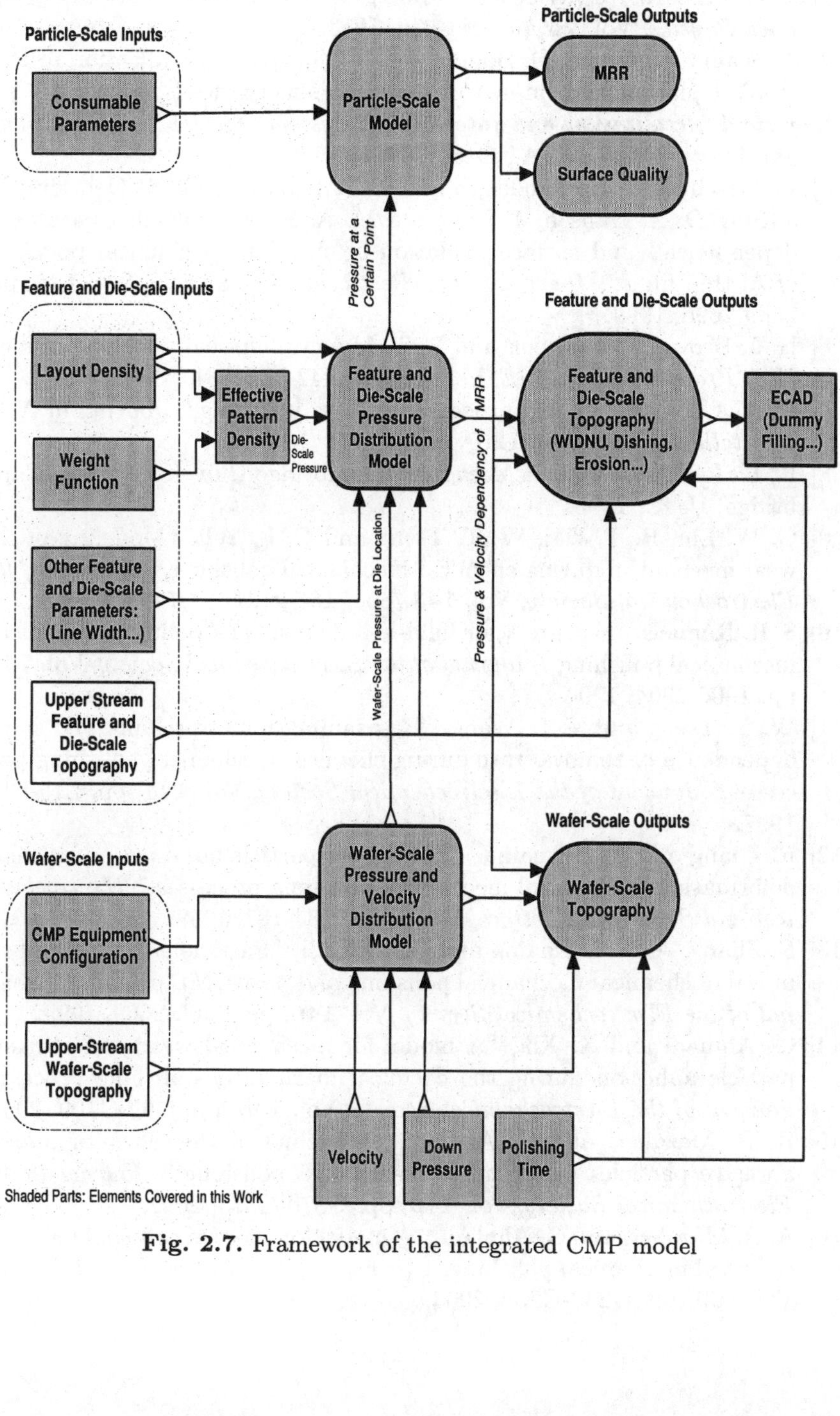

Fig. 2.7. Framework of the integrated CMP model

[2.2] F. Preston, "The theory and design of plate glass polishing machines," *Journal of the Society of Glass Technology*, Vol. **11**, pp. 214-256, 1927.

[2.3] J. F. Archard, "Contact and rubbing of flat surfaces," *Journal of Applied Physics*, Vol. **24**, pp. 981-985, 1953.

[2.4] A. Maury, D. Ouma, D. Boning and J. Chung, "A modification to Preston's equation and impact on pattern density effect modeling," *Advanced Metalization and Interconnect Systems for ULSI Applications*, San Diego, CA, U. S. A., Oct. 1997.

[2.5] P. Wrschka, J. Hernandez, Y. Hsu, T. S. Kuan, G. S. Oehrlein, H. J. Sun, D. A. Hansen, J. King and M. A. Fury, "Polishing parameter dependencies and surface oxidation of chemical mechanical polishing of Al thin films," *Journal of the Electrochemical Society*, Vol. **146**, pp. 2689-2696, 1999.

[2.6] N. J. Brown, P. C. Baker and R. T. Maney, "Optical polishing of metals," *Proceedings of SPIE*, Vol. **306**, pp. 42-57, 1981.

[2.7] L. M. Cook, "Chemical processes in glass polishing," *Journal of Non-Crystalline Solids*, Vol. **120**, pp. 152-171, 1990.

[2.8] K. L. Johnson, *Contact Mechanics*, Cambridge University Press, Cambridge, U. K., 1985.

[2.9] C. W. Liu, B. T. Dai, W. T. Tseng and C. F. Yeh, "Modeling of the wear mechanism during chemical-mechanical polishing," *Journal of the Electrochemical Society*, Vol. **143**, pp. 716-721, Feb., 1996.

[2.10] S. R. Runnels, "Feature-scale fluid-based erosion modeling for chemical-mechanical polishing," *Journal of the Electrochemical Society*, Vol. **141**, pp. 1900-1904, 1994.

[2.11] W. T. Tseng and Y. L. Wang, "Re-examination of pressure and speed dependence of removal rate during chemical mechanical polishing processes," *Journal of the Electrochemical Society*, Vol. **144**, pp. L15-L17, 1997.

[2.12] F. Zhang and A. Busnaina, "The role of particle adhesion and surface deformation in chemical mechanical polishing processes," *Electrochemical and Solid-State Letters*, Vol. **1**, pp. 184-187, 1998.

[2.13] F. Zhang, A. A. Busnaina and G. Ahmadi, "Particle adhesion and removal in chemical mechanical polishing and post-CMP cleaning," *Journal of the Electrochemical Society*, Vol. **146**, pp. 2665-2669, 1999.

[2.14] G. Ahmadi and X. Xia, "A model for mechanical wear and abrasive particle adhesion during the chemical mechanical polishing process," *Journal of the Electrochemical Society*, Vol. **148**, pp. G99-G109, 2001.

[2.15] A. R. Mazaheri and G. Ahmadi, "Modeling of the effect of bumpy abrasive particles on chemical mechanical polishing," *Journal of the Electrochemical Society*, Vol. **149**, pp. G370-G375, 2002.

[2.16] A. R. Mazaheri and G. Ahmadi, "A model for effect of colloidal forces on chemical mechanical polishing," *Journal of the Electrochemical Society*, Vol. **150**, pp. G233-G239, 2003.

[2.17] Y-T Su, "Investigation of removal rate properties of a floating polishing process," *Journal of the Electrochemical Society*, Vol. **147**, pp. 2290-2296, 2000.

[2.18] Y. Moon, "Mechanical aspects of the material removal mechanism in chemical mechanical polishing (CMP)," *Ph.D. Dissertation, Department of Mechanical Engineering, University of California at Berkeley*, Berkeley, CA, U. S. A., 1999.

[2.19] J. Z. Zhang, S. K. Huang, W. S. Toh, W. S. Tay, F. Chen and B-B. Zhou, "Optimization of pad conditioning for stable oxide CMP process," *1997 CMP-MIC Conference*, pp. 315-321, Santa Clara, CA, U. S. A., Feb. 1997.

[2.20] T. Yu, C. Yu and M. Orlowski, "A statistical polishing pad model for chemical-mechanical polishing," *IEEE International Electron Devices Meetings*, pp. 865-868, Dec. 1993.

[2.21] J. M. Steigerwald, S. P. Murarka and R. J. Gutmann, *Chemical Mechanical Planarization of Microelectronic Materials*, John Wiley & Sons., New York, U. S. A., 1997.

[2.22] B. Zhao and F. G. Shi, "Chemical mechanical polishing in IC processes: new fundamental insights," *1999 CMP-MIC Conference*, pp. 13-22, Santa Clara, CA, U. S. A., Feb. 1999.

[2.23] J. Luo and D. A. Dornfeld, "Material removal mechanism in chemical mechanical polishing: theory and modeling," *IEEE Transaction: Semiconductor Manufacturing*, Vol. **14**, pp. 112-133, 2001.

[2.24] J. Luo and D. A. Dornfeld, "Effects of abrasive size distribution in chemical-mechanical planarization: modeling and verification," *IEEE Transaction: Semiconductor Manufacturing*, Vol. **16**, pp. 469-476, 2003.

[2.25] J. Luo and D. A. Dornfeld, "Material removal regions in chemical mechanical planarization for sub-micron integrated circuit fabrication: coupling effects of slurry chemicals, abrasive size distribution, and wafer-pad contact area," *IEEE Transaction: Semiconductor Manufacturing*, Vol. **16**, pp. 45-56, 2003.

[2.26] G. Fu, A. Chandra, S. Guha and G. Subhash, "A plasticity-based model of material removal in chemical-mechanical polishing (CMP)," *IEEE Transaction: Semiconductor Manufacturing*, Vol. **14**, pp. 406-417, 2001.

[2.27] K. Osseo-Asare, "Surface chemical processes in chemical mechanical polishing: relationship between silica material removal rate and the point of zero charge of the abrasive material," *Journal of the Electrochemical Society*, Vol. **149**, pp. G651-G655, 2002.

[2.28] F. B. Kaufman, D. B. Thompson, R. E. Broadie, M. A. Jaso, W. L. Guthrie, D. J. Pearson and M. B. Small, "Chemical-mechanical polishing for fabricating patterned W metal features as chip interconnects," *Journal of the Electrochemical Society*, Vol. **138**, pp. 3460-3465, 1991.

[2.29] E. Paul, "A model of chemical mechanical polishing," *Journal of the Electrochemical Society,* Vol. **148**, pp. G355-G358, 2001.

[2.30] Y. Zhao, L. Chang and S. H. Kim, "A mathematical model for chemical mechanical polishing based on formation and removal of weakly bonded molecular species," *Wear*, Vol. **254**, pp. 332-339, 2003.

[2.31] C. L. Borst, D. G. Thakurta, W. N. Gill and R. J. Gutmann, "Surface kinetics model for SiLK chemical mechanical polishing," *Journal of the Electrochemical Society*, Vol. **149**, pp. G118-G127, 2002.

[2.32] D. Castillo-Mejia, S. Gold, V. Burrows and S. Beaudoin, "The effect of interactions between water and polishing pads on chemical mechanical polishing removal rates," *Journal of the Electrochemical Society*, Vol. **150**, pp. G76-G82, 2003.

[2.33] Y. Xie and B. Bhushan, "Effects of particle size, polishing pad and contact pressure in free abrasive polishing," *Wear*, Vol. **200**, pp. 281-295, 1996.

[2.34] Y. Zhao and L. Chang "A micro-contact and wear model for chemical-mechanical polishing of silicon wafers," *Wear*, Vol. **252**, pp. 220-226, 2002.

[2.35] A. Bastawros, A. Chandra, Y. J. Guo and B. Yan, "Pad effects on material-removal rate in chemical-mechanical planarization," *Journal of Electronic Materials*, Vol. **31**, pp. 1022-1031, 2002.

[2.36] B. Stine, D. Ouma, R. Divecha, D. Boning and J. Chung, "Rapid characterization and modeling of pattern dependent variation in chemical mechanical polishing," *IEEE Transaction: Semiconductor Manufacturing,* Vol. **11**, pp. 129-140, 1998.

[2.37] B. Stine, D. Ouma, R. Divecha, D. Boning, J. Chung, D. L. Hetherington, I. Ali, G. Shinn, J. Clark, O. S. Nakagawa and S. Y. Oh, "A closed-form analytic model for ILD thickness variation in CMP processes," *1997 CMP-MIC Conference*, pp. 266-273, Santa Clara, CA, U. S. A., Feb. 1997.

[2.38] D. O. Ouma, D. S. Boning, J. E. Chung, W. G. Easter, V. Saxena, S. Misra, and A. Crevasse, "Characterization and modeling of oxide chemical mechanical polishing using planarization length and pattern density concepts," *IEEE Transaction: Semiconductor Manufacturing*, Vol. **15**, pp. 232-244, 2002.

[2.39] E. Tseng, C. Yi and H. C. Chen, "A mechanical model for DRAM dielectric chemical-mechanical polishing process," *1997 CMP-MIC Conference*, pp. 258-265, Santa Clara, CA, U. S. A., Feb. 1997.

[2.40] T. H. Smith, S. J. Fang, D. S. Boning, G. B. Shinn and J. A. Stefani, "A CMP model combining density and time dependencies," *1999 CMP-MIC Conference*, pp. 97-104, Santa Clara, CA, U. S. A., Feb. 1999.

[2.41] J. Grillaert, M. Meuris, E. Vrancken, N. Heylen, K. Devriendt, W. Fyen and M. Heyns, "The use of a semi empirical CMP model for the optimization of the STI module," *1999 CMP-MIC Conference*, pp. 105-112, Santa Clara, CA, U. S. A., Feb. 1999.

[2.42] E. Tseng, M. Meng and S. C. Peng, "Modeling and discussions for STI CMP process," *1999 CMP-MIC Conference*, pp. 113-119, Santa Clara, CA, U. S. A., Feb. 1999.

[2.43] T. Tugbawa, T. Park, D. Boning, T. Pan, P. Li, S. Hymes, T. Brown and L. Camilletti, "A mathematical model of pattern dependencies in Cu CMP processes," *CMP Symposium, Electrochemical Society Proceedings*, Vol. **99-37**, pp. 605-615, Honolulu, HA, U. S. A., Oct. 1999.

[2.44] C. Ouyang, K. Ryu, L. Milor, W. Maly, G. Hill and Y-K Peng, "An analytical model of multiple ILD thickness variation induced by interaction of layout pattern and CMP process," *IEEE Transaction: Semiconductor Manufacturing*, Vol. **13**, pp. 286-292, 2000.

[2.45] O. G. Chekina, L. M. Keer and H. Liang, "Wear-contact problems and modeling of chemical mechanical polishing," *Journal of the Electrochemical Society*, Vol. **145**, pp. 2100-2106, 1998.

[2.46] O. S. Nakagawa, K. Rahmat, N. Chang, S-Y Oh, P. Nikkel and D. Crook, "Impact of CMP ILD thickness variation on interconnect capacitance and circuit performance," *1997 CMP-MIC Conference*, pp. 251-257, Santa Clara, CA, U. S. A., Feb. 1997.

[2.47] A. Kahng, G. Robins, A. Singh and A. Zelikovsky, "Filling algorithms and analyzes for layout density control," *IEEE Transaction: Computer Aided Design of Integrated Circuits and Systems*, Vol. **18**, pp. 445-462, 1999.

[2.48] R. Tian, D. F. Wong and R. Boone, "Model-based dummy feature placement for oxide chemical-mechanical polishing manufacturability," *IEEE Transaction: Computer-Aided Design of Integrated Circuits and Systems*, Vol. **20**, pp. 902-910, 2001.

[2.49] Y. Chen, A. B. Kahng, G. Robins, and A. Zelikovsky, "Area fill synthesis for uniform layout density," *IEEE Transaction: Computer-Aided Design of Integrated Circuits and Systems,* Vol. **21**, pp. 1132-1147, 2002.

[2.50] H. Hocheng, H-Y. Tsai and L-J. Chen, "A kinematic analysis of CMP based on velocity model," *1997 CMP-MIC Conference,* pp. 277-280, Santa Clara, CA, U. S. A., Feb. 1997.

[2.51] D. Wang D, J. Lee, K. Holland,T. Bibby, Beaudoin S and T. Cale, "Von Mises stress in chemical-mechanical polishing processes," *Journal of the Electrochemical Society*, Vol. **144**, pp. 1121-1127, 1997.

[2.52] W-T. Tseng, Y-H. Wang and J-H. Chin, "Effects of film stress on the chemical mechanical polishing process," *Journal of the Electrochemical Society*, Vol. **146**, pp. 4273-4280, 1999.

[2.53] G. Fu and A. Chandra, "A model for wafer scale variation of material removal rate in chemical mechanical polishing based on viscoelastic pad deformation," *Journal of Electronic Materials*, Vol. **31**, pp. 1066-1073, 2002.

[2.54] G. Fu and A. Chandra, "A model for wafer scale variation of removal rate in chemical mechanical polishing based on elastic pad deformation," *Journal of Electronic Materials*, Vol. **30**, pp. 400-408, 2001.

[2.55] J. Tichy, J. A. Levert, L. Shan and S. Danyluk, "Contact mechanics and lubrication hydrodynamics of chemical mechanical polishing," *Journal of the Electrochemical Society*, Vol. **146**, pp. 1523-1528, 1999.

[2.56] D. G. Thakurta, C. L. Borst, D. W. Schwendeman, R. J. Gutmann and W. N. Gill, "Pad porosity, compressibility and slurry delivery effects in chemical-mechanical planarization: modeling and experiments," *Thin Solid Films*, Vol. **366**, pp. 181-190, 2000.

[2.57] D. G. Thakurta, D. W. Schwendeman, R. J. Gutmann, S. Shankar, L. Jiang and W. N. Gill, "Three-dimensional wafer-scale copper chemical-mechanical planarization model," *Thin Solid Films*, Vol. **414**, pp. 78-90, 2002.

[2.58] D. G. Thakurta, C. L. Borst, D. W. Schwendeman, R. J. Gutmann and W. N. Gill, "Three-dimensional chemical mechanical planarization slurry flow model based on lubrication theory," *Journal of the Electrochemical Society*, Vol. **148**, pp. G207-G214, 2001.

[2.59] C. H. Cho, S. S. Park and Y. Ahn, "Three-dimensional wafer scale hydrodynamic modeling for chemical mechanical polishing," *Thin Solid Films*, Vol. **389**, pp. 254-260, 2001.

[2.60] J. Seok, C. P. Sukam, A. T. Kim and J. A. Tichy and T. S. Cale, "Multiscale material removal modeling of chemical mechanical polishing," *Wear*, Vol. **254**, pp. 307-320, 2003.

[2.61] J. Luo and D. A. Dornfeld, "Optimization of CMP from the viewpoint of consumable effects," *Journal of the Electrochemical Society*, Vol. **150**, pp. G807-G815, 2003.

[2.62] T. Tugbawa, T. Park and D. Boning, "Integrated chip-scale simulation of pattern dependencies in copper electroplating and copper chemical mechanical polishing processes," *International Interconnect Technology Conference,* San Francisco, CA, U. S. A., June 2002.

[2.63] J. F. Luo, Y. J. Liu and E. J. Berger, "Analysis of two-dimensional thin structures (from micro- to nano-scales) using the boundary element method," *Computational Mechanics*, Vol. **22**, pp. 404-412, 1998.

[2.64] J. F. Luo, Y. J. Liu and E. J. Berger, "Interfacial stress analysis for multi-coating systems using an advanced boundary element method," *Computational Mechanics* , Vol. **24**, pp. 448- 455, 2000.

[2.65] Y. J. Liu, N. Xu and J. F. Luo, "Modeling of interphases in fiber-reinforced composites under transverse loading using the boundary element method," *ASME Transaction: Journal of Applied Mechanics*, Vol. **67**, pp. 41-49, 2000.

[2.66] A. T. Kim, J. Seok, J. A. Tichy and T. S. Cale, "A multiscale elasto-hydrodynamic contact model for CMP," *Journal of the Electrochemical Society*, Vol. **150**, pp. G570-G576, 2003.

[2.67] T. Yoshida, "Three-dimensional chemical-mechanical polishing process model by BEM," *Electrochemical Society Proceedings*, Vol. **99-37**, pp. 593-604, 1999.

[2.68] T. Yoshida, "Three-dimensional wafer process model for nanotopography," *Material Research Society Symposium*, Vol. **767**, 2003.

[2.69] C-H. Yao, D. L. Feke, K. M. Robinson and S. Meikle, "The influence of feature-scale surface geometry on CMP processes," *Journal of the Electrochemical Society*, Vol. **147**, pp. 3094-3099, 2000.

[2.70] R. Tian, X. Tang and M. D. F. Wong, "Dummy-feature placement for chemical-mechanical polishing uniformity in a shallow-trench isolation process," *IEEE Transactions: Computer-Aided Design of Integrated Circuits and Systems*, Vol. **21**, pp. 63-71, 2002.

[2.71] E. Chagarov and J. B. Adams, "Molecular dynamics simulations of mechanical deformation of amorphous silicon dioxide during chemical mechanical polishing," *Journal of Applied Physics*, Vol. **94**, pp. 3853-3861, 2003.

[2.72] Y. Ye, R. Biswas, A. Bastawros and A. Chandra, "Simulation of chemical mechanical planarization of copper with molecular dynamics," *Applied Physics Letters*, Vol. **81**, pp. 1875-1877, 2002.

[2.73] Y. Y. Ye, R. Biswas, J. R. Morris, A. Bastawros and A. Chandra, "Molecular dynamics simulation of nanoscale machining of copper," *Nanotechnology*, Vol. **14**, pp. 390-396, 2003.

[2.74] P. H. Chen, H. C. Shih, B-W. Huang and J-W. Hsu, "Catalytic-pad chemical kinetics model of CMP," *Electrochemical and Solid-State Letters*, Vol. **6**, pp. G140-G142, 2003.

[2.75] B. Lee, "Modeling of chemical mechanical polishing for shallow trench isolation," *Ph.D. Dissertation, Department of Electrical Engineering and Computer Science, MIT*, Cambridge, MA, U. S. A., 2002.

[2.76] Y-R. Jeng and H-J. Tsai, "Improved model of wafer/pad powder slurry for CMP," *Journal of the Electrochemical Society*, Vol. **150**, pp. G348-G354, 2003.

[2.77] Y-R. Jeng, P-Y. Huang and W-C. Pan, "Tribological analysis of CMP with partial asperity contact," *Journal of the Electrochemical Society*, Vol. **150**, pp. G630-G637, 2003.

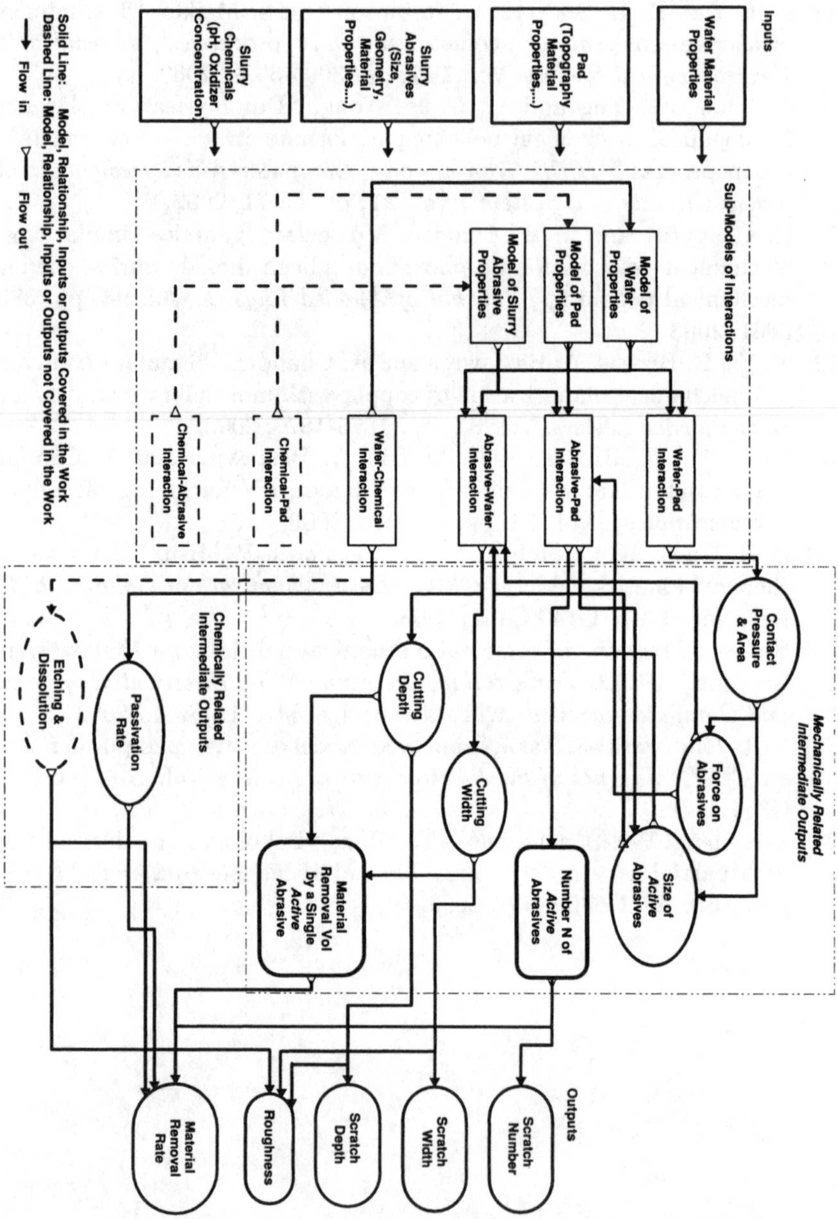

Fig. 2.A. Detailed framework of the comprehensive particle-scale material removal model

3

MATERIAL REMOVAL MECHANISM IN CMP: A COMPREHENSIVE MODEL OF ABRASIVE PARTICLE, PAD ASPERITY AND WAFER INTERACTIONS

3.1 Introduction

The CMP process has been widely accepted in the semiconductor industry for oxide dielectric and metal layer planarization [3.1]. It is used to insure that the interconnects between multi-layer chips are achieved reliably and that the thickness of dielectric materials is uniform and sufficient. During CMP processes, a wafer is rotated about its axis while being pressed face-down by a carrier and a carrier film against a polishing pad/belt covered with colloidal silica slurry with specific chemical properties. The abrasive slurry, composed of, for example, 50-70 nanometer fused silica in an aqueous solution with pH between 8.5-11, plays an important role in the material removal mechanism in CMP. Microscopic observations of polished surfaces have shown that material removal in CMP occurs as a consequence of a combination of chemical reaction of the slurry chemicals with the wafer surface materials and the repeated sliding, rolling or indentation of the abrasive particles against the wafer surface [3.2]. Most analyses of the mechanism of CMP treat the chemical effects and mechanical effects of the slurry separately. In this chapter, we focus on studying the mechanical effect.

The detailed mechanical interaction between the wafer, pad and slurry has been the subject of research for some time [3.3-3.9]. Most of the earlier models were at the macro-scale and concentrated on significant input variables to the CMP process such as slurry flow rate, pressure, velocity, friction force, lubrication, pad and wafer geometry, and output variables including polish rate, planarization rate, polish rate uniformity and surface quality. The abrasive and abrasion mechanism in CMP need further investigation. The abrasion/wear-based Preston's equation $MRR= K_e PV$ [3.3] has been widely used in CMP process control and consumable development for IC fabrication and manufacturing. However, Preston's equation only reflects the influence of process parameters including down pressure and relative velocity. Experimental results have shown that in CMP more factors, especially the consumables such as the polishing pad made of polymer materials and slurry composed of nanoscale abrasives, have large influences on the material removal rate [3.10-3.13]. For example, it is found that a softer and rougher polishing pad will yield a larger material removal. This cannot be explained by Preston's equation. Furthermore, experimental results show that the pressure dependence of the removal rate for CMP with soft pads satisfies a nonlinear relationship [3.14-3.15]. Although a roughly linear curve with a non-zero initial value at zero pressure can be fitted for some nonlinear experimental data, leading to a linear revised Preston's equation $MRR{=}K_ePV{+}\ MRR_0$, no physical or experimental evidence supports the initial material removal rate, MRR_0. Actually, it is found that two separate linear curves must be used to fit the experimental data for MRR under higher pressures and MRR under lower pressures, if the non-linearity of the experimental data is apparent [3.16].

In the last several years, the observed discrepancy between Preston's equation and experimental observations in CMP has stimulated efforts to modify Preston's equation [3.17-3.21]. The revised Preston's equations concentrated on different elements of CMP. For example, Zhang and Busnaina [3.20] proposed an equation $MRR{=}\ K_e\sqrt{PV}$ taking into account the normal stress and shear stress acting on the contact area between abrasive particles and wafer surfaces. However, most of the models are quite rough, and, similar to Preston's equation, the only process parameters included are pressure and velocity. An all-purpose parameter, K_e, representing both chemical effects and consumable effects has been introduced. Although some researchers such as Liu [3.21] took the deformation over the wafer-abrasive interface into account and developed a more complicated model reflecting the influence of abrasive and wafer hardness, the interactions between the abrasive and pad and those between wafer and pad are neglected. Although the models may agree well with some experimental results, they are not very robust.

The latest work to consider the effects of the pad hardness and the contact between wafer and pad is by Zhao and Shi [3.17-3.18]. In their work the nonlinear pressure dependence of the material removal rate is shown due to the nonlinear pressure dependence of contact area between wafer and pad. They argue that the number of abrasives involved in material removal will increase with the contact area and, since the MRR is linearly related to the number of abrasives, the material removal rate will be nonlinearly dependent on the pressure. They proposed an equation $MRR{=}\ K_eP^{2/3}V$ to describe this. Many important factors are not considered in this model. For example, the model neglects the decrease of the force applied on abrasives with the increase of contact area, which may lead to a smaller *MRR*. Based on this view of the relationship between contact area and number of abrasives involved in material removal, the observation that *MRR* decreases when using a smooth pad, which contacts the wafer with a larger area, cannot be explained.

Zhao and Shi [3.17] recognized the limitation of the model and introduced a threshold pressure P_{th}, arguing that only when the down pressure is larger than the threshold pressure will material removal occur. A revised equation, $MRR = K_e V(P^{2/3} - P_{th}^{2/3})$, was proposed to include this. Experimental results agree well with the equation. So far, no physical or experimental results support the proposed threshold pressure [3.43]. In addition, many factors related to the consumables such as pad hardness and distribution of abrasive size are not considered in this model. Therefore what is included in the all-purpose parameter K_e is unclear.

Given the rapid integration and scaling up of CMP processes in IC manufacturing and the increasingly strict process requirements in manufacturing current and future generation IC chips, a more detailed and accurate model to understand the fundamental mechanism involved in the CMP processes is in need.

CMP is quite different from the conventional lapping and polishing processes. A detailed analysis shows that the linear Preston's equation is only applicable to polishing with a pad whose hardness is similar to or harder than that of the abrasive particles and the polished wafer [3.17]. In CMP, polishing pads are typically made of polyurethane polymers which are much softer than the polishing abrasives and wafer surface. The wafer and polishing pad contact each other directly in CMP while they are separated by abrasives in the conventional hard pad polishing processes. Therefore, the Young's modulus of the pad, hardness of the pad and the contact area between wafer and pad have large influences on the material removal. Moreover, the size of the nano-scale polishing abrasives used in CMP is much smaller than the micro-scale height of the pad asperities. Thus, the pad topography and the distribution of the abrasive size play a significant role in CMP.

In this chapter, we propose a detailed model to describe the interactions among the wafer, pad and abrasives considering the above unique properties of CMP. This model is based on assumptions of plastic contact over the wafer-abrasive interface and pad-abrasive interface, a normal distribution of abrasive size and a periodic rough surface of the polishing pad. *MRR* prediction formulations and the theoretical basis and analysis on which they are based are presented. The chemical effect and mechanical effect are separated in the formulations while a number of important input parameters and their influences on material removal are integrated into the model.

3.2 Solid-Solid Contact Mode in CMP

Before further discussions, let us review the two typical contact modes in CMP, namely the hydro-dynamical contact mode and the solid-solid contact mode. Figs. 3.1 (a) and (b) show the two contact modes schematically. As shown in Fig. 3.1 (a), when the down pressure applied on the wafer surface is small and the relative velocity of the wafer is large, a thin fluid film (polishing slurry) with *micro-scale* thickness [3.11] will be formed between the wafer and pad surface. A typical property of CMP is that the size of the abrasive particles (*nano-scale*) are much smaller than the thickness of the slurry film (*micro-scale*), and therefore a lot of abrasive particles are inactive even if their sizes are relative large compared with the average size of all particles. Almost all material removal is due to three-body abrasion by some of the abrasives floating in the slurry and the chemical etching by slurry chemicals, Fig. 3.1 (a).

The accurate prediction of the number of abrasive particles involved in abrasion is important in investigating the material removal rate. However, in the hydro-dynamical contact mode it is almost impossible to predict how many particles are engaged in three-body abrasion or inactive. The complex mechanics and dynamics associated with three-body abrasion and the effect of slurry flows present additional difficulties for modeling of CMP material removal under the hydro-dynamical contact mode [3.42].

In this chapter, only the material removal mechanism in the solid-solid contact mode, when the down pressure applied on the wafer surface is large and the relative velocity of the wafer is small, is investigated. In this mode, the wafer and pad asperity contact each other and both two-body abrasion and three-body abrasion occur, Fig. 3.1 (b). In two-body abrasion the abrasive particles are embedded in the pad asperities (i.e., sliding without rotation) and almost all effective material removals happen due to these abrasions. The number of abrasives involved in two-body abrasion, based on the distribution of the abrasive sizes and the topography of the pad surface can be determined and this will be shown in detail in section 3.3.

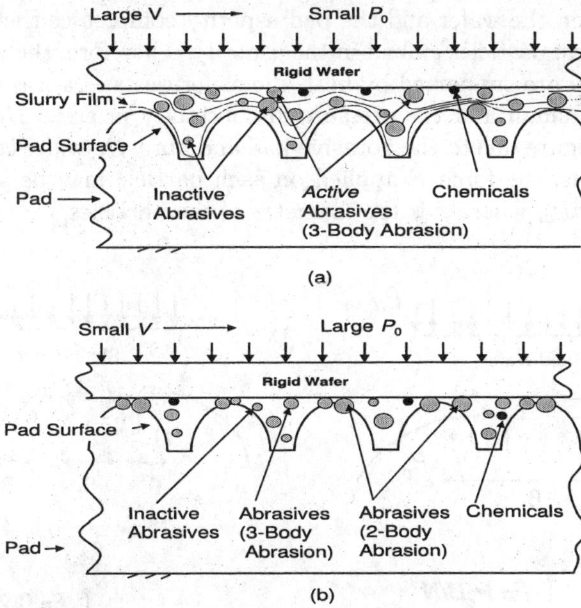

Fig. 3.1. Two contact modes of CMP (a) hydro-dynamical contact mode and (b) solid-solid contact mode

Although the mechanism of two-body abrasion in the solid-solid contact mode is relatively clear, there are still two fundamental differences between the chemical mechanical polishing process, where the polishing pad is soft, and other polishing processes, where the polishing pad is hard. These differences are related to how the abrasive particles are held against the wafer surface and how the force is applied to the pad asperity and the abrasive particles. Figure 3.2 (a) and (b) show schematically the contact between a wafer, abrasive particles, and an asperity of a polishing pad for a polishing process with a hard pad and a soft pad, respectively. It is assumed that there are N active abrasives between the wafer surface and the pad asperity. As shown in the figure, for the hard pad polishing, the force applied on the asperity is supported totally by the N *active* abrasive particles. The mean force on each particle is $F = P_c B/N$, where P_c is the contact pressure and B the contact area of the asperity. This force is independent of the abrasive size. The B/N is the *effective* asperity area occupied by a single abrasive. Since not all the area of the asperity is occupied by abrasives, this effective area is larger than the abrasive size. Note here that the pressure P_c on the asperity is dependent on the *real* contact area between wafer and pad and not equal to the down pressure P_0 applied on the wafer top surface. The influence of pad surface on pressure P_c will be discussed in section 3.3. In Fig. 3.2 (b), however, the contact between the wafer and the pad asperity compresses the pad asperity and the abrasive particles embed in the asperity. Therefore, the forces applied on the asperity are supported by both the asperities and abrasives. The particles are very small and the pressure discontinuity in areas B_1, B_2 and B_3 shown in the figure due to the abrasive-pad and abrasive-wafer contact can be neglected. Thus, the force F applied on each particle may be approximated as $F = 0.25\pi x^2 P_c$, where x is the diameter of the particles.

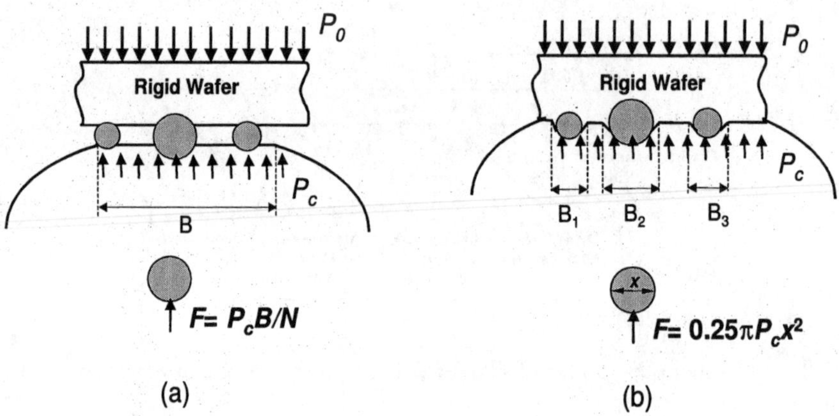

Fig. 3.2. Schematic of the solid-solid contact between the wafer, abrasive particles and a pad asperity of (a) a hard polishing pad and (b) a soft polishing pad

Obviously, in CMP with a soft polishing pad the force applied on the abrasives is smaller than that in polishing with a hard pad because the abrasive size $0.25\pi x^2$ is usually smaller than the *effective* area B/N occupied by a single abrasive. Moreover, in polishing with a hard pad, the force is dependent on the number of active particles in the contact area. In CMP with a soft pad, however, the force is only determined by the pressure and abrasive size and independent of the abrasive number.

Another fundamental difference between CMP and other polishing processes in solid-solid contact mode is that in CMP most material removal is by abrasives on the pad asperities that directly contact the wafer. As mentioned before, only parts of the wafer-pad interface (*real* contact surface) are in direct contact due to the roughness or waviness of the pad surface, Figs. 3.3 (a) and (b). In conventional polishing processes, the abrasive size is on the same order (*micro-scale*) as the asperity of the polishing pad. So, even abrasives embedded in the lower areas of the pad, which do not contact the wafer, can be involved in material removal and the abrasion due to such abrasives are two-body, Fig. 3.3 (a). However, in CMP because the abrasive size (*nano-scale*) is much smaller than the scale of the pad roughness, the abrasives in the area without wafer-pad direct-contact will not embed in the pad, Fig. 3.3 (b), and therefore they are either inactive or act in three-body abrasion. The material removed by three-body abrasion is negligible compared with the two-body abrasion happening in the wafer-pad contact area.

In this chapter, all material removal is considered due to the two-body abrasion. However, it is noteworthy that not all abrasives on the contact area are involved in material removal either in CMP or other polishing processes [3.22-3.24]. The number of *active* abrasives involved in material removal in the contact areas is determined by the distribution of abrasive size and the hardness of the wafer and pad. Details on the calculation of the number of *active* abrasives in CMP will be presented in section 3.3 and Chapter 4. In addition, the *active* abrasives are actually not always embedded in the polishing pad but abrade the pad also. Thus, while the action of the active abrasives is one of two-body abrasion, they are not always fixed but sometimes move relative to the polishing pad.

3.3 Assumptions for Model Development

Figure 3.4 shows an abrasive involved in material removal. The wafer surface and the pad asperity are separated by a distance g, which depends on the force applied on the abrasive, F, the hardness of wafer, H_w and the hardness of pad, H_p. The penetration depth of the abrasive into the wafer and pad surfaces is not only related to the material removal by the abrasive but also the likelihood of other abrasives on the asperity to being involved in the material removal process. The pressure P_c applied on the abrasive, the penetration depth, the volume removed by a single particle and the number of *active* abrasives are the four most important parameters in determining material removal in our model. In the following, we discuss how to evaluate each of them. Correspondingly, assumptions about the roughness of pad surface, distribution of abrasive size, geometry of the abrasive and deformation mode on the wafer-abrasive and pad-abrasive surface will be stated.

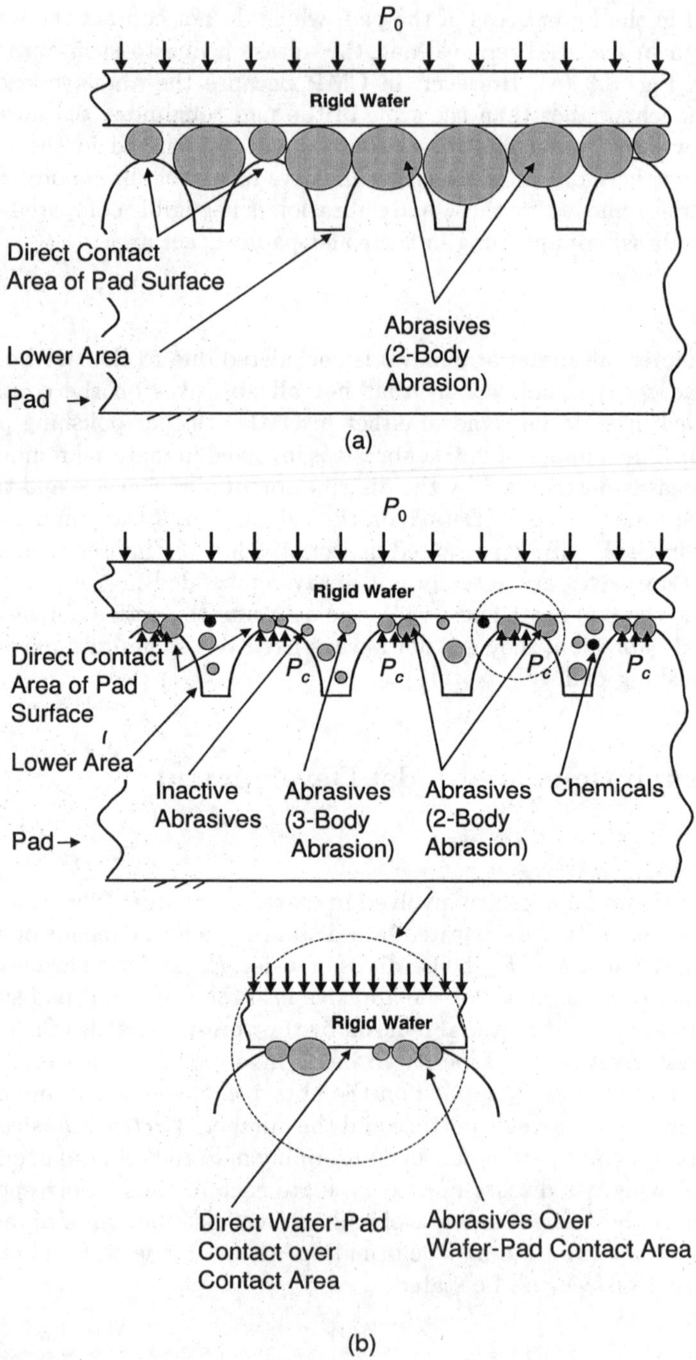

Fig. 3.3. Schematic of the solid-solid contact in a polishing process with (a) micro-scale abrasives and (b) nano-scale abrasives

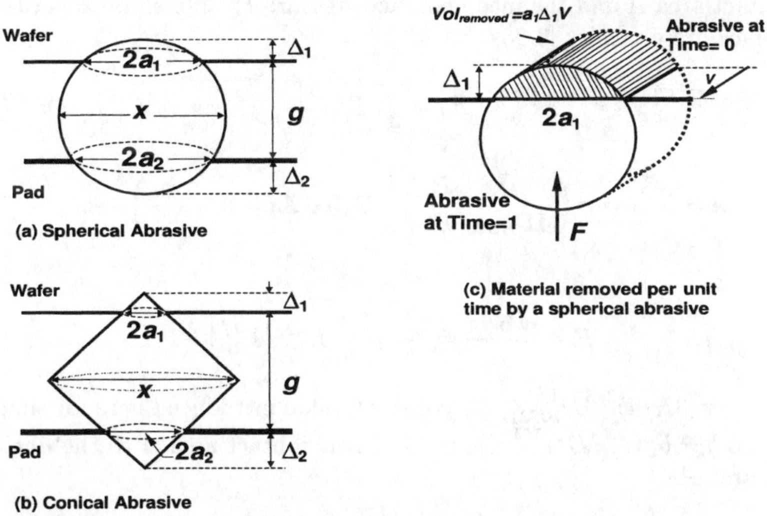

Fig. 3.4. Model of the plastic contact between an abrasive particle, wafer and pad

3.3.1 Periodic Roughness of Pad Surface

In solid-solid contact mode, we neglect the force applied on the wafer surface by the fluid flow and assume that the contact is purely solid-solid contact. Contact mechanics [3.25] shows that the contact between solid surfaces is discontinuous and the *real* area of contact is a small fraction of the *apparent* contact area, Fig. 3.3. The real area of contact is determined by the down pressure and the topography of the contact surfaces of the two solids. Greenwood and Williamson [3.26] developed one of the first models for the contact of two flat elastic planes, one of which is rough and the other of which is smooth. In their model, the rough surface is presumed to be covered with local high spots or asperities whose summits are all spherical in shape, Fig. 3.5. A uniform distribution of the summits over the rough surface with a known density D_{SUM} of summits per unit area is assumed. All the summits are presumed to have the same radius R and the asperity height are assumed to follow a normal distribution. In our model, the heights of asperities are assumed to follow a periodic distribution, meaning that the height of asperities are equal or close to each other so that all asperities contact the wafer and deform under down pressure. This assumption is reasonable if the standard deviation of the normal distribution is small or the elastic modulus of asperities is small. Based on this assumption, when the down pressure applied on the wafer top surface is P_0 and the diameter of wafer is D, the *apparent* contact area A_0, *real* contact area A and the mean contact pressure P_c applied on asperities of pad surface are given by [3.25]

$$A_0 = \frac{1}{4}\pi D^2 \tag{3.1}$$

$$A = bA_0 = \pi \left(\frac{3R}{4D_{SUM}}\frac{P_0}{E^*}\right)^{\frac{2}{3}} D_{SUM}A_0 = b_1 \left(\frac{P_0}{E^*}\right)^{\frac{2}{3}} A_0 \tag{3.2}$$

and

$$P_c = \frac{P_0 A_0}{A} = \frac{P_0}{b} = \frac{1}{b_1}E^{*2/3}P_0^{1/3} \tag{3.3}$$

where $b_1 = \pi(3R/4)^{2/3}D_{SUM}^{1/3}$ is a constant value introduced here for simplification and $b = b_1\,(P_0/E^*)^{2/3}$ the ratio of *real* contact area A to the *apparent* contact area A_0.

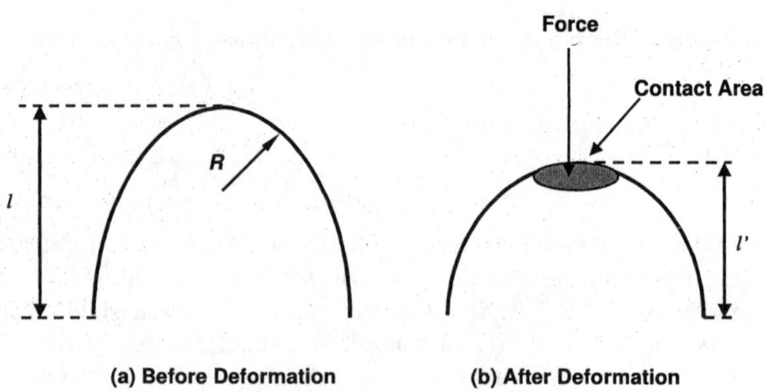

Fig. 3.5. Spherical capped asperity in contact

In the above equations, the equivalent modulus of elasticity, E^*, is equal to $1/[(1-\upsilon_w^2)/E_w + (1-\upsilon_p^2)/E_p]$ [3.25] where E_w and υ_w are the Young's modulus and Poisson's ratio of the wafer, respectively, and E_p and υ_p the Young's modulus and Poisson's ratio of the pad, respectively. Because the Young's modulus of the wafer E_w (for example, 2.9×10^6 psi (or 2×10^{10}Pa) for a silicon wafer) is much larger than the Young's modulus of the pad E_p (for example, 725psi (or 5×10^6Pa) for pad material such as IC 1000) and the

Poisson's ratio of pad v_p is close to 0.5 for polymer materials [3.27-3.28], the E^* is approximately equal to $\frac{4}{3}E_p$. In the following discussions, $\frac{4}{3}E_p$ instead of E^* is used. The mean pressure P_c on an asperity, which is dependent on the down pressure P_0, Young's modulus of pad E_p and topography of pad surface, as discussed in section 3.2, is approximately the pressure applied on the whole abrasive body when the polishing pad is soft. Therefore the force F applied on an abrasive will be $0.25\pi P_c x^2$=$0.25\pi(1/b_1)(4/3)^{2/3}E_p^{2/3}P_0^{1/3}x^2$ if the diameter of the abrasive is x. Note that the down pressure dependence of the force is nonlinear.

The influence of *real* contact area on CMP has been studied by Moon [3.11] and Zhao and Shi [3.17] in earlier work. Revised Preston's equations were developed by Zhao and Shi based on the consideration of the *real* contact area. However only qualitative discussions were presented and the nonlinear down pressure dependence of the contact pressure was not observed. This nonlinear dependence of the applied contact pressure P_c is shown to be important in determining the material removal rate in CMP in later sections.

3.3.2 Plastic Deformations Over Wafer-Particle and Pad-Particle Interfaces

Understanding the deformation mode over the particle-wafer interface is important in determining material removal. If the deformation is elastic (Hertizian elastic deformation), little or no abrasion occurs and the material removal should be mainly by chemical etching. Zhang et al. [3.29] have shown theoretically that the deformation over the particle-wafer interface is not purely elastic because the calculated Hertzian contact pressure is much larger than the yield stress of the wafer material. In this chapter, the deformation over the particle-wafer interface due to the sliding indentation of a half-space by a hard indenter is considered. This kind of deformation has been assumed by many investigators in modeling of grinding, lapping and polishing processes.

For CMP, some may argue that the machining of brittle materials, such as silicon, should be due mainly to micro-fracture. However, studies have shown that machining at very small depths of cut ($< 1\mu$m) often causes brittle materials to machine in a ductile manner [3.30]. Due to the small size of the abrasive, the cutting depth (0.1-1nm) in CMP is smaller than the nano-scale size of abrasive. Therefore, the assumption of plastic deformation over the particle-wafer interface is reasonable. A possible problem associated with the assumption is that because the cutting depth/roughness of the surface finish (0.1-1nm) is close to the size of several layers of atoms, the validation of idealized aspects of material behaviors, such as plasticity and indentation-sliding models for material removal, introduced in macroscopic mechanics of materials, may be questionable. Recently, a molecular dynamics (MD) simulation has been done to study the effect of tool geometry in nanometric cutting [3.41]. The cutting depths in the simulation range from 0.363 to 2.172nm. It is concluded from the simulation that when considering machining brittle materials with tools of large edge radius relative to the depth of cut, which is the case in CMP processes, the indentation-sliding model appears more appropriate than in macroscopic scale [3.41]. This implies the validation of the fully plastic deformation assumption over the wafer-abrasive interface and the indentation-sliding model to be used in this chapter.

The hardness of the pad is much less than that of the wafer and the abrasives are almost embedded statically in the pad. Thus, a fully developed plastic deformation (quasi-static indentation) can be assumed over the particle-pad interface.

Besides the mode of deformation, another factor influencing the magnitude of the deformation is the geometry of abrasives. The scanning electron microscope (SEM) photograph [3.11] [3.31], Fig. 3.6, shows that the geometry of most of the abrasives used in CMP is close to a sphere. In this chapter, two different geometries of abrasives, cone and sphere, are analyzed, Fig. 3.4, to see the influence of abrasive geometry on our model. It will be shown that the *MRR* predictions based on a spherical geometry agree well with the experimental *MRR* results while those based on conical abrasives do not agree. In the following discussion, for simplicity, all the assumptions are based on a spherical geometry. Similar processes can be used to determine the corresponding formulations for the conical geometry.

As discussed above, the contact between an abrasive particle and the wafer is considered to be a sliding indentation of a half-space by a hard indenter and the contact between polishing pad/pad asperity and abrasive particle is considered to be a quasi-static indentation of a half-space by a hard indenter. A plastic zone (with a semi projected-circle for sliding indentation and a full projected-circle for semi-static indentation) forms around this contact and the mean contact pressures applied by the abrasive on the wafer and pad surfaces

Fig. 3.6. SEM photographs of silica abrasives in the Nalco2352 slurry [3.31]

are equal to their hardness H_w and H_p respectively. Assume the force applied on the abrasive is F. Then the radii a_1 and a_2 of the projected circle of contact between the spherical indenter and the wafer and pad surfaces, Fig. 3.4, are given by

$$a_1 = \sqrt{2F/(\pi H_w)} \tag{3.4}$$

and

$$a_2 = \sqrt{F/(\pi H_p)}\ . \tag{3.5}$$

Since, typically, the penetration depths Δ_1 and Δ_2 of the spherical indenter into the wafer and pad surfaces are much smaller than the diameter of the abrasives x ($\Delta_1 << x$ and $\Delta_2 << x$), we can say $x\Delta_1 = a_1^2$ and $x\Delta_2 = a_2^2$. So the depths Δ_1 and Δ_2 are given by

$$\Delta_1 = \frac{2F}{\pi x H_w} \tag{3.6}$$

and

$$\Delta_2 = \frac{F}{\pi x H_p} \tag{3.7}$$

where F=$0.25\pi P_c x^2$.

The overall deformation of the two contact points is

$$\Delta = \Delta_1 + \Delta_2 = \frac{F}{\pi x}\left(\frac{2}{H_w} + \frac{1}{H_p}\right) = 0.25 P_c x \left(\frac{2}{H_w} + \frac{1}{H_p}\right) \tag{3.8}$$

From this equation, it is seen that the deformation is larger when the pad and wafer are softer.

The mean value of cutting depth Δ_1 in Eq. 3.6 is approximately equal to the final roughness R_a of the polished wafer. In CMP, R_a is around 0.1~1nm, which can be measured readily using an atomic force microscope (AFM). Once knowing the measured roughness of the polished wafer and the average sizes of *active* abrasives, we can estimate the pressure P_c applied on the polishing pad and the ratio b of contact area using Eq. 3.6 as follows:

$$P_c = \frac{\Delta_1 H_w}{0.5 x_{avg-a}} = \frac{R_a H_w}{0.5 x_{avg-a}} \tag{3.9}$$

and

$$b = \frac{P_0}{P_c} = \frac{0.5 P_0 x_{avg-a}}{R_a H_w}, \tag{3.10}$$

where x_{avg-a} is the average size of *active* abrasives. The above relationships are important. In the following discussion, they are used to estimate the ratio b of contact area, which is difficult to measure directly. Moreover, according to our model, there is $R_a \propto P_0^{1/3}$. Experimental results from Zhang et al. [3.29] support this relationship although they did not specifically note this. They also support our assumptions of $A \propto P_0^{2/3}$ and $P_c \propto P_0^{1/3}$ in the last section.

For a polishing process with down pressure P_0, relative velocity V (or sliding distance in unit time) and mean size x_{avg-a} of active abrasives, the mean volume removed by a single abrasive per unit time is, Fig. 3.4 [3.23-3.24],

$$\begin{aligned} \overline{Vol_{removed}} &= \Delta_1 a_1 V = \frac{2\sqrt{2}F}{\pi x_{avg-a} H_w} \sqrt{F/(\pi H_w)}V \\ &= \frac{\sqrt{2}}{4} x_{avg-a}^2 \left(\frac{P_c}{H_w}\right)^{\frac{3}{2}} V = \frac{\sqrt{2}}{3} \frac{x_{avg-a}^2 E_p}{(b_1 H_w)^{\frac{3}{2}}} \sqrt{P_0} V \end{aligned} \tag{3.11}$$

where $P_c = (1/b_1)\left(\frac{4}{3}E_p\right)^{2/3} P_0^{1/3}$ (see Eq. 3.3) is the pressure at the locations of abrasives. The mean size x_{avg-a} of *active* abrasive will change with the down pressure P_0. It will be discussed next how the pressure influences the average size. Note that the role of velocity in CMP should be more complicated than that shown in our model. Besides the material removed by a single abrasive in unit time, its influence should be mainly related to the fluid flow. The model here is concentrated on the effect of the pressure.

Obviously the volume removed by a single abrasive is dependent on abrasive size, wafer hardness, down pressure, Young's modulus of the polishing pad, topography of the polishing pad and the relative velocity. The Young's modulus of the pad, the topography of the polishing pad and down pressure influence the volume through the force applied on the abrasives. Wafer hardness and abrasive size influence the volume through the cutting depth.

3.3.3 Normal Distribution of Abrasive Particle Size

As discussed in section 3.2, due to the small size of abrasive particles, the particles which are not in the contact area will not be involved in two-body abrasion and the material removed by them is negligible. However, not all of the particles on the contact area will be involved in material removal.

Let us review the contact between the wafer and a pad asperity. At first it is assumed that there is no load applied on the wafer. The wafer and the asperity are separated by abrasives in fluid with a separating distance g, where g is equal to (or greater than) the diameter of largest abrasives packed between the wafer and the asperity. When a load is applied, the wafer and pad surface contact the largest abrasive particles first, and then in the neighborhood of the largest abrasive particles, an initial deformation Δ occurs and a separating distance $g = x_{max} - \Delta$ is introduced, where x_{max} is the diameter of the largest particles. (A "neighborhood" is referred to here because not all parts of the wafer-pad interface are separated by the gap. As discussed in section 3.2, most of the contact in CMP is direct contact between the wafer and pad asperity without abrasives present and, therefore, the F applied on abrasives could be approximated by pressure times the abrasive size.) More particles will be engaged in the removal process, as shown in Fig. 3.7, in the neighborhood of the initial deformation area, if their diameters are larger than the gap $g = x_{max} - \Delta$. Note that the increase in the number of particles involved in abrasion will not lead to a decrease of the initial gap since the force F applied on each particle is independent of particle numbers. The portion of particles involved in two-body abrasion is referred to as *active* abrasive particles. The number of *active* abrasives N increases with the decrease of the separating distance or the increase of the deformation Δ, Fig. 3.7.

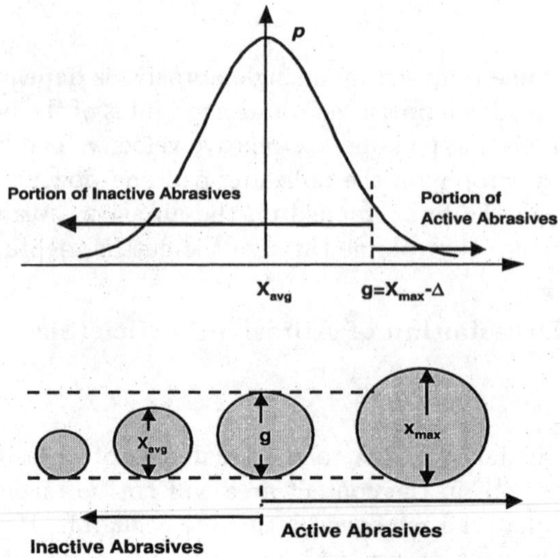

Fig. 3.7. Schematic showing the relationship between the deformation Δ in the neighborhood of large particles and the probability of *active* abrasive particles

In most situations the distribution of abrasive particles sizes (i.e. diameters), x, satisfies normal probability density functions as shown in earlier studies [3.22-3.24]

$$p\{x = x_a\} = p(x_a) = \frac{1}{\sqrt{2\pi}} e^{-\frac{1}{2}(\frac{x_a - x_{avg}}{\sigma})^2}$$

and

$$p\{x \leq x_a\} = \Phi(\frac{x_a - x_{avg}}{\sigma}) = \frac{1}{\sqrt{2\pi}} \int_{-\infty}^{\frac{x_a - x_{avg}}{\sigma}} e^{-\frac{1}{2}t^2} dt \qquad (3.12)$$

where x_{avg} is the mean abrasive diameter and σ the standard deviation. Then the number of the *active* particles N over the wafer-pad interface can be given by

$$\begin{aligned} N &= n\left[\Phi\left(\frac{x_{max} - x_{avg}}{\sigma}\right) - \Phi\left(\frac{x_{max} - \Delta - x_{avg}}{\sigma}\right)\right] \\ &= n\left\{\Phi\left(\frac{x_{max} - x_{avg}}{\sigma}\right) - \Phi\left[\frac{x_{max} - x_{avg} - \frac{F}{\pi x_{max}}\left(\frac{1}{H_p} + \frac{2}{H_w}\right)}{\sigma}\right]\right\} \end{aligned} \qquad (3.13)$$

where n is the number of all particles including *active* particles and *inactive* particles over the wafer-pad contact area. In most situations, $(x_{max} - x_{avg})/\sigma$

can be approximated as 3 and $\Phi[(x_{max} - x_{avg})/\sigma]$ is 1.0. The F in Eq. 3.13 should be approximately equal to the pressure P_c times the size x_{max} of the largest particles. So Eq. 3.13 can be further written as

$$N \approx n\left\{1-\Phi\left[3-\frac{0.25x_{max}\left(\frac{1}{H_p}+\frac{2}{H_w}\right)}{\sigma}P_c\right]\right\}$$
$$= n\left\{1-\Phi\left[3-\frac{0.25\left(\frac{4}{3}\right)^{\frac{2}{3}}(x_{avg}+3\sigma)\left(\frac{1}{H_p}+\frac{2}{H_w}\right)}{\sigma}\frac{E_p^{\frac{2}{3}}}{b_1}P_0^{\frac{1}{3}}\right]\right\} \tag{3.14}$$

Similarly, the average size x_{avg-a} of the *active* abrasive*Average size of active abrasive* can be deduced as

$$x_{avg-a} = x_{avg} + \frac{\sigma\, p\left(\frac{x_{max}-\Delta-x_{avg}}{\sigma}\right)}{1-\Phi\left(\frac{x_{max}-\Delta-x_{avg}}{\sigma}\right)}$$
$$= x_{avg} + \frac{\sigma p\left[3-\frac{0.25\times\left(\frac{4}{3}\right)^{\frac{2}{3}}(x_{avg}+3\sigma)\left(\frac{1}{H_p}+\frac{2}{H_w}\right)}{\sigma}\frac{E_p^{\frac{2}{3}}}{b_1}P_0^{\frac{1}{3}}\right]}{1-\Phi\left[3-\frac{0.25\times\left(\frac{4}{3}\right)^{\frac{2}{3}}(x_{avg}+3\sigma)\left(\frac{1}{H_p}+\frac{2}{H_w}\right)}{\sigma}\frac{E_p^{\frac{2}{3}}}{b_1}P_0^{\frac{1}{3}}\right]} \tag{3.15}$$

It can be seen that the average size of the *active* abrasives changes with the down pressure. However, it will be shown in later sections that this change is quite small when the standard deviation σ is small. Therefore the average size of active abrasives can be taken as constant, kx_{avg}. The gap Δ is usually much smaller than the abrasive size x, therefore the average size of *active* abrasives x_{avg-a} is approximately equal to $x_{avg}+3\sigma$, Eq. 3.15, and k is approximately equal to $1+3\sigma/x_{avg}$. A more detailed discussion of how the abrasive size distribution comes into the model and affects the material removal process will be given in Chapter 4.

The number n of abrasives captured over the wafer-pad contact area is independent of down pressure in our model. If we assume that the volume concentration (number of abrasives per unit volume of the slurry) of the abrasives in the slurry is G_1 and all abrasives flow out the wafer-pad interface along the velocity direction, the number n_{all} of abrasives over the wafer-pad interface will be $G_1 A_0 l'$ where l' is the gap between wafer and pad, or the average height of the asperities after deformation, Fig. 3.8. If the concentration of abrasives in the fluid over wafer-pad interface is kept as G_1, the number n_f of abrasives in the fluid is equal to $G_1\ (A_0 - A'_a)\ \ l'$, where A'_a is the mean area of all asperities after deformation. Thus, the number n of abrasives captured by the asperities on the contact area is $n_{all} - n_f = G_1 A'_a l' = G_1\, Vol'$ where Vol' is the volume of the asperities. For polymer materials, the volume Vol' is approximately constant (independent of down pressure) because the Poisson's ratio of polymer materials is close to 0.5 [3.27-3.28], so n is equal to $G_1 A_0 D_{SUM}\, al$ where a is the mean area of a single asperity and l the mean height of a single asperity. In the model, n is dependent on the roughness of the pad (asperity volume) but independent of the down pressure P_0. The concentration G_1 of abrasives can be calculated using the following equation:

$$G_1 = \frac{d_s \rho_s m_{s-a}}{\rho_a \overline{Vol}_a} = \frac{d_s \rho_s m_{s-a}}{\rho_a \frac{\pi}{6}\, x_{avg}^3} \tag{3.16}$$

where d_s is the dilution ratio of slurry to DI water (volume of slurry/(volume of slurry + DI water)), ρ_s the density of the slurry before dilution, m_{s-a} the concentration of slurry before dilution (weight of abrasive in slurry / weight of liquid in slurry), ρ_a the density of abrasive and $\overline{Vol}_a$ the average volume of a single abrasive. (Note that m_{s-a} times d_s is the weight concentration C of slurry abrasives in the diluted slurry, which is to be used in Chapters 4 and 5.) Therefore,

$$n = G_1 D_{SUM} A_0 al = \frac{d_s \rho_s m_{s-a} A_0}{\rho_a \frac{\pi}{6}\, x_{avg}^3} D_{SUM} al$$

and Eq. 3.14 can be written in the following form:

$$N \approx \frac{d_s \rho_s m_{s-a} A_0}{\rho_a \frac{\pi}{6} x_{avg}^3} D_{SUM} al \times \left\{ 1 - \Phi \left[3 - \frac{0.25 \times \left(\frac{4}{3}\right)^{\frac{2}{3}} (x_{avg} + 3\sigma) \left(\frac{1}{H_p} + \frac{2}{H_w}\right) E_p^{\frac{2}{3}}}{\sigma} \frac{}{b_1} P_0^{\frac{1}{3}} \right] \right\} \tag{3.17}$$

Eq. 3.17 is important since it links the hardness of the wafer H_w and pad H_p with the number of *active* particles, which is proportional to the material removal rate. According to Eq. 3.17, when the ratio b of the real contact to the apparent contact area is the same, and the pad is softer, the number of active abrasives is larger and there is a higher material removal rate. The higher material removal rate by a softer pad is a phenomenon observed extensively in experiments. Eq. 3.17 offers some quantitative explanation of this phenomenon.

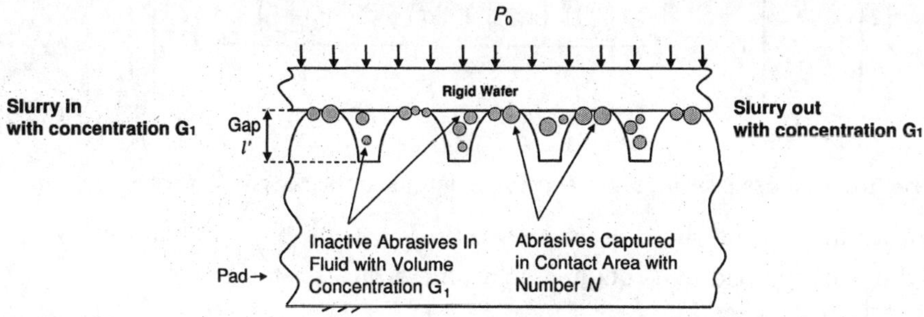

Fig. 3.8. The number of abrasive particles captured over wafer-pad interface

3.4 Formulations of Material Removal Rate

Based on the above discussions, the mass material removal rate can be written as

$$MRR_{mass} = \rho_w N \overline{Vol}_{removed} + C_0 \tag{3.18}$$

where ρ_w is the density of the wafer material, N the number of *active* abrasive particles, $\overline{Vol}_{removed}$ the volume removed by a single abrasive in unit time and C_0 material removal due to chemical etching. Basically, the chemical affects the material removal in two respects, one, the direct chemical removal, or C_0 in Eq. 3.18, and the other, the enhancing effects of chemicals on mechanical removal, which is to be addressed later. In solid-solid contact mode, C_0 is small compared with the mechanical removal $\rho_w N \overline{Vol}_{removed}$ in the silicon CMP process [3.11]. However, note that the mechanical removal and chemical removal influence each other. C_0 may be enhanced by mechanical abrasion, fluid flow and other mechanical factors. So it cannot be measured directly from the static etching of the wafer. Based on similar reasoning, the influence of chemical etching on mechanical parameters, such as the hardness of the wafer surface, cannot be measured from the static passivation of the wafer either.

Substituting Eq. 3.11, Eq. 3.17 and $x_{avg-a} = kx_{avg}$ into Eq. 3.18 and ignoring the direct chemical etching yields

$$MRR_{mass} = \rho_w A_0 k^2 \frac{2\sqrt{2} d_s \rho_s m_{s-a} D_{SUM} a l}{\rho_a \pi x_{avg}} \frac{E_p}{(b_1 H_w)^{\frac{3}{2}}}$$

$$\times \left\{ 1 - \Phi \left[3 - \frac{0.25 \times \left(\frac{4}{3}\right)^{\frac{2}{3}} (x_{avg} + 3\sigma) \left(\frac{1}{H_p} + \frac{2}{H_w}\right) E_p^{\frac{2}{3}}}{\sigma} \frac{}{b_1} P_0^{\frac{1}{3}} \right] \right\} \sqrt{P_0} V \tag{3.19}$$

Here, for discussions in later sections, let us express the above equation in another form using the contact area ratio $b = b_1 \left(\frac{P_0}{\frac{4}{3}E_p} \right)^{\frac{2}{3}}$ instead of Young's modulus of the pad and roughness parameter b_1. Then there is,

$$MRR_{mass} = \frac{3\sqrt{2}}{2} k^2 \rho_w A_0 \frac{d_s \rho_s m_{s-a} D_{SUM} al}{\rho_a \pi x_{avg}} \frac{1}{(bH_w)^{\frac{3}{2}}}$$
$$\times \left\{ 1 - \Phi \left[3 - \frac{0.25(x_{avg} + 3\sigma) \left(\frac{1}{H_p} + \frac{2}{H_w}\right)}{\sigma} \frac{P_0}{b} \right] \right\} P_0^{\frac{3}{2}} V \tag{3.20}$$

Note that in Eq. 3.20 P_0 in the probability density function Φ and $P_0^{\frac{3}{2}}$ outside of it do not represent the pressure dependence of *MRR* since the parameter b is a function of pressure P_0. The real dependence is reflected in Eq. 3.19.

For simplicity Eq. 3.19 can be written in the following form:

$$MRR_{mass} = C_1 \left[1 - \Phi \left(3 - C_2 P_0^{\frac{1}{3}} \right) \right] \sqrt{P_0} V \tag{3.21}$$

where

$$C_1 = \rho_w k^2 A_0 \frac{2\sqrt{2} d_s \rho_s m_{s-a} D_{SUM} al}{\rho_a \pi x_{avg}} \frac{E_p}{(b_1 H_w)^{\frac{3}{2}}}$$

is a value representing the effect of slurry chemicals, slurry abrasives, wafer size, wafer density, wafer hardness, pad material and pad topography, and

$$C_2 = \frac{0.25 \times \left(\frac{4}{3}\right)^{\frac{2}{3}} (x_{avg} + 3\sigma) \left(\frac{1}{H_p} + \frac{2}{H_w}\right)}{\sigma} \frac{E_p^{\frac{2}{3}}}{b_1}$$

is a value representing the effect of slurry abrasives (average size and size distribution), wafer and pad hardness and pad roughness. The values C_1 and C_2 are independent of the down pressure P_0 and the relative velocity V.

In most situations, the thickness removed per unit time is used to approximate the material removal rate. The following equation can be derived for *MRR* based on thickness removed:

$$
\begin{aligned}
MRR_{thickness} &= \frac{MRR_{mass}}{\rho_w A_0} \\
&= \frac{C_1\left[1 - \Phi\left(3 - C_2 P_0^{\frac{1}{3}}\right)\right]\sqrt{P_0}V}{\rho_w A_0} \\
&= C_3\left[1 - \Phi\left(3 - C_2 P_0^{\frac{1}{3}}\right)\right]\sqrt{P_0}V
\end{aligned} \tag{3.22}
$$

where

$$
C_3 = \frac{C_1}{\rho_w A_0} = \frac{2\sqrt{2}d_s k^2 \rho_s m_{s-a} D_{SUM} al}{\rho_a \pi x_{avg}} \frac{E_p}{(b_1 H_w)^{\frac{3}{2}}}
$$

is a constant independent of pressure P_0 and relative velocity V. In this work, when not explained explicitly, *MRR* indicates the thickness removal rate.

Similarly, the thickness removal rate formulation for a conical abrasive can be derived as follows:

$$
MRR_{thickness} = C_4\left[1 - \Phi\left(3 - C_5 P_0^{\frac{1}{6}}\right)\right] P_0^{\frac{1}{3}} V \tag{3.23}
$$

where C_4 and C_5 are two parameters independent of pressure and velocity. Due to the different geometry of abrasives, the pressure dependence of *MRR*, $P_0^{\frac{1}{6}}$ in the probability density function Φ and $P_0^{\frac{1}{3}}$ outside of the function are different from those for spherical abrasives.

The enhancing effects of chemical reaction on the mechanical material removal are reflected by density ρ_w and hardness H_w in Eqs. 3.19 and 3.20 and C_1 and C_3 in Eqs. 3.21 and 3.22. (Note that the density does not have influence on the thickness removal rate.) Generally, it is believed that a softened/hardened layer with material properties different from those of the wafer substrate will be formed continuously on the wafer surface [3.32]. In CMP, it is this layer that is removed mechanically by abrasives. In Chapter 5 of this work, a detailed model is developed to explain how the slurry chemicals affect/enhance the material removal process.

An interesting point deserving mention here is that when other conditions of slurry are the same, larger abrasives may yield smaller C_3 and therefore smaller *MRR*. This has been observed experimentally [3.10]. The reason for small abrasives to yield more material removal is that the number of cutting tools (abrasives) over wafer-pad interface is larger (see equation of n in section 3.3.). This effect of abrasive size will be addressed in Chapter 4.

In addition, the rate of slurry flow (volume/time) is not included in our model and therefore does not influence the material removal rate. This is observed by TEOS CMP experiments [3.33] with flowing rates $50\times10^3 \sim 150\times10^3 \text{mm}^3/\text{min}$. This indicates that our assumption of steady-state condition of slurry dispensing in section 3.3.3 is reasonable.

Before verification of the above equations, several points about the *MRR* equations deserve further mention:

i. The pressure dependency of removal rate is nonlinear. One part of the non-linearity is from the probability density function of the abrasive size distribution. The other part is from the elastic deformation of the pad. The pressure influences the material removal rate through both the active abrasive number and the volume removed by a single abrasive. When pressure increases, the material removal increases.
ii. The pad hardness influences the *MRR* by the active abrasive number. Since the hardness of the pad is in general much smaller than wafer hardness (see the equation of constant C_2), it has a larger influence on the *active* abrasive number than the wafer hardness. If the ratio of *real* contact area to the *apparent* contact area b is the same (see Eq. 3.20) (a softer and rougher pad could yield the same contact area ratio b as a harder and smoother pad), a softer pad yields larger *MRR*.
iii. Based on our model, for constant pad hardness, H_p, a larger contact ratio b (smoother pad) yields smaller material removal. This is because when the contact ratio b is larger the force applied on each abrasive will become smaller and therefore the volume removed by a single abrasive is smaller. It is known that a smoother pad, which contacts the wafer with a larger contact area, yields a smaller material removal rate. The analysis in this chapter helps to explain this phenomenon.
iv. Wafer hardness affects the *MRR* through both the *active* abrasive number and the removal volume by a single abrasive. The influence of wafer hardness on the active abrasive number is much smaller than that of the pad hardness. A softer wafer has a larger *MRR*.
v. The roughness parameters D_{SUM}, a, l and R of the pad have a large influence on the material removal rate. When a, l and D_{SUM} are larger, the material removal rate is larger.
vi. The *MRR* formulations are different for different abrasive geometry. Especially, the pressure dependence of *MRR* is different.
vii. An optimal abrasive size exists for maximizing material removal rate. Under the same dilution ratio d_s and concentration m_{s-a} of slurry before dilution, larger abrasive size does not necessarily yield larger material removal rate due to the smaller number of abrasives over the wafer-pad

interface. If the abrasive size is optimized, a lower dilution ratio of slurry, which is one of the most expensive consumables in CMP, may be used to generate a sufficiently large material removal.

3.5 Model Verification

In this section, the predictions of material removal rate by Eqs. 3.20 and 3.22 are verified. Due to the complicated form of Eq. 3.20 and the unavailability of accurate values of some of the input parameters, such as pad hardness, the validity of Eq. 3.20 can only be estimated at this stage by comparing the order of the *MRR* prediction with that of experimental results. The full verification of Eq. 3.20 relies on the measurement of pad hardness, contact area ratio and distribution of abrasive size. The parameters C_2 and C_3 in Eq. 3.22 are constant when only process parameters P_0 and V change. Therefore Eq. 3.22 can be verified by calculating the C_2 and C_3 using partial experimental *MRR* results, and then comparing experimental results with those predicted using the calculated C_2 and C_3 values.

3.5.1 An Estimation of Material Removal Rate

First, let us estimate the mass *MRR* for a typical silicon CMP process using Eq. 3.20 to see whether the predicted *MRR* is in correct order. It is noteworthy that Preston's equation and revised Preston's equations are not useful for estimating the order of the material removal rate due to the existence of the all-purpose parameter K_e. Experimental *MRR* results of a silicon CMP process from Moon et al. [3.12] are used to verify the equation. Table 1 lists the range of the parameters in Eq. 3.20. Most of these values are from those used in experiments. However, some of the parameters are difficult to measure directly and have to be estimated as follows.

For pad materials, only shore hardness is available, which cannot be converted directly to hardness with unit Pa. In material handbooks [3.34] it is shown that a type of polymer material - unplasticized polyvinylchloride - has both shore hardness D65-85 and Rockwell hardness M115. The Rockwell hardness M115 corresponds approximately to Vickers hardness $30{\sim}80\times10^6$Pa. Since pad materials such as IC60 used by Moon et al. have shore hardness D65-85, it is approximately estimated that the hardness of the three polishing pads used in the experiment are 30, 50 and 100×10^6Pa, respectively. Clearly, obtaining accurate hardness values of pads is a challenge in CMP modeling.

The hardness of a polished silicon wafer is measured to be around 12×10^9Pa. However, the hardness H_w in our model should be lower than this value considering the effect of chemical reaction. The hardness of a silicon wafer after 4 hours of static passivation and etching is measured to be around $6\sim7\times10^9$Pa, much smaller than the value of 12×10^9Pa without passivation. However, the value of H_w, as a fitting parameter in the model, cannot be measured directly from static chemical passivation and etching due to the enhancing effect of the mechanical abrasion and fluid flow on the chemical reaction. Therefore, an indirect method must be used. Moon [3.11] did experiments to investigate the chemical effect on mechanical material removal. Two silicon CMP experiments, one with both slurry chemicals and slurry abrasives, the other with slurry abrasives only, were conducted. It was found that the material removal of silicon wafers with slurry chemicals is about twice that without using slurry chemicals. According to the model, the hardness of the softened layer should be around 0.62 times the hardness of the silicon wafer (12×10^9Pa) to explain this *MRR* difference (see Eq. 3.20). Thus, it is concluded that the hardness H_w is around 7.6×10^9Pa for the silicon CMP process. Note that here the small value of chemical etching C_0 in Eq. 3.18 has been neglected for silicon CMP (see Eq. 3.19) and therefore all *MRR* differences can be attributed to the fitting parameter H_w.

Table 3.1. Ranges of parameters in Eq. 3.20

Parameters	Approximate Ranges of Values
Dilution Ratio d_s (volume of slurry/(volume of slurry + DI water))	0.1
Slurry concentration m_{s-a}(weight of abrasives/weight of fluid) (%)	4.5
Density ρ_s of diluted slurry (g/mm^3)	1.00×10^{-3}
Density ρ_a of abrasive (silica) (g/mm^3)	1.08×10^{-3}
Density ρ_w of wafer (silicon) (g/mm^3)	2.33×10^{-3}
Area A_0 of wafer (mm^2)	7.85×10^3
Average size x_{avg} of abrasive (mm)	80×10^{-6}
Hardness H_w of silicon wafer (Pa)	7.6×10^9
Hardness H_p of polishing pad (Pa)	30×10^6, 50×10^6, 100×10^6
Ratio of real contact area and apparent contact area b	1/6631
Relative velocity V of wafer (mm/hour)	518400
Down pressure P_0 (Pa)	1924
Height l of asperities (mm)	58×10^{-3}
k	1.56

One single hardness value H_w is used in this chapter to represent the influences of slurry chemicals on the wafer surface and material removal process. It is noted that more complicated interactions between slurry chemicals and wafer surface may exist and these will be covered in Chapter 5.

Nalco2352 slurry is used as the polishing slurry. The mean size, 80nm, of the abrasives is from specifications provided by the abrasive supplier. According to the SEM photograph of the abrasives, Fig. 3.6, it is assumed that the standard deviation σ of the abrasive size distribution is 15nm. The k is estimated to be $1+3\sigma/x_{avg}= 1.56$. Chapter 4 will discuss the effects of abrasive size distributions on the material removal process in detail.

The ratio b of the contact area is difficult to measure directly. However, it can be estimated using Eq. 3.10. The hardness of H_w is about 7.6×10^9Pa. The roughness R_a of the polished wafer is around 0.1nm. By substituting H_w, R_a, $P_0= 1924$Pa and $x_{avg}= 80$nm into Eq. 3.10, the value of b can be calculated to be about 1/6331. Thus, the real pressure P_c applied on the asperities is about 6,331 times larger than the down pressure P_0.

SEM photographs of top views of polishing pads, Fig. 3.9, from Moon [3.11] are used to estimate the volume $A_0 D_{SUM} al$ of asperities in Eq. 3.20. The asperity height l is about 58μm for polishing pads. From the SEM photograph, the mean area of total asperities is estimated to be around $0.005A_0$ and, therefore, the volume is around $0.005A_0l = 2.2765\text{mm}^3$. The structures of the surface asperities are quite different for the three pads, one with fibers, the other two with isolated pores and solid walls. However, the different structures do not affect the results predicted by the model if the volumes of the asperities (mean area times mean height l) are the same. Note that the asperity heights are close to each other, although not exactly the same, and, therefore, it is assumed here that all asperities contact the wafer and deform under pressure so that the contact pressure model in section 3.3.1 is applicable.

By substituting values in Table 1 into Eq. 3.20, MRR values 0.045~0.24 g/hour for three pad hardness 100, 50 and 30×10^6Pa are obtained. These values are quite reasonable compared with experimental results of 0.010~0.035 g/hour for three different kinds of pads (IC60, SUBA 500 and UR 100) in [3.12]. Note that the unit g/hour for MRR is used because the thickness change cannot be measured directly for a silicon wafer. The value that can be directly measured is the mass. Using thickness removed per unit time to approximate the *MRR*, the predicted removal rates are 160~880nm/min.

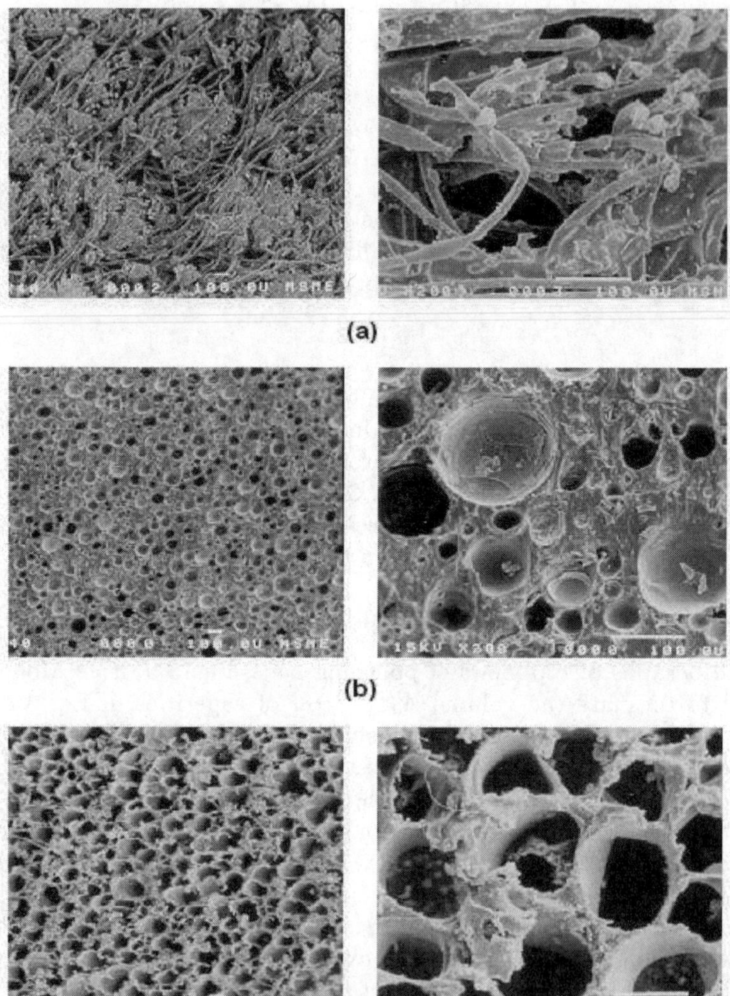

Fig. 3.9. Top view of (a) SUBA500, (b) IC60, and (c) UR100 polishing pads at two magnifications (×40 and ×200) [3.11]

It is helpful to see the influences of the above input parameters on *MRR*. Figs. 3.10 and 3.11 show the variation of material removal rate with the hardness and the contact area. Obviously, a softer pad yields a larger *MRR* when the contact area is kept constant. Larger contact area yields smaller *MRR* when the hardness of the pad is kept constant. This agrees well with the 2nd and 3rd points mentioned in the last section. To realize the same large *MRR* as a softer pad, a harder pad must contact the wafer with a smaller contact area. This explains why in a real situation, where a harder pad does contact the wafer with a smaller area, the range of material removal by the pad, whether it is hard or soft, does not change much in order (for example, 0.01~0.035g/hour *MRR* using three different pads in Moon et al. [3.12]).

According to Figs. 3.10 and 3.11 if a pad becomes harder, and the contact area between pad and wafer increases, the material removal rate will reduce. Practical experience supporting this is that if a pad is used in a polishing process for a long time, it becomes harder and smoother (loading and aging), and the *MRR* of polishing reduces. Based on the model, if no conditioning of the pad is done, the material removal rate will be close to a constant value, but not equal to zero. Fig. 3.12 shows typical polishing rate curves for a pad without conditioning and with non-optimized conditioning [3.35]. The trend of the polishing rate when using a pad without conditioning is the same as that predicted. For polishing with pad conditioning, because the hardness of pad and contact area do not change much, the material removal rate keeps constant, Fig. 3.12. Although the decrease of *MRR* with no conditioning might be due to the poor slurry dispensing on the 'glazed' pad surface [3.32], the modeling effort in this chapter provides an alternative explanation.

An important point is that the probability of active abrasive presence ($1-\Phi(x)$), say 0.0139 for pad hardness 50MPa, contact ratio 10^{-4} and standard deviation 15nm, plays a significant role in the order of the predicted *MRR*. If all abrasives are involved in material removal, the order of predicted *MRR* will be about 100 times higher. A detailed look at Eq. 3.20 shows that the order of the material removal rate is primarily the integrated effect of wafer hardness (cutting volume by a single abrasive), pad hardness & topography (probability of active abrasives) and down pressure (probability of *active* abrasive and cutting volume). The resulting correctness of the estimation of the order shows that the proposed equation is reasonable, and especially, the concept of *active* abrasives is reasonable. More accurate information on distribution of abrasive size, the number of abrasives, pad hardness and pad topography will be critical to further verify the equation.

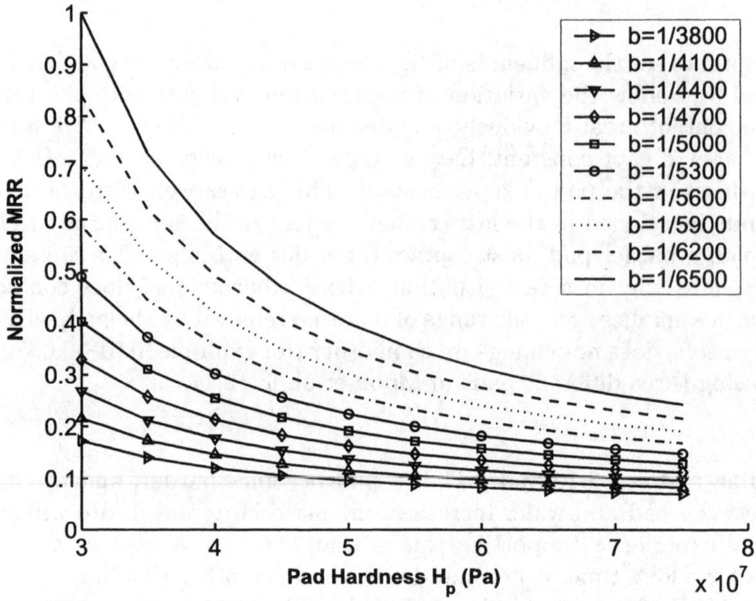

Fig. 3.10. Estimation of material removal rate as a function of pad hardness (P_0= 1924Pa, x_{avg} =80nm and σ= 15nm)

3.5.2 Down Pressure Dependency of *MRR*

The full verification of the *MRR* estimation in Eq. 3.20 is difficult at this stage due to the lack of accurate information on abrasive size, abrasive number, pad hardness and pad roughness. However, the pressure dependence of material removal rate in Eqs. 3.20 and 3.22 can be fully verified by comparing the predicted values with the experimental results using the following method. Assuming that two sets of down pressures and velocities, say, P_1, V_1, P_2 and V_2, and the corresponding material removal rates, say, MRR_1 and MRR_2, are known, then by substituting these values into Eq. 3.22, the values of C_3 and C_2 can be calculated. Since C_3 and C_2 are considered constant when only down pressures and velocities change, substituting the calculated values of C_3 and C_2 into Eq. 3.22 should allow an accurate prediction of material removal rate under other pressures and velocities if the *MRR* equation is correct. Based on this idea, four sets of experimental *MRR* data [3.15] [3.36] are used to verify the developed pressure dependence of *MRR*. In all four experimental sets, the velocities are kept constant when pressures change. Note that due to the complicated form of the function $\Phi(x) = \frac{1}{\sqrt{2\pi}} \int_{-\infty}^{x} e^{-\frac{1}{2}t^2} dt$, an iteration method of trial and error has been used to solve for values of C_3 and C_2.

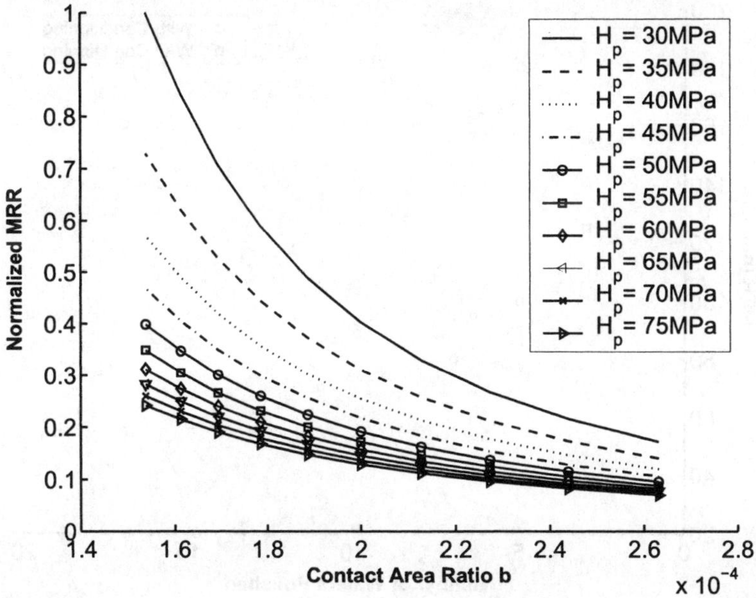

Fig. 3.11. Estimation of material removal rate as a function of contact area ratio b (P_0 =1924Pa, x_{avg}= 80nm and σ =15nm)

Fig. 3.13 shows two sets of experimental data for the thickness removal rate of a PECVD SiO_2 CMP. One data set (slurry 1) is for CMP using an alkaline/cerium-oxide slurry [3.15] and the other (slurry 2) is for CMP using a conventional alkaline/silica slurry [3.36]. In both cases the same kind of polishing pad IC1400 is used and the down pressure is changed from 2psi to 12psi to obtain four experimental data points for each set. For the case of slurry 1, as discussed, by substituting the *MRR* value 280nm/min under pressure P_0= 2psi and 455nm/min under pressure P_0= 4psi into Eq. 3.22, the values of VC_3 and C_2 can be solved to be 84148.76 and 0.137, respectively, where V is the relative velocity of the wafer in the experiment. Thus the *MRR* prediction will be $MRR = (84148.76/V)\left[1 - \Phi(3 - 0.137\sqrt[3]{P_0})\right]\sqrt{P_0}$. By substituting pressure P_0= 6psi and P_0= 8psi into the above equation, *MRR* values of 612.39nm/min and 764.90nm/min are obtained, respectively. Compared with the experimental data, 610nm/min and 765 nm/min, the relative errors of predicted values are negligible. Line 1 in Fig. 3.13 shows the *MRR* predicted by Eq. 3.22. A similar process is used to check the validity of the formulation for data set 2. Very accurate prediction results are also obtained. Line 2 in Fig. 3.13 shows the prediction.

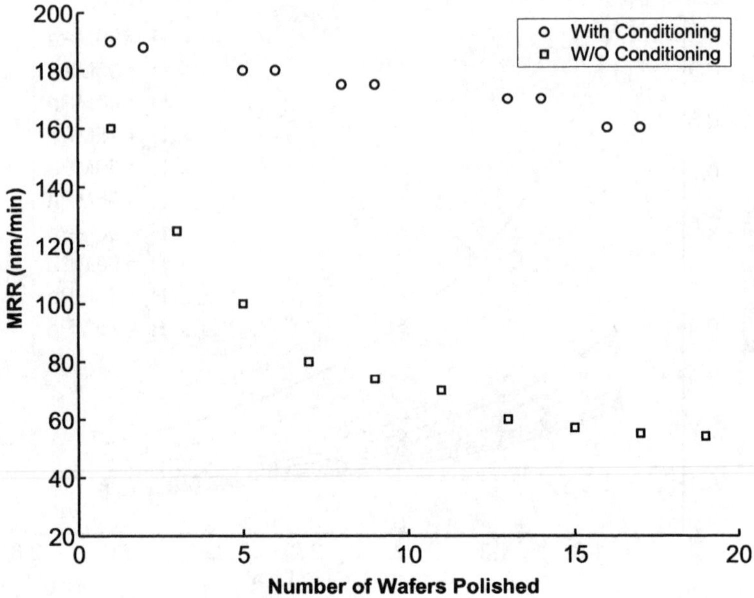

Fig. 3.12. The typical polishing rate curves for pad without conditioning and with non-optimized conditioning [3.35]

The probability of *active* abrasives under different pressures are around 0.2%∼ 1.5%. These values are obtained by solving Eq. 3.22 using experimental down pressures and the corresponding *MRR* values. In the last section the probability is evaluated by directly substituting typical CMP input values into the function

$$1-\Phi\left[3-\frac{0.25(x_{avg}+3\sigma)\left(\frac{1}{H_p}+\frac{2}{H_w}\right)}{\sigma}\frac{P_0}{b}\right].$$

Values around 0.9∼ 1.6%, close to the 0.2%∼1.5%, are obtained. The closeness of the probability values obtained using the two completely unrelated methods is strong evidence supporting the model.

By substituting value of C_2 into Eq. 3.15, it is seen that the mean size of active abrasive is $x_{avg}+$ 3.03011σ for pressure P_0= 2psi (13.78kPa) and x_{avg}+3.12821σ for pressure P_0 =8psi (55.12kPa). The change in the average size of *active* abrasive is quite small under the two down pressures, for typical x_{avg} value of 80nm and standard deviation σ= 5∼15nm. This means that it is

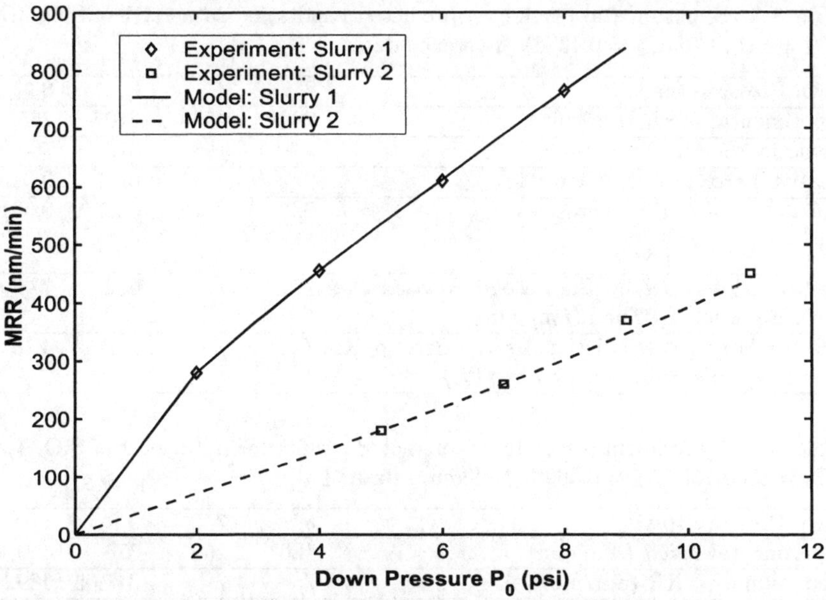

Fig. 3.13. Two sets of experimental material removal rates vs. predictions of the model for SiO_2 CMP processes

reasonable to take the mean size of active abrasive as a constant independent of the down pressure.

In Tables 3.2 and 3.3, the experimental *MRR* vs. the prediction of *MRR* and the relative errors of the predictions are shown. The probabilities of *active* abrasives 1-$\Phi(x)$ are important parameters and are also listed in the table. In the caption of each table, the calculated parameters VC_3 and C_2 are given. The relative errors of the predictions for data set 2 are larger than those for data set 1. Note that the experimental results at down pressures 5psi (34.45kPa) and 7psi (48.23kPa) are taken as references and their errors are assumed to be zero. If considering errors exist in the experimental *MRR* results for pressures 5psi (34.45kPa) and 7psi (48.23kPa) and a line is used based on Eq. 3.22 to fit all the 4 experimental data, the prediction errors for pressures 9psi (62.01kPa) and 11psi (75.79kPa) will be smaller. The current errors, 6.2% and 2.31%, however, are still small considering the measurement error and other factors influencing the experimental results.

It is helpful to review here the predictions by Preston's equation [3.3] and Zhao and Shi's Equation [3.17-3.18]. Lines 1 and 2 in Fig. 3.14 are the predictions of

Table 3.2. Experimental results vs. predicted results for data set 1 of SiO_2 CMP ($VC_3 = 84148.76$, $C_2 = 0.1370$) *reference data

Down Pressure (psi)	2*	4*	6	8
Experimental *MRR* (nm/min)	280*	455*	610	765
Prediction of *MRR* (nm/min)	/	/	612	764.9
Relative Error of Prediction (%)	/	/	0.33	0.013
Probability of Active Abrasives $1\text{-}\Phi\left(3 - C_2 P_0^{1/3}\right)$ (%)	0.2353	0.2704	0.2971	0.3214
Prediction of *MRR* by the Second Revised Preston's Equation by Zhao (nm/min)	/	/	602	733
Relative Error of Prediction by the Second Revised Preston's Equation by Zhao (%)	/	/	1.31	4.18

Table 3.3. Experimental results vs. predicted results for data set 2 of SiO_2 CMP ($VC_3 = 8989.155$, $C_2 = 0.3698$) *reference data

Down Pressure (psi)	5*	7*	9	11
Experimental *MRR* (nm/min)	180*	260*	370	450
Prediction of *MRR* (nm/min)	/	/	347.05	439.57
Relative Error of Prediction (%)	/	/	6.2	2.31
Probability of Active Abrasives $1\text{-}\Phi\left(3 - C_2 P_0^{1/3}\right)$ (%)	0.8955	1.0929	1.2869	1.4744
Prediction of *MRR* by the Second Revised Preston's Equation by Zhao (nm/min)	/	/	331.5	398.4
Relative Error of Prediction by the Second Revised Preston's Equation by Zhao (%)	/	/	10.4	11.5

material removal rates based on Preston's equation. *MRR* at down pressure 2psi (13.78kPa) is used as the reference for data set 1 and *MRR* at down pressure 5psi (34.45kPa) is used as the reference for data set 2. Obviously, Preston's equation does not accurately predict the *MRR*. It was noted that the experimental data could be roughly fitted with a linear line, as shown by lines 3 and 4 in the figure. This leads to a revised Preston's equation $MRR = K_e P_0 V + MRR_0$, where MRR_0 is the material removal when down pressure is zero. However, no theoretical basis supports MRR_0

It has been argued that, as in the parameter C_0 introduced in Eq. 3.18, the initial material removal rate MRR_0 is due to the static chemical etching. However, in oxide CMP, using a typical alkaline-based slurry, the etching of the oxide is minimal while the initial value obtained from the line fitted to the experimental data is quite large; see line 3 in the Fig. 3.14 ($MRR_0 =$ 120nm/min). Moreover, in some of the oxide CMP data, such as that shown by line 4 (MRR_0= -50nm/min), the initial value from the fitted line is negative. This cannot be explained by static chemical etching.

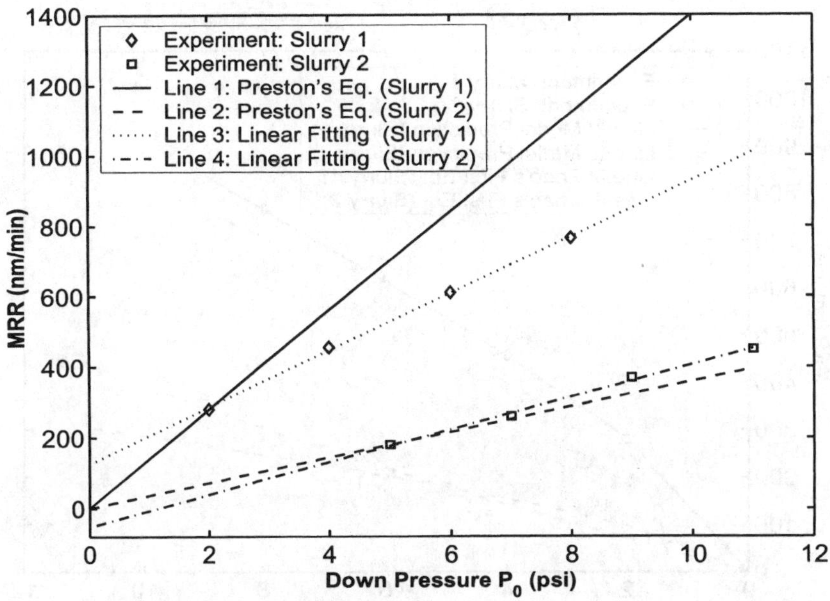

Fig. 3.14. Two sets of experimental material removal rates vs. predictions of Preston's Equation for SiO_2 CMP processes

Zhao and Shi [3.17-3.18] tried to explain the above observations and proposed two revised Preston's equations $MRR= K_e VP_0^{2/3}$ and $MRR= K_eV(P_0^{2/3} - P_{th}^{2/3})$ (P_{th} is a threshold pressure). Lines 3 in Figs. 3.15 and 3.16 show the *MRR* prediction for slurry 1 based on their first equation and second equation, respectively. Lines 4 in Figs. 3.15 and 3.16 show the same predictions for slurry 2. The predictions based on the model proposed here are shown with lines 1 and 2 in both figures. Zhao and Shi argued that because the real contact area between an asperity and the wafer is given by $A \propto P_0^{2/3}$, the number of active abrasives N is $\propto P_0^{2/3}$. Thus yields the first revised Preston's equation. Zhao and Shi only note that the number of *active* abrasives changes with the real contact area. They did not notice that the mean pressure applied on the asperity also changes, and, therefore, the volume cut by a single *active* abrasive changes. The second equation fits the experimental data well (see lines 3 and 4 in Fig. 3.16) while the first equation does not (see lines 3 and

4 in Fig. 3.15). For comparison, the predictions and relative errors based on the second equation are listed in Tables 3.2 and 3.3. The relative errors are larger than those using the model developed here.

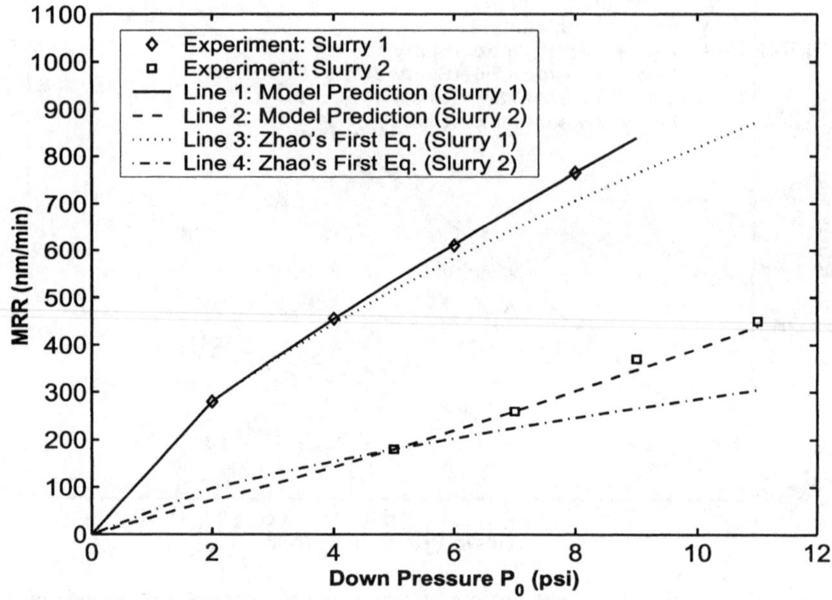

Fig. 3.15. Predictions of the model vs. predictions of Zhao's first equation for SiO_2 CMP processes

Eq. 3.23, based on conical abrasive shapes, was also used to predict the material removal rate. The values of parameters C_4 and C_5 cannot be obtained for data set 2. Therefore, the assumption of conical abrasive shape is incorrect. This agrees with the SEM photograph of abrasives, Fig. 3.6, and the observation by Xie and Bhushan [3.37] that smaller abrasives become more round.

Fig. 3.17 and Tables 3.4 and 3.5 show the prediction of *MRR* vs. two sets of experimental *MRR* for a Tungsten (W) CMP process [3.38]. The relative velocity for data set 1 is 40rpm and the relative velocity for data set 2 is 20rpm. The *MRR* value at pressures 7psi and 8psi can be predicted by using

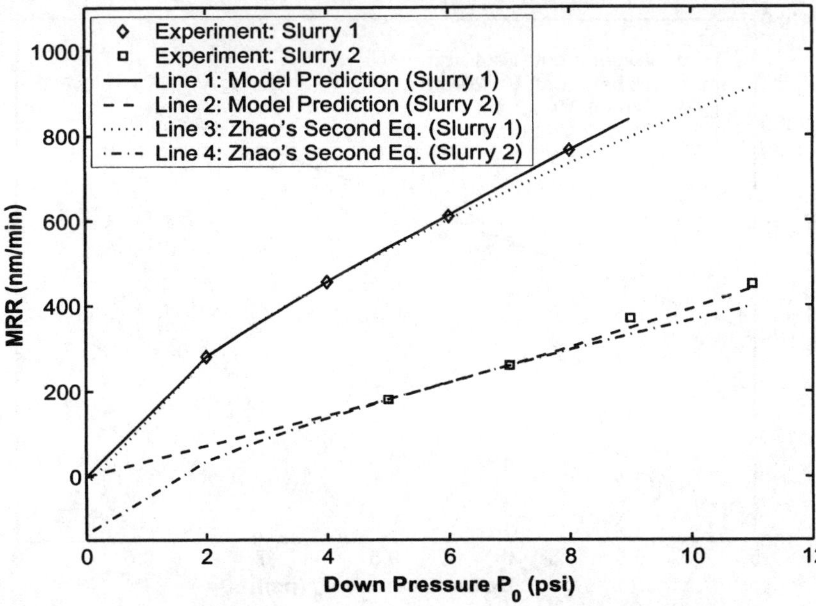

Fig. 3.16. Predictions of the model vs. predictions of Zhao's second equation for SiO_2 CMP processes

the *MRR* at pressures 5psi and 6psi as reference. The relative errors of 0.487% and 2.94% are small. Since the same pad and slurry are used for data set 2 and only the relative velocity changes, the C_2 calculated from data set 1 can be used to predict the *MRR* values of set 2. The relative error in the prediction is quite small for down pressures 6psi and 8psi although it is relatively large (7.87%) for a down pressure of 7psi.

Table 3.4. Experimental results vs. predicted results for data set 2 of tungsten (W) CMP when relative velocity V = 40rpm (VC_3= 9965.32, C_2= 0.37) *reference data

Down Pressure (psi)	5*	6*	7	8
Experimental *MRR* (nm/min)	200*	243.7*	287.5	326.5
Prediction of *MRR* (nm/min)	/	/	288.9	336.1
Relative Error of Prediction (%)	/	/	0.487	2.94
Probability of active abrasives $1\text{-}\Phi\left(3 - C_2P_0^{1/3}\right)$ (%)	0.8975	0.9984	1.0958	1.1924

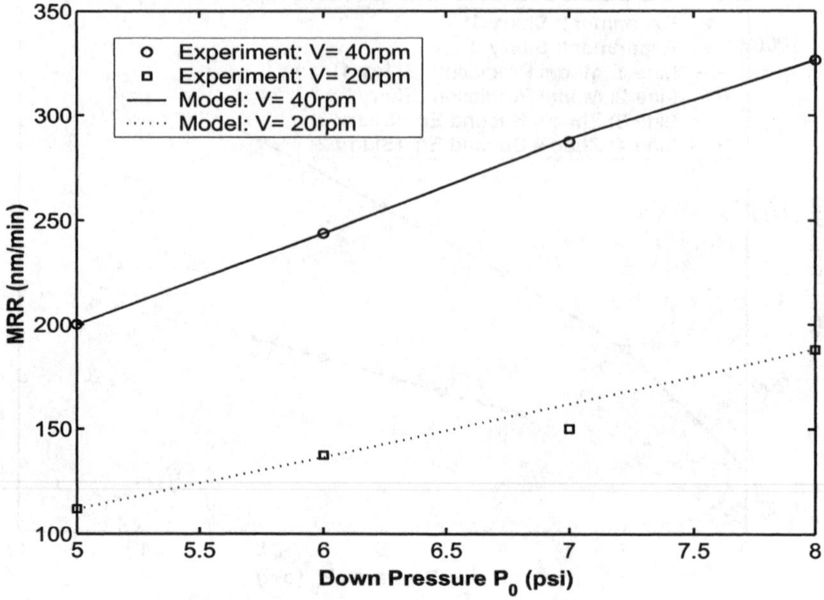

Fig. 3.17. Two sets of experimental material removal rates vs. predictions for Tungsten (W) CMP processes

The same procedures are used to predict the *MRR* of aluminum (Al) and copper (Cu) CMP processes based on reported data [3.39-3.40] and accurate results are also obtained.

Table 3.5. Experimental results vs. predicted results for data set 2 of tungsten (W) CMP when relative velocity V = 20rpm (VC_3= 5580.58, C_2= 0.37) *reference data

Down Pressure (psi)	5*	6	7	8
Experimental *MRR* (nm/min)	112.0*	137.5	150.0	188.0
Prediction of *MRR* (nm/min)	/	136.5	161.8	188.2
Relative Error of Prediction (%)	/	0.727	7.87	0.106
Probability of active abrasives $1\text{-}\Phi\left(3 - C_2 P_0^{1/3}\right)$ (%)	0.8975	0.9984	1.0958	1.1924

3.6 Discussion

A model has been developed to investigate the material removal mechanism in the CMP process. The model relies on an estimation of the number of *active* abrasive particles and the forces acting on the abrasive particles. The number of active abrasives is calculated based on the normal distribution of the abrasive size while the force is calculated based on down pressure, the real contact area and the assumption that the force applied on a single particle is equal to the pressure times the abrasive size. It is found that the pressure dependence of the material removal rate is related to a probability density function due to the assumed normal distribution of abrasive size. The material removal rate is estimated based on an estimation of input parameters. The results agree well with published experimental data. The nonlinear pressure dependence of the polishing rate obtained is verified by experimental results as well.

The key idea of the model is that the sources of material removal can be separated into two parts, one the number of the active abrasive, and the other the material removal by a single active abrasive. The different roles of consumables and the influences of fluid and the chemical elements on mechanical material removal can be attributed to one of these two parameters. A number of input parameters, including pad hardness, pad elastic modulus, pad roughness, wafer material and abrasive size, are shown to be important in the model. The pad hardness, the volume of pad asperities and distribution of abrasive size are critical since they determine the number of active abrasives and, therefore, the order of the material removal rate. The topography of the pad also plays an important role due to its influence on the contact ratio b, which affects the *active* abrasive number and the magnitude of the pressure applied on the asperities and abrasives. The interactions between chemical removal, fluid flow and abrasion are important. In this chapter, an interface between the chemical effect and the mechanical effect has been constructed through a fitting parameter H_w, the hardness value of the wafer surface. A detailed model covering the interactions of chemicals, wafer surface and other mechanical elements will be discussed in Chapter 5. The fluid effect in the current model is attributed to the number of active abrasives. For static fluid flow, it is assumed in the model that the number n of abrasive on the wafer-pad interface is independent of the down pressure.

The modeling work in this chapter may provide a quantitative tool for the design of consumables including slurries and pads, whose effects on material removal are mainly reflected in the parameters C_3 and C_2. Note that C_3 and C_2 actually determine the slope and the initial material removal rate in the revised linear Preston's equation. By adjusting C_3 and C_2, the slope that reflects the sensitivity of material removal on process parameters can be controlled. Fig. 3.18 shows the variance of pressure dependence of *MRR* under different values of C_2. When C_2 is larger, down pressure has a larger influence on *MRR*.

The model developed here is also helpful for CMP process control. The material removal rate affects a number of output parameters including the within-wafer non-uniformity (WIWNU), the selectivity of the material removal rate in polishing different wafer materials, and the wafer-wafer non-uniformity (WWNU). The pressure and velocity distribution over the wafer-pad interface determine the WIWN of the material removal rate. The adjustment of consumable parameters (C_3 and C_2) can reduce the sensitivity of *MRR* to the pressure distribution, leading to a smaller WIWNU when the distribution of pressure and velocity remains the same. Chapter 8 of this work discusses in detail how to optimize the non-uniformity from the viewpoint of consumable effects using the model developed in this chapter. The selectivity of the material removal rate for different wafer materials can also be optimized by selecting optimal down pressures and relative velocities based on the model. The wafer-wafer non-uniformity is determined by the pad hardness and contact area change during planarization. An on-line change of the down pressure with an on-line measurement of the hardness and contact area may be helpful to improve the WWNU.

One final point worth mentioning is that from the model development there is no limitation to apply the *MRR* Eqs. 3.19∼3.21 to each single wafer point. The material removal rate is equal to the number of *active* abrasives times the material removed by a single abrasive. After normalizing the mass *MRR* equation to an equation of height reduction rate, the area term A_0 disappears; Eqs. 3.19∼3.21.

3.7 Conclusion

Based on the assumptions of plastic contact over wafer-abrasive and pad-abrasive interfaces, the normal distribution of abrasive size and the periodic

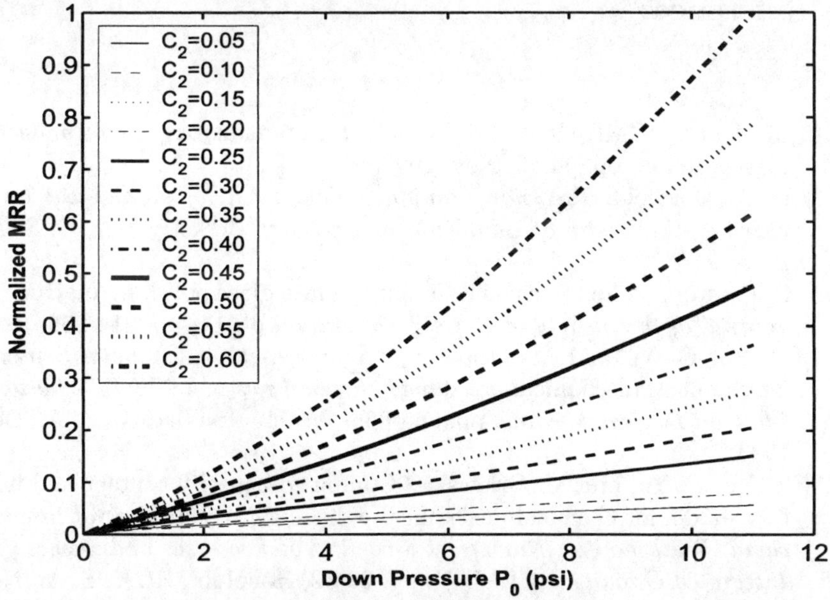

Fig. 3.18. Influence of C_2 on pressure dependence of MRR

roughness of the pad surface, a novel model was developed for predicting the material removal in CMP. The basic idea of the model is that $MRR = \rho_w N Vol_{removed}$, where ρ_w is the density of wafer, N the number of active abrasives and $Vol_{removed}$ the volume of material removed by a single abrasive. Compared with previous approaches at modeling, such as Preston's equation, the model proposed here integrates not only the process parameters pressure and velocity, but also other important input parameters including the wafer hardness, pad hardness, pad roughness, abrasive size and abrasive geometry into the same formulation to predict the material removal rate. An interface between the chemical effect and mechanical effect has been constructed through a fitting parameter H_w in the model. It reflects the influence of chemicals on the mechanical material removal. The fluid effect in the current model is attributed to the number of active abrasives. The nonlinear down pressure dependence of the material removal rate is related to a probability density function of the abrasive size and the elastic deformation of the pad. Compared with experiment results, the proposed model predicts the MRR accurately. By further verification of the model, a better understanding of the fundamental mechanism involved in CMP processes can be obtained.

3.8 References

[3.1] R. Dejule, "CMP challenges below a quarter micron," *Semiconductor International*, pp. 54-60, Nov. 1997.

[3.2] D. A. Dornfeld, *Precision Manufacturing, Dept. of Mechanical Engineering, University of California at Berkeley*, Berkeley, CA, U. S. A., 1999.

[3.3] F. Preston, "The theory and design of plate glass polishing machines," *Journal of the Society of Glass Technology*, Vol. **11**, pp. 214-256, 1927.

[3.4] T. Yu, C. Yu and M. Orlowski, "A statistical polishing pad model for chemical mechanical polishing," *Proceedings of IEEE International Electron Devices Meeting*, pp. 865-868, Washington, DC, U. S. A., Dec. 1993.

[3.5] T. Yu, C. Yu, and M. Orlowski, "Combined asperity contact and fluid flow model for chemical mechanical polishing," *Proceedings of International Workshop on Numerical Modeling of Processes and Devices for Integrated Circuits*: NUPAD V, pp. 29-32, Honolulu, HI, U. S. A., New York, NY, U. S. A., June 1994.

[3.6] S. R. Runnels and L. M. Eyman, "Physical modeling of CMP fluid flow analysis," *Sematech Report*, 1993.

[3.7] S. R. Runnels, "Tribology analysis of chemical mechanical polishing," *Journal of the Electrochemical Society*, Vol. **141**, pp. 1698-1701, 1994.

[3.8] D. Wang, J. Lee, K. Holland, T. Bibby, S. Beaudoin and T. Cale, "Von mises stress in chemical-mechanical polishing processes," *Journal of the Electrochemical Society*, Vol. **144**, pp. 1121-1127, 1997.

[3.9] J. Tichy, J. A. Levert, L. Shan and S. Danyluk, "Contact mechanics and lubrication hydrodynamics of chemical mechanical polishing," *Journal of the Electrochemical Society,* Vol. **146**, pp. 1523-1528, 1999.

[3.10] R. Xu, G. Smart and M. Zhang, "Particle characteristics and removal rate in CMP process", *1999 CMP-MIC Conference*, pp. 253-255, Santa Clara, CA, U. S. A., Feb. 1999.

[3.11] Y. Moon, "Mechanical aspects of the material removal mechanism in chemical mechanical polishing (CMP)," *Ph.D. Dissertation, Department of Mechanical Engineering, University of California at Berkeley*, Berkeley, CA, U. S. A., 1999.

[3.12] Y. Moon, I. Park and D. A. Dornfeld, "Mechanical properties and relationship to process performance of the polishing pad in chemical mechanical polishing (CMP) of silicon," *Proceedings of ASPE Spring Topical Meeting on Silicon Machining*, pp. 83-87, April 1998.

[3.13] M. C. Pohl and D. A. Griffiths, "The importance of particle size to the performance of abrasive particles in the CMP processes," *Journal of Electronic Materials*, Vol. **25**, pp. 1612-1616, 1996.

[3.14] W. T. Tseng and Y. L. Wang, "Re-examination of pressure and speed dependence of removal rate during chemical-mechanical polishing processes," *Journal of the Electrochemical Society*, Vol. **144**, pp. L15-L17, 1997.

[3.15] K. Pak, Y. R. Park, U-I. Chung, Y. B. Koh and M. Y. Lee, "A CMP process using a fast oxide slurry," *1997 CMP-MIC Conference*, pp. 299-306, Santa Clara, CA, U. S. A., Feb. 1997.

[3.16] D. O. Ouma, "Modeling of chemical mechanical polishing for dielectric planarization," *Ph.D. Dissertation, Department of Electrical Engineering and Computer Science, MIT*, Cambridge, MA, U. S. A., 1999.

[3.17] B. Zhao, and F. G. Shi, "Chemical mechanical polishing in IC processes: new fundamental insights," *1999 CMP-MIC Conference*, pp. 13-22, Santa Clara, CA, U. S. A., Feb. 1999.

[3.18] F. G. Shi and B. Zhao, "Modeling of chemical-mechanical polishing with soft pads," *Applied Physics A*, Vol. **67**, pp. 249-252, 1998.

[3.19] G. Wu and L. Cook, "Mechanism of copper damascene CMP," *1998 CMP-MIC Conference*, pp. 150-156, Santa Clara, CA, U. S. A., Feb. 1998.

[3.20] F. Zhang, A. A. Busnaina, J. Feng and M. A. Fury, "Particle adhesion force in CMP and subsequent cleaning processes," *1999 CMP-MIC Conference*, pp. 61-64, Santa Clara, CA, U. S. A., Feb. 1999.

[3.21] C. W. Liu, B. T. Dai, W. T. Tseng and C. F. Yeh, "Modeling of the wear mechanism during chemical-mechanical polishing," *Journal of the Electrochemical Society,* Vol. **143**, pp. 716-721, 1996.

[3.22] R. Chauhan, Y. Ahn, S. Chandrasekar and T. N. Farris, "Role of indentation fracture in free abrasive machining of ceramics," *Wear*, Vol. **162-164**, pp. 246-257, 1993.

[3.23] V. H. Bulsara, Y. Ahn, S. Chanraseka and T. N. Farris, "Polishing and lapping temperatures," *ASME Transaction*: *Journal of Tribology*, Vol. **119**, pp. 163-170, 1997.

[3.24] Z. B. Hou and R. Komanduri, "Magnetic field assisted finishing of ceramics-part II: on the thermal aspects of magnetic float polishing (MFP) of ceramic balls," *ASME Transaction: Journal of Tribology*, Vol. **120**, pp. 652-659, 1998.

[3.25] K. L. Johnson, *Contact Mechanics*, Cambridge University Press, Cambridge, U. K., 1985.

[3.26] J. A. Greenwood and J. B. Williamson, "Contact of nominally flat surfaces," *Proceedings of the Royal Society of London*, Vol. **A295**, pp. 300-319, 1966.

[3.27] I. M. Ward and D. W. Hadley, *An Introduction to Mechanical Properties of Solid Polymers*, John Wiley & Sons., New York, U. S. A., 1993.

[3.28] I. M. Ward, *Mechanical Properties of Solid Polymers*, John Wiley & Sons., New York, U. S. A., 1983.

[3.29] F. Zhang, A. A. Busnaina and G. Ahmadi, "Particle adhesion and removal in chemical mechanical polishing and post-CMP cleaning," *Journal of the Electrochemical Society*, Vol. **146**, pp. 2665-2669, 1999.

[3.30] T. Bifano, T. A. Dow and R. O. Scattergood, "Ductile-regime grinding: a new technology for machining brittle materials," *ASME Transaction: Journal of Engineering Industry*, Vol. **113**, pp. 184-189, 1991.

[3.31] Y. Moon, A. Chang, J. F. Luo, K. Bevans and D. A. Dornfeld, "Small future reproducibility, a focus on chemical mechanical polishing (CMP)", *UC-SMART Major Program Award First Annual Review, University of California at Berkeley*, Berkeley, CA, U. S. A., May 1999.

[3.32] J. M. Steigerwald, S. P. Murarka and R. J. Gutmann, *Chemical Mechanical Planarization of Microelectronic Materials*, John Wiley & Sons., New York, U. S. A., 1997.

[3.33] A. Maury and V. Czitrom, "Optimization of CMP process for TEOS based oxide films using a designed experiment," *1996 CMP-MIC Conference*, pp. 285-290, Santa Clara, CA, U.S. A., Feb. 1996.

[3.34] W. Bolton, *Engineering Materials Pocket Book*, CRC Press, Boca Raton, Florida, U. S. A., 1989.

[3.35] J. Z. Zhang, S. K. Huang, W. S. Toh, W. S. Tay, F. Chen and B-B. Zhou, "Optimization of pad conditioning for stable oxide CMP process," *1997 CMP-MIC Conference*, pp. 315-321, Santa Clara, CA, U. S. A., Feb. 1997.

[3.36] S-Y. Shih and L-J. Chen, "Thermal characteristics study of CMP low dielectric constant material," *1998 CMP-MIC Conference*, pp. 305-312, Santa Clara, CA, U. S. A., Feb. 1998.

[3.37] Y. Xie and B. Bhushan, "Effects of particle size, polishing pad and contact pressure in free abrasive polishing," *Wear*, Vol. **200**, pp. 281-295, 1996.

[3.38] C. W. Liu, W. T. Tseng, B. T. Dai, C. Y. Lee and C. F. Yeh, "A perspective of the wear mechanism during chemical-mechanical polishing of tungsten thin films," *1996 CMP-MIC Conference*, pp. 138-144, Santa Clara, CA, U. S. A., Feb. 1996.

[3.39] C. G. Kallingal, D. J. Duquette and S. P. Murarka, "An investigation of slurry chemistry used in chemical mechanical planarization of aluminum", *Journal of the Electrochemical Society,* Vol. **145**, pp. 2074-2081, 1998.

[3.40] Z. Stavreva, D. Zeidler, M. Plotner, G. Grasshoff and K. Drescher, "Chemical-mechanical polishing of copper for interconnect formation," *Microelectronic Engineering*, Vol. **33**, pp. 249-257, 1997.

[3.41] R. Komanduri, N. Chandrasekaran and L. M. Raff, "Effect of tool geometry in nanometric cutting: a molecular dynamics simulation approach," *Wear*, Vol. **219**, pp. 84-97, 1998.

[3.42] Y-S. Su, "Investigation of removal rate properties of a floating polishing process," *Journal of the Electrochemical Society,*Vol. **147**, pp. 2290-2296, 2000.

[3.43] D. J. Stein and D. L. Hetherington, "Review and experimental analysis of oxide CMP models," *Electrochemical Society Proceedings*, Vol. **99-37**, pp. 217-233, 1999.

4

EFFECTS OF ABRASIVE SIZE DISTRIBUTION IN CMP

4.1 Introduction

The understanding of the material removal mechanism in CMP should be based on understanding the roles of the cutting tools, namely, the abrasives, and their interactions with other important input values such as the pad, chemical and wafer materials. The effect of abrasive size distribution in chemical-mechanical planarization has long been observed [4.1-4.7] [4.13] [4.14]. For example, experimental results show that there is an inverse proportional relationship between the abrasive size and the material removal rate [4.2-4.3]. Connections between size distribution and the scratching of the wafer surface have also been observed and reported [4.5-4.6]. Beside the experimental research, however, there is a general lack of models which can accurately predict the performance of consumables, and specially, the abrasive particles. This limits the application of the experimental results for the optimization of the CMP process.

In Chapter 3, a comprehensive model has been proposed to explain the fundamental mechanism and interactions of consumable parameters in solid-solid contact of CMP, including abrasive size and size distribution. It is principally based on three assumptions: a regularly periodic pad topography, normal distribution of abrasive size and plastic deformations over the wafer-abrasive and pad-abrasive interfaces (two-body abrasion). The key idea of the model is that the effects of the consumables on material removal can be attributed to two parameters. One is the number of active abrasives, and the other the material removed by a single abrasive, which is a function of abrasive size. Based on the model, only part of the abrasives, or namely, the *active* abrasives, are involved in the material removal process, Fig. 3.7. An '*active*' abrasive must satisfy two conditions. First, it should be located on the contact area between wafer and pad. The number of abrasives there is proportional to the weight concentration of abrasives in the slurry. Second, the abrasive should be large enough. The wafer will contact the largest abrasives first when a force is applied on the wafer top surface. A gap is formed between the pad and wafer surface in the neighborhood of the larger abrasives. Only abrasives larger than this gap can participate in material removal. Apparently, the distribution function affects both the size of active abrasives, and therefore the volume removed by a single abrasive, and the number of active abrasives.

In Chapter 3, a material removal formulation as a function of the process parameters including the down pressure and velocity as well as consumable parameters including the abrasive size distribution was proposed. The model predictions correlate with experimental results of *MRR* as a function of down pressure.

To fully verify the roles of the abrasive size distributions proposed in the material removal model, however, experimental results of *MRR* as a function of abrasive size distribution are needed. Recently, tungsten (W) CMP experiments have been done by Bielmann et al. [4.3] using five different kinds of abrasive size distributions, which were measured using dynamic light scattering technology. Five different material removal rates were obtained. This chapter discusses the correlation of the model prediction with these experimental results. A detailed view of wafer-abrasive-pad contact is proposed to explain the roles of the abrasive size distribution. Material removal rate formulation as a function of abrasive size distribution is developed and verified.

4.2 Detailed View of Wafer-Abrasive-Pad Contact and Material Removal Rate as a Function of Abrasive Size Distribution

The following detailed view of wafer-pad-abrasive contact, which has been discussed briefly in Chapter 3, is proposed. First, it is proposed that only part of the pad contacts the wafer directly. This part of the pad is called the contact area. This view of wafer-pad contact is shown schematically in Fig. 4.1 (a). The relationship between the down pressure P, contact pressure P_c and contact area A can be obtained based on a contact model as discussed in Chapter 3. The contact area A and contact pressure P_c are functions of the down pressure, pad topography and pad materials. The abrasive geometry is assumed to be spherical. The *active* abrasives involved in the two-body abrasion will locate on the contact area, Fig. 4.1 (b)-(e). When there is no abrasive, the wafer and pad contact area A is a function of the wafer-pad topography and wafer-pad materials. Since the nano-scale *active* abrasives are much smaller than the micro-scale features of the pad topography and the pad surface is quite soft related to the abrasive and wafer surface, the final *effective* contact area A and contact pressure P_c in the situations where abrasives are embedded into the contact area, should be approximately equal to those without abrasives, Fig. 4.1 (e). The trend of this contact, which makes the effective contact area between the wafer and pad as close as possible to that without abrasives, is called 'stable contact'. There are four stages, Figs. 4.1 (b)-(e), from the point that the wafer begins to contact the pad and abrasives to where the final 'stable contact' is realized. When there is no down pressure applied on the wafer top surface, the abrasives will disperse on the pad contact area randomly, Fig. 4.1 (b). The number of abrasives located on the contact area is proportional to the abrasive weight concentration in the slurry; see Chapter 3. The wafer and pad are separated by abrasives and the gap g between them is equal to the size of the largest abrasives, Fig. 4.1 (b). Once down pressure is applied on the wafer top surface, the wafer will initially be supported by abrasives only, Fig. 4.1 (c). The *effective* contact area is approximately equal to $0.25\pi x_{avg}^2 n$, where n is the number of abrasives and x_{avg} the average size of abrasives. The forces applied on each single abrasive are quite large, and all the abrasives are embedded deeply into the pad. A very small gap is formed between the pad and wafer, Fig. 4.1 (c), which is a function of the pad hardness and force applied on the abrasives. The trend for wafer and pad to contact directly will push the abrasives to agglomerate. In this stage, some part of the contact area is due to the direct contact between the wafer and pad, while another part of contact area is occupied by abrasives with closer relative locations, Fig. 4.1 (d). The *effective* contact area, equal to the direct contact area between the wafer and pad plus the area $0.25n\pi x_{avg-a}^2$ occupied by the abrasives, becomes closer to that without abrasives, Fig. 4.1 (d). The force applied on each single abrasive becomes smaller and therefore the gap g' between the wafer and pad becomes larger. Abrasives smaller than this gap g' are pushed off the contact area, indicating the number n of abrasives is decreasing. The process increasing the direct contact between wafer and pad will continue until all the abrasives are finally packed closely, as shown in Fig. 4.1 (e). The *effective* contact area, $A_1 + A_2$, between wafer and pad in this stage is approximately equal to that without abrasives. The direct contact area A_1 will not increase and a stable gap g is finally formed, Fig. 4.1 (e).

From the above discussion, it can be seen that only a portion of the abrasives is involved in the material removal process and the portion of these *active*

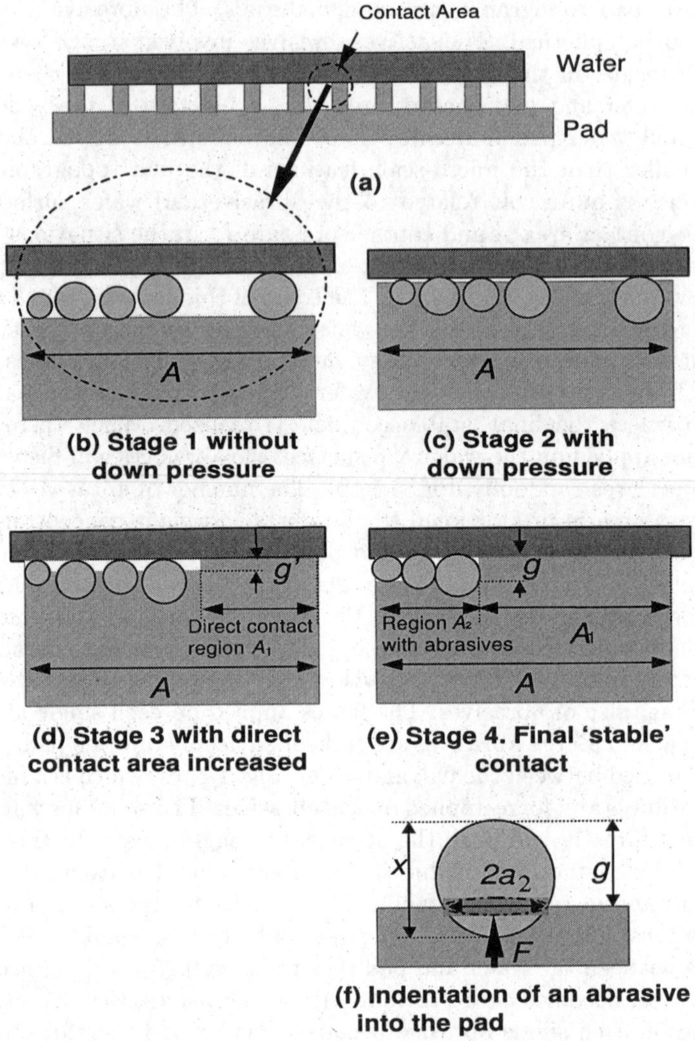

Fig. 4.1. Detailed view of abrasive-pad interaction

abrasives is determined by the gap g, which is the lower bound of the *active* abrasive size. In most situations the distribution of abrasive particles sizes (i.e. diameters), x, satisfies a normal probability density function, Chapter 3,

$$p\{x = x_a\} = p(x_a) = \frac{1}{\sqrt{2\pi}} e^{-\frac{1}{2}(\frac{x_a - x_{avg}}{\sigma})^2}$$

and

$$p\{x \leq x_a\} = \Phi(\frac{x_a - x_{avg}}{\sigma}) = \frac{1}{\sqrt{2\pi}} \int_{-\infty}^{\frac{x_a - x_{avg}}{\sigma}} e^{-\frac{1}{2}t^2} dt \tag{4.1}$$

where x_{avg} is the mean abrasive diameter and σ the standard deviation. Then the area occupied by the abrasives in Fig. 4.1 (e), is approximately equal to

$$\begin{aligned} A_2 &= 0.25n \int_g^{\infty} p(x)\pi x^2 dx \\ &= 0.25n\pi \int_g^{\infty} p(x)x^2 dx \\ &\approx 0.25n[1 - \Phi(g)]\pi x_{avg-a}^2(g) \end{aligned} \tag{4.2}$$

where n is the number of abrasives on the pad asperities before down pressure is applied (i.e. number of abrasives in Fig. 4.1 (b)), and $x_{avg-a}(g)$, a function of g, is the average size of *active* abrasives which is larger than the gap g. The total force F_{total} supported by the abrasives should be equal to the contact pressure P_c times the area A_2, or

$$F_{total} = 0.25n[1 - \Phi(g)]\pi {x_{avg-a}}^2(g)P_c \tag{4.3}$$

Note that n[1-$\Phi(g)$] in Eq. 4.3 is actually the number of *active* abrasives larger than g.

The *active* abrasives are indented into the pad surface under forces F, Fig. 4.1 (f). The indentation depth Δ into the pad surface is $x - g$ for a single active abrasive with size x. The indentation depth of abrasive into the wafer surface has been neglected here since it is much smaller than that in the pad surface, Chapter 3. Then the projected area of the indentation into the pad surface is approximately equal to

$$\pi {a_2}^2 = \pi x(x - g) \tag{4.4}$$

where a_2 is the radius of the projected area of the indentation. Note here that the indentation depth Δ is much smaller than the abrasive size x. Assuming the indentation leads to a fully plastic deformation, Chapter 3 [4.8], the force applied on the single abrasive with size x can be obtained as

$$F(x) = \pi {a_2}^2 H_p = \pi x(x - g)H_p \tag{4.5}$$

where H_p is the hardness of the pad. From Eq. 4.5 it is seen that the force on an abrasive with size g is zero. The force supported becomes larger for larger abrasives.

Once the force applied on each single abrasive is known, the total force supported by all abrasives larger than gap g can be calculated as

$$\begin{aligned} F_{total} &= n\int_g^\infty p(x)F(x)dx \\ &= nH_p\pi\left[\int_g^\infty p(x)x^2dx - g\int_g^\infty p(x)xdx\right] \\ &\approx n[1-\Phi(g)]H_p\pi[x^2_{avg-a} - gx_{avg-a}] \end{aligned} \tag{4.6}$$

From Eqs. 4.3 and 4.6, it is obtained that

$$\begin{aligned} g &= \frac{H_p - 0.25P_c}{H_p}x_{avg-a}(g) \\ &= \frac{H_p - 0.25P_c}{H_p}\left[x_{avg} + \frac{\sigma\, p\,(g)}{1-\Phi\,(g)}\right] \end{aligned} \tag{4.7}$$

For a specified abrasive size distribution, the gap g can be obtained by solving Eq. 4.7. The gap is a function of pad hardness, contact pressure (function of down pressure, pad topography and pad materials) and abrasive size distribution.

Based on the material removal model developed in Chapter 3, the material removal rate can be written as the product of the *active* abrasive number $N= n[1\text{-}\Phi(g)]$ and the volume removed by a single abrasive $cx^2_{avg-a}V$ per unit time, where V is the relative velocity of wafer to pad and c a constant independent of the abrasive sizes; see Chapter 3:

$$\begin{aligned} MRR &= \underbrace{\underbrace{n}_{part1}\left[1-\Phi\left(\frac{g-x_{avg}}{\sigma}\right)\right]}_{part2}\, c\underbrace{x^2_{avg-a}}_{part3}V \\ &= cn\left[1-\Phi\left(\frac{g-x_{avg}}{\sigma}\right)\right]\left[x_{avg} + \frac{\sigma\, p\left(\frac{g-x_{avg}}{\sigma}\right)}{1-\Phi\left(\frac{g-x_{avg}}{\sigma}\right)}\right]^2 V \end{aligned} \tag{4.8}$$

From Eq. 4.8, it could be seen that the role of the gap g on material removal is twofold. When the g is increased, Fig. 4.1, the number of active abrasives is decreased, while the size of active abrasives is increased. Gap g is a function of the contact pressure, pad hardness, and abrasive size distribution, Eq. 4.7. Therefore, the roles of contact pressure, pad hardness, and abrasive size distribution on material removal are also twofold.

It is noted that it is difficult to obtain an accurate analytical solution of Eq. 4.7. An approximate determination of gap g is preferred. The contact pressure should be much smaller than the pad hardness so that the indentation depth $x - g$ into the pad is much smaller than the abrasive size x. Therefore, the *active* abrasive size larger than g is usually located in the tail of the distribution function. For a small change of gap g in this region, the average size of *active* abrasives $x_{avg-a}(g)$ may be taken approximately as a constant value of $x_{avg}+3\sigma$. Then the gap g is approximately

$$g = \frac{H_p - 0.25P_c}{H_p}(x_{avg} + 3\sigma) \tag{4.9}$$

Substituting the approximate solution of gap g (Eq. 4.9) into Eq. 4.8 and considering that the number n of abrasives on the contact area before down pressure is applied is proportional to the weight concentration of the abrasives over the mean volume of a single abrasive, we obtain

$$MRR = \underbrace{\frac{C_6}{x_{avg}^3}}_{part\ 1} \underbrace{\left[1 - \Phi\left(3 - C_7\frac{x_{avg}+3\sigma}{\sigma}\right)\right]}_{part\ 2} \underbrace{\left[x_{avg} + \frac{\sigma p\left(3 - C_7\frac{x_{avg}+3\sigma}{\sigma}\right)}{\Phi\left(3 - C_7\frac{x_{avg}+3\sigma}{\sigma}\right)}\right]^2}_{part\ 3} \tag{4.10}$$

where the C_6 and $C_7 = 0.25P_c/H_p$ are two parameters representing the effects of other consumable parameters including the weight/volume concentration of the abrasives, and process parameters including the down pressure and velocity. The first part of the equation represents the effect of the abrasive size on the total number of abrasives. The x_{avg}^3 is proportional to the average volume of a single abrasive and C_6 includes a term for the weight concentration of abrasives. For the same weight concentration, when the average abrasive size is larger, the total amount of abrasives on the contact area is small, leading to a small material removal rate. It is noteworthy that there is an upper limit for the weight concentration. When the weight concentration is so large that the total contact area is occupied by abrasives, the material removal will saturate and the above formulation will need revision. This is discussed in detail in Chapter 5 and [4.9].

The second part of Eq. 4.10 is a proportional function, representing the effect of abrasive size distribution on the number of *active* abrasives. The value of the proportion is smaller than 1, indicating that only part of the abrasives on contact area is *active*. The third part represents the effect of the average size of *active* abrasives on the material removal. (Note that the material removal is proportional to the square of the average size of abrasives. This square term is introduced by the indentation model in Chapter 3.) The average size of the *active* abrasives is larger than the average size x_{avg} of all abrasives, including *active* and *inactive* abrasives. If the size distribution function of abrasive size is not considered, the *MRR* can be written as a function of the average size x_{avg}:

$$MRR = C_6/x_{avg} = C_6 x_{avg}^{-1} \tag{4.11}$$

Here, part 2 in Eq. 4.10 is equal to one, and part 3 of *active* abrasive size is equal to x_{avg}^2, indicating that all abrasives on the contact area are involved in the material removal process.

It is noted that recently Bastawros et al. [4.10] have proposed three contact modes between the pad and abrasive particles, namely, the full contact mode, the partial contact mode, and the non-contact mode. The partial contact mode and the non-contact mode can be found in Fu et al. [4.13] as well. The contact mode proposed in this paper, which is from the concept of 'stable contact', i.e., the *effective* contact area between the wafer and pad with abrasives approximately equals that without abrasives, does not fall in these three modes. In the proposed partial contact and non-contact mode, similar to that in stage 3 of Fig. 4.1 (d), there is still room for abrasives to be packed close to each other so that the '*effective*' contact area is approximately equal to that without abrasives. An extreme case of the 'stable contact mode' is that the total contact area is occupied by the abrasives and therefore there is no direct wafer-pad contact; see Chapter 5 [4.9]. This is different from the 'non-contact mode' by Bastawros et al. [4.10] and Fu et al. [4.13]. In the 'non-contact mode' proposed in [4.10] & [4.13], there is no need for abrasives to be closely packed to each other, which may be the case for a hard pad due to the small pad deformation [4.13]. In our extreme case, the abrasives should be closely packed next to form a mono-abrasive layer; see Chapter 5. The contact mode proposed by us separates the contact area into two different regions: one, the direct wafer-pad contact region, and the other, the closely packed region of abrasives. It is necessary to point out that the wafer-abrasive-pad contact shown in Fig. 4.1 (e) is just one representative part of this contact mode. The top view of the contact between wafer, abrasives and a single pad asperity with these representative regions dispersing randomly on the contact area is shown schematically in Fig. 4.2 (a) as a comparison with the partial contact mode and full contact mode of Bastawros et al. [4.10] and Fu et al. [4.13], Fig. 4.2 (b). The 'full contact mode', in which each single abrasive is fully wrapped by the pad material as proposed in [4.10], is an extreme case of the contact mode proposed here.

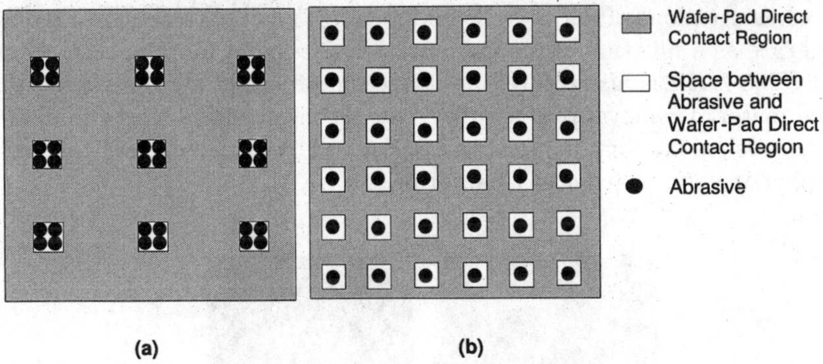

Fig. 4.2. (a). Top view of contact between wafer, abrasive and one single pad asperity in our contact mode (space is approximately equal to zero and effective contact area is approximately equal to that without abrasives.) and (b). Top view of partial contact mode [4.10] [4.13] (effective contact area is smaller than the contact area without abrasives; full contact mode [4.10], where the space is approximately equal to zero and effective contact area is approximately equal to that without abrasives is an extreme case of the contact mode in (a))

It is also worthy to point out that when developing the material removal rate formulations (15) and (25) in [4.13], Fu et al [4.13] does not consider that the number N of abrasives may be a function of abrasive size, which is proportional to $n[1\text{-}\Phi(g)] \propto (1/x_{avg}^3)\Phi(x_{avg}, \sigma)$ in our model. Therefore, while the model on the whole is quite reasonable, any experimental verification of material removal rate formulations (15) and (25) in [4.13] as a function of the abrasive size might lead to some confusion.

4.3 Experimental Verification

Bielmann et al. [4.3] did extensive tungsten CMP experiments using five different distributions of abrasive size. This experimental data is used for model verification. Figure 4.3 shows SEM/TEM pictures of three of the five abrasives [4.3]. The size distribution is measured using dynamical light scattering technology and is shown schematically in Figure 4.4 [4.3]. The average sizes of the five kinds of abrasives (AKP50, AKP30, AKP15, AA07 and AA2) used by Bielmann [4.3] were 0.29μm, 0.38μm, 0.60μm, 0.88μm and 2μm, respectively. By assuming that the size distribution satisfies a normal distribution, the standard deviation of the size distribution can be obtained using the measured data in Fig. 4.4 and are listed in Table 4.1. Bielmann [4.3] changed the weight concentrations of abrasives from 2% to 15% and obtained the material removal rate as a function of median abrasive size under four concentrations, Fig. 4.5. The *MRR* is also plotted as a function of weight concentration, Fig. 4.6. The material removal rate saturates for a concentration larger than 10% for abrasive sizes 0.29μm, 0.38μm and 0.60μm. The explanation of this saturation phenomenon is given in Chapter 5.

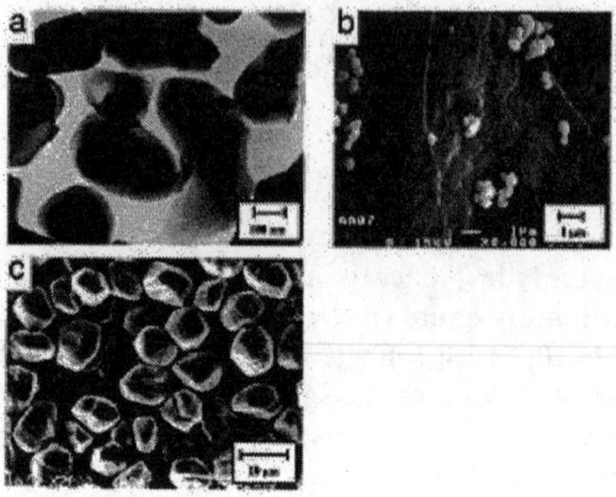

Fig. 4.3. SEM/TEM pictures of abrasive particles [4.3]

It is concentrated here on the linear region, concentration smaller than 10%. Figure 4.7 shows the perfect correlation between the experimental results (average values of the experimental data under the three concentrations 2%, 5% and 10%) and the model predictions of Eq. 4.10. It is found that with an increase in abrasive size, the decrease in material removal rate can be fit with a power function. From Eq. 4.10 we can see this is due to the exponent -3 in the term x_{avg}^3 of part 1 and coefficient C_7 in parts 2 and 3. The C_7 value is a function of consumable values including pad hardness, pad topography and down pressure, but independent of the weight concentration. This implies that the weight concentration does not affect size dependence of the material removal rate. Experimental results support this conclusion. Three values of the exponent– -0.6764, -0.6853 and –0.7756, essentially equivalent, were obtained for slurry concentrations 10%, 5% and 2%. When the weight concentration is 15%, the value –0.4985 of the exponent is much larger. This is because C_6 representing the effect of weight concentration is no longer a constant. Its value for concentration 15% is smaller for the smaller abrasive sizes (0.29, 0.38 and 0.6μm) than for the larger abrasive sizes (0.88 and 2μm); see Chapter 5. This contributes to the increase of the fitted exponent value. Moreover, it is noted that the distribution function of abrasives does play a significant role in the material removal from the fitted exponent values. Otherwise, the values should be equal or close to –1 as indicated by Eq. 4.11, instead of –0.67~ -0.78.

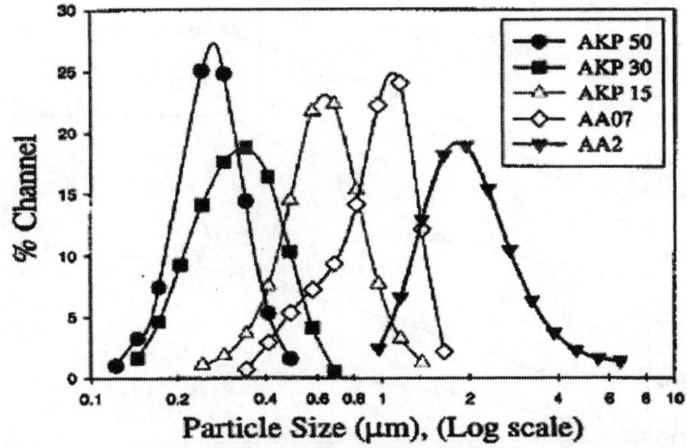

Fig. 4.4. Abrasive size distribution [4.3]

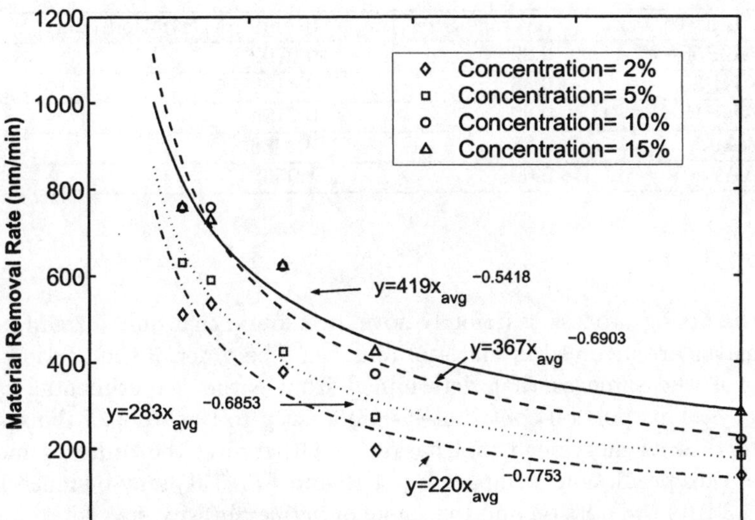

Fig. 4.5. Material removal rate as a function of abrasive size distribution [4.3]

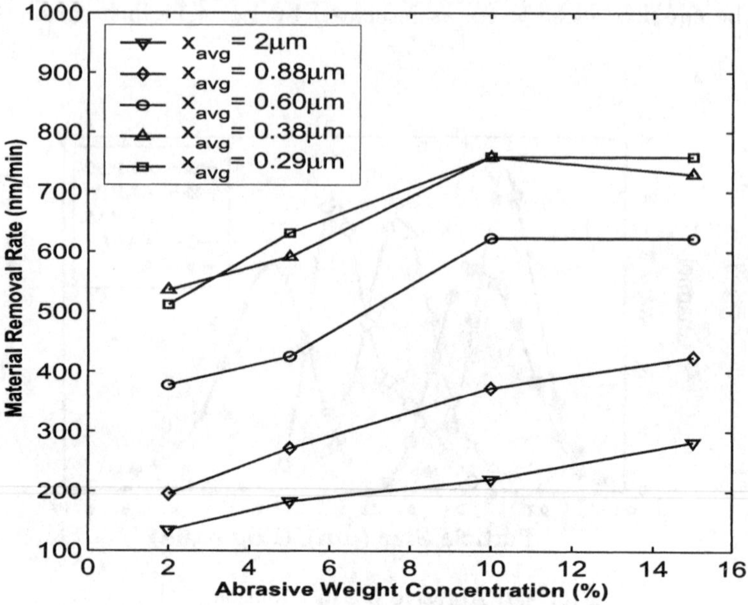

Fig. 4.6. Material removal rate as a function of abrasive weight concentration [4.3]

Table 4.1. The average size and standard deviation of the abrasive size distributions

Abrasive Type	Average Size (μm)	Standard Deviation (μm)
AKP50	0.29	0.0702
AKP30	0.38	0.1189
AKP15	0.60	0.2106
AA07	0.88	0.2888
AA2	2.00	1.0562

From the fitting process described above, it is found that only a small portion of abrasives are involved in material removal. The order of the portion (0.1% - 0.2%) is the same as that determined from earlier experimental data in Chapter 3. This portion contributes significantly to the order of the material removal rate as mentioned in Chapter 3. Otherwise, the order of material removal rate prediction (compare Eq. 4.10 and Eq. 4.11) may be much larger. Table 4.2 lists the portion and the range of *active* abrasive size determined by fitting to the experimental data. The portion of *active* abrasives is close for the five size distributions in these experiments and the average sizes of active abrasives are close to $x_{avg}+3\sigma$, as assumed in approximate estimation of the gap g in Eqs. 4.9 and 4.10. Therefore, a simplified relationship between the material removal rate and abrasive size distribution may be written as

$$MRR \propto \frac{(x_{avg} + 3\sigma)^2}{x_{avg}^3} \tag{4.12}$$

This formulation also fits the experimental results quite well, Fig. 4.8.

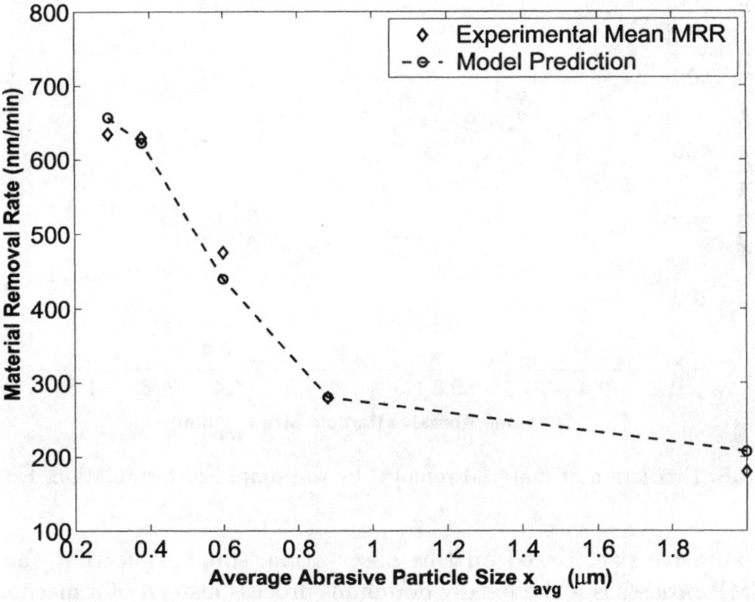

Fig. 4.7. Material removal rate prediction as a function of abrasive size distribution

Table 4.2. The fraction of active abrasive and the range of *active* abrasive size

Abrasive Type	Fraction of *Active* Abrasives (%)	Range of *Active* Abrasive Size (nm)	Average Size of *Active* Abrasives (nm)
AKP50	0.1663	> 496	516
AKP30	0.1619	> 730	764
AKP15	0.1603	> 1221	1281
AA07	0.1612	> 1731	1814
AA2	0.1559	> 5122	5424

The model predicts that the cutting depth of the abrasives into the wafer is proportional to the abrasive size, Chapter 3. It is noted from experimental results in [4.3] that the tungsten wafer roughness after polishing is independent

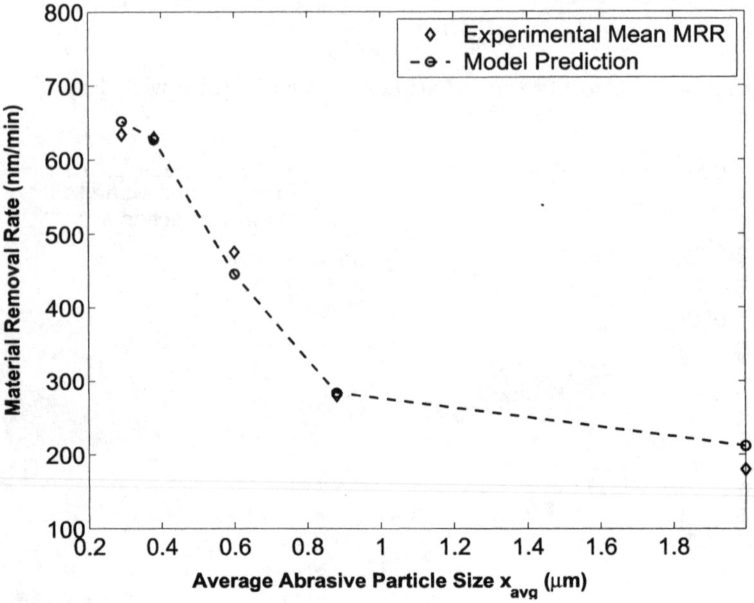

Fig. 4.8. Prediction of material removal by the simplified formulation, Eq. 4.12

of the abrasive size. Based on this observation, some argue that the tungsten CMP process is a chemically dominant process instead of a mechanically dominant process as implied by the model. The roughness due to the chemical reactions in tungsten CMP, including etching, dissolution and re-deposition, which are independent of the abrasive size, may be larger that that due to the abrasion. In another word, the final roughness of a polished tungsten wafer is due to the chemical reaction but not exclusively the mechanical abrasive removal. However, the wafer roughness is mainly a parameter related to the material removal by a single abrasive or chemical reaction. The total material removal is equal to the number of abrasives/chemical reactions times the material removed per single abrasive/chemical reaction. This implies that even though the roughness due to the chemicals is larger than that due to the abrasion (or independent of abrasive particle size), the material removal process is not necessarily a chemical dominant process.

It is also noteworthy that smaller abrasive sizes do not always benefit the material removal rate. The material removal rate is equal to the number of *active* abrasives times the material removed by a single *active* abrasive. When the cutting depth predicted by the model is smaller than 0.1nm, the material removed by a single abrasive may be minimal. In this case, even if the abrasive number is large, the total material removal rate may be still negligible.

Therefore, the cutting depth, a function of abrasive size, wafer-pad contact area and down pressure, sets the lower boundary for abrasive sizes under a certain combination of down pressure and polishing pads. Recently, experimental evidences supporting this have been reported by Zhou et al. [4.14].

4.4 Discussions on Standard Deviation

The work in this chapter shows that the distribution of abrasive size plays an important role in the material removal process. The effect of the standard deviation on material removal can be predicted using Eq. 4.10. Fig. 4.9 shows the prediction for five mean abrasive sizes schematically. It is found that for each single size, a unique value of standard deviation exists. On the right side of the figure, the material removal rate increases with the standard deviation. And on the left side, the material removal rate decreases with the standard deviation. The right side region is called the 'size dominant region', since on this side, the average size of the active abrasives (part 3 in Eq. 4.10) increases with the standard deviation and the increase is much faster than the decrease of the *active* abrasive number (part 2 in Eq. 4.10). The left side region is called the 'number dominant region', since in this region the decrease of the number of active abrasives is much faster than the increase of the active abrasive size. In normal cases, the standard deviation will fall in the 'size dominant region' since it is quite difficult to manufacture fine abrasives with small standard deviations falling in the 'number dominant region'. This means that decreasing the standard deviation, or using fine abrasives, may not lead to a larger material removal rate, if the standard deviation is not in the number dominant region.

In practice, filtering is sometimes used to remove the large abrasives to minimize scratching. While the number of surface scratches is reduced, it is usually found that the material removal is reduced as well [4.15]. Based on the model, this is due to the reduction of the standard deviation in the size domain region. An optimal standard deviation, may be obtained based on the model, satisfying the requirements for minimum surface scratching, and large material removal rate simultaneously.

4.5 Conclusion

The abrasive size distribution plays an important role in the material removal in CMP. Basically, it influences the material removal from two aspects: the

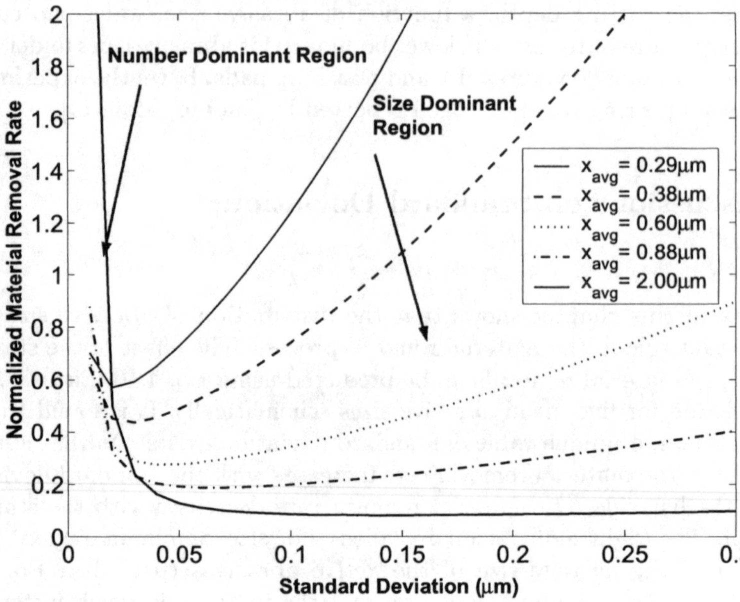

Fig. 4.9. Material removal rate as a function of standard deviation of abrasive size distribution

number of *active* abrasives, and the size of the active abrasives. In this chapter, a model for material removal rates explaining the effects of abrasive size distribution has been proposed and verified. In the future, the application of the model for process optimization, for example, for improving non-uniformity by changing the standard deviation or minimizing the surface scratching and maintaining the material removal rate by using optimal standard deviation, may be attempted.

4.6 References

[4.1] M. C. Pohl and D. A. Griffith, "The importance of particle size to the performance of abrasive particles in the CMP processes," *Journal of Electronic Materials*, Vol. **25**, pp. 1612-1616, 1996.

[4.2] R. Xu, G. Smart and M. Zhang, "Particle characteristics and removal rate in CMP process", *1999 CMP-MIC Conference*, pp. 253-255, Santa Clara, CA, U.S. A., Feb. 1999.

[4.3] M. Bielmann, U. Mahajan and R. K. Singh, "Effect of particle size during tungsten chemical mechanical polishing," *Electrochemical and Solid-State Letters*, Vol. **2**, pp. 401-403, 1999.

[4.4] R. Jairath, M. Desai, M. Stell, R. Tolles and D. Scherber-Brewer, "Consumables for the chemical mechanical polishing (CMP) of dielectrics and conductors," *Material Research Society Symposium Proceedings*, Vol. **337**, pp. 121-131, 1994.

[4.5] G. B. Basim, J. J. Adler, U. Mahajan, R. K. Singh, and B. M. Moudgil, "Effect of particle size of chemical mechanical polishing slurries for enhanced polishing with minimal defects," *Journal of the Electrochemical Society*, Vol. **147**, pp. 3523-3528, 2000.

[4.6] J. Huang, H. C. Chen, J. Y. Wu and W. Lur, "Investigation of CMP micro-scratch in the fabrication of sub-quarter micron VLSI circuits," *1999 CMP-MIC Conference*, pp. 77-79, Santa Clara, CA, U.S. A., Feb. 1999.

[4.7] K. Nicholes, R. Singh, D. Grant and M. Litchy, "Measuring particles in CMP slurries," *Semiconductor International*, Vol. **24**, pp. 201-206, 2001.

[4.8] J. Luo and D. A. Dornfeld, "Material removal mechanism in chemical mechanical polishing: Theory and modeling," *IEEE Transaction: Semiconductor Manufacturing,* Vol. **14**, pp. 112-133, 2001.

[4.9] J. Luo and D. A. Dornfeld, "Material removal regions in chemical mechanical planarization for sub-micron integrated circuit fabrication: coupling effects of slurry chemicals, abrasive size distribution, and wafer-pad contact area," *IEEE Transaction: Semiconductor Manufacturing*, Vol. **16**, pp. 45-56, 2003.

[4.10] A. Bastawros, A. Chandra, Y. J. Guo and B. Yan, "Pad effects on material-removal rate in chemical-mechanical planarization," *Journal of Electronic Materials*, Vol. **31**, pp. 1022-1031, 2002

[4.11] J. Luo and D. A. Dornfeld, "Improvement of non-uniformity in chemical mechanical polishing from the viewpoint of consumable effects based on a developed material removal model: a research proposal," *ESRC Reports (01-2001), University of California at Berkeley*, Berkeley, CA, U. S. A, Jan. 2001.

[4.12] J. Luo and David A. Dornfeld, "Optimization of CMP from the viewpoint of consumable effects," *Journal of the Electrochemical Society*, Vol. **150**, pp. G807-G815, 2003.

[4.13] G. Fu, A. Chandra, S. Guha and G. Subhash, "A plasticity-based model of material removal in chemical-mechanical polishing (CMP)," *IEEE Transaction: Semiconductor Manufacturing*, Vol. **14**, pp. 406-417, 2001.

[4.14] C. H. Zhou, L. Shan, J. R. Hight, S. Danyluk, S. H. Ng and A. J. Paszkowski, "Influence of colloidal abrasive size on material removal rate and surface finish in SiO2 chemical mechanical polishing," *Tribology Transactions*, Vol. **45**, pp. 232-238, 2002.

[4.15] Y. Ahn, C-W. Baek and Y-K. Kim, "Chemical mechanical polishing by colloidal silica-based slurry for micro-scratch reduction," *Wear*, submitted for publication, 2003.

5

MATERIAL REMOVAL REGIONS IN CMP: COUPLING EFFECTS OF SLURRY CHEMICALS, ABRASIVE PARTICLE SIZE DISTRIBUTION AND WAFER-PAD CONTACT AREA

5.1 Introduction

The material removal rate in the solid-solid contact mode of CMP usually increases linearly with the abrasive weight concentration. This is observed experimentally [5.3-5.6] [5.38]. However, this linear increase only holds for a limited range of abrasive weight concentrations. Two exceptions exist. First, when there are no or few abrasives in the slurry, the material removal is usually close to zero [5.8-5.11]. This has been observed for various wafer materials including copper [5.8], aluminum [5.9] [5.42], tungsten [5.10] and silicon [5.11], and slurry recipes. This material removal, mainly due to the chemical etching and dissolution, is much smaller than that due to chemical-mechanical removal with abrasives. This is reasonable. Otherwise, both high and low features on the wafer surface will be removed as aggressively as in isotropic wet etching, and therefore no "planarization" can be realized. A small increase of the abrasive concentration in this region close to the pure chemical removal usually leads to a rapid increase of the material removal, Fig. 5.1. The observed linear increase is much slower than this rapid increase, and usually does not cross zero at zero weight concentration, Fig. 5.1. Second, when the concentration of abrasives is larger than a certain value, say, C_s, the material removal rate will stop increasing but keep constant [5.3], [5.6-5.7], [5.41]. This phenomenon is called material removal saturation. A qualitative explanation of this is that the total contact area between the wafer and pad surface is occupied by the *active* abrasives when saturation occurs, and a further increase in concentration cannot increase the number of abrasives on the contact area. This leads to the material removal saturation since MRR is supposed to be proportional to the *active* abrasive number (Chapters 3 and 4 [5.1-5.2]). This idea is shown schematically in Fig. 5.2.

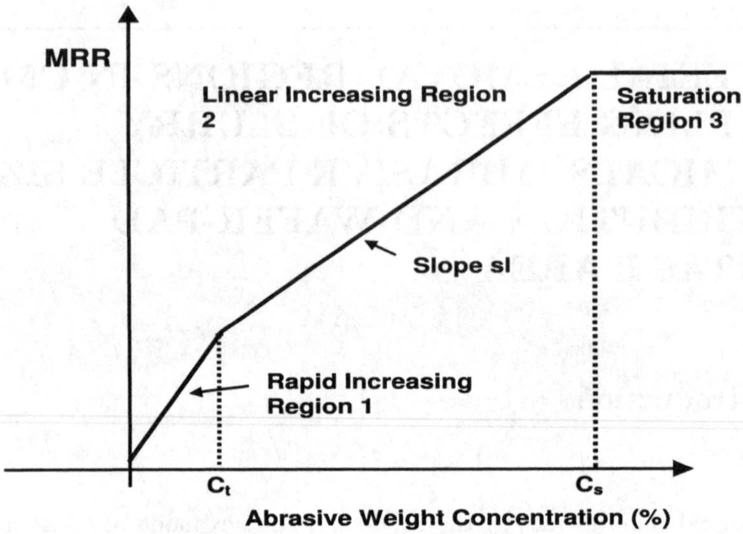

Fig. 5.1. Three regions of *MRR* with the increase of abrasive weight concentration

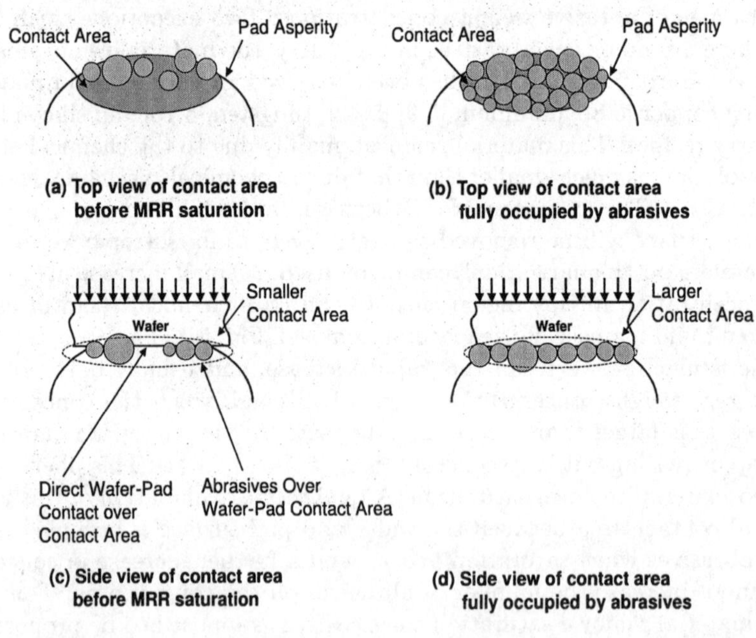

Fig. 5.2. Schematic of two contact modes with different abrasive weight concentrations

Therefore, there are two transitions of material removal regions with the increase of the abrasive weight concentration. First, a transition from a rapid increase region at small abrasive concentrations to a slower linear increase region. And second, a transition from the linear increase region to the saturation region at large abrasive concentrations. This chapter discusses the extension of the material removal model proposed in Chapter 3 to explain these transitions quantitatively. Three regions of material removal are proposed. First, a chemically dominant and rapidly increasing region, whose range is determined by the generation/passivation rate and hardness of the surface passivation layer. Second, a mechanically dominant linear region, where material removal increases proportionally with the weight concentration of abrasives. The material removal in both of the above regions increases linearly with abrasive weight concentrations, although with different slopes. It is proposed that this transition is due to a transition from a wafer surface covered with a single softer material to a surface covered with both softer and harder materials. And, third, a mechanically dominant saturation region, where the material removal saturates because the total contact area is fully occupied by the abrasives. The range of the first region, slopes of the rapid increasing and slower linear increasing region, the two transition concentrations and the saturation material removal rate are proposed to be functions of the chemicals, abrasive size distribution and wafer-pad contact area. Formulations of material removal rate as a function of the abrasive weight concentration are proposed for the above three regions. They agree with the previous model in Chapters 3 and 4. Experimental evidence supporting the existence of these regions is presented. The understanding of the proposed coupling effects of slurry chemicals, abrasive weight concentration, abrasive size distribution and wafer-pad contact area may help to understand the fundamental mechanism in CMP and optimize the process in the future.

5.2 Transition from the First Region to the Second Region: Effects of Bi-layer Property of the Passive Film

It is clear that the slurry chemicals must play an important role in the rapid increase of material removal in Fig. 5.1. The roles of chemicals in CMP may be attributed to a wear-accelerated corrosion [5.12] or a combined corrosion-wear [5.13] depending on the polished materials, slurry composition/environment and the applied mechanical load. If this understanding is correct, CMP is basically a chemical process, which is enhanced by mechanical elements including abrasion. Another understanding of the roles of chemicals, however, is based on Kaufman's model for tungsten CMP [5.10]. Since oxidants such as

H_2O_2, $KMnO_4$, KIO_3, NH_4OH and $K_3Fe(CN)_6$ are frequently used in CMP slurry, Kaufman et al. [5.10] proposed that the chemical action of the slurry is responsible for continuously oxidizing the metal surface to form a thin passive layer (usually several to ten nanometers) which is immediately removed by the slurry abrasives. The fresh metal surface exposed due to the abrasion is then rapidly repassivated and removed. This passivation-abrasion-repassivation process continues until the desired metal thickness is realized. Kaufman's model implies that the CMP is basically a mechanical process enhanced by chemical actions, whose effects on *MRR* are attributed to the material and mechanical properties of this passive surface layer, such as the hardness value, and its generation rate. This passivation-abrasion-repassivation mechanism of material removal has also been applied to explain the metal removal and planarization in copper and aluminum CMP [5.14-5.17]. For silicon and silicon oxide CMP, a near surface change was observed [5.11] [5.18-5.19]. Cook [5.20] proposed that a surface hydrated softer layer (≡SiOH) is formed under chemical attack for silica CMP and it is this layer that is removed. Therefore, a similar mechanism of surface modification, abrasion and then re-modification seems to be at work for silicon and silicon oxide CMP as well.

Based on different slurry chemicals used, Carpio [5.21] pointed out that both dissolution-type chemistry and passivation-type chemistry may exist in CMP, specially, in copper CMP. For example, while ammonium hydroxide (NH_4OH) based slurry passivates the copper surface, nitric acid (HNO_3) may dissolve it easily with the aid of abrasion to expose the copper surface [5.21]. A corrosion inhibitor such as BTA has to be used to protect the recessed areas on the wafer surface from the dissolution. Whether CMP is dissolution-dominant or passivation dominant is determined by chemical parameters such as the pH value. For example, Wang and Tseng found that in aluminum CMP, when the slurry pH value is smaller than 4, the process is dissolution-dominant, when it is larger than 4, it is passivation dominant [5.40]. The phenomena that occur in the dissolution-based CMP may not be able to be explained by the model proposed here. In this chapter, Kaufman's model is extended to explain that the rapid increase of material removal in the region of small abrasive weight concentration is due to a bilayer property of the passive films.

Various micrographic and microchemical examinations of passive films on many metals and alloys have shown that they form as bilayers, consisting of a compact, harder barrier layer underlying a porous, precipitated, hydrated and softer upper layer [5.22-5.24], Fig. 5.3. Recently, X-ray photoelectron spectroscopy (XPS) analysis has demonstrated that in the slurry modified copper surface a Cu(OH) and CuO bilayer may exist [5.25]. The formation mechanism of the passive layer can be attributed to the diffusion of the metal cation and oxygen anions through the passive films and the generation and annihilation of vacancies at the metal/bottom film and upper film/bottom film interfaces. The growth rate of the bilayer on the metal surface including the upper and bottom layers, may be either diffusion-controlled when the films are thick or reaction-controlled when the films are thin. This property of bilayer structure with different microstructures and hardness has also been observed in near surface modification of the silicon and silica structure in the CMP [5.18-5.19]. Trogolo and Rajan [5.18] found in silica CMP the existence of a 2 nm surface layer with lower density than the bulk, below which the density increases to a value greater than the bulk, gradually returning to the bulk density at a depth of 15-20 nm below the silica surface. Although Trogolo and Rajan [5.18] explain the formation of the bilayer structure from a viewpoint of material science, a similar diffusion and reaction mechanism as that in metal, however, may underlie this phenomenon. The detailed discussion on the dynamical film growth mechanism is beyond the scope of this book. Here, a parameter GR is simply introduced to represent the generation rate of the upper and bottom layers. It may be a coupling function of slurry chemical and mechanical elements. Two hardness parameters, H_{w1} and H_{w2}, are used to represent the different material properties of the upper and bottom layers, respectively.

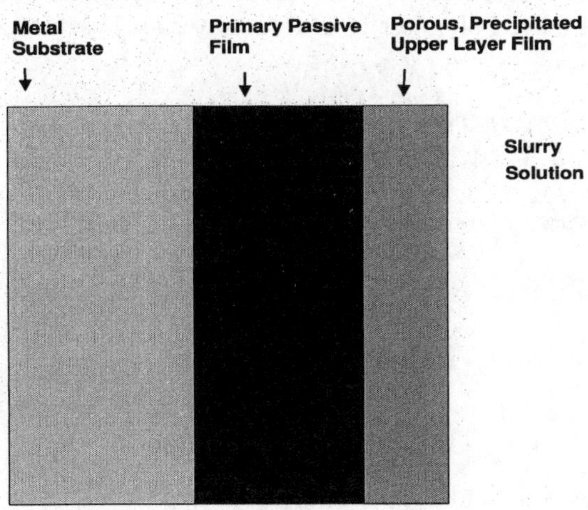

Fig. 5.3. The bilayer-structure of the passive layers [5.22-5.24]

Due to the bilayer nature of the metal passive film, two regions of materials removed with the increase of abrasive weight concentrations are proposed as follows. When the weight concentration of slurry abrasives is small, the (MRR) is close to zero as mentioned before and probably smaller than the growth rate

(GR) of the upper porous layer. This implies that the wafer surface is fully covered with the upper porous softer layer during the polishing, as shown in Fig. 5.4 (a). The material removal rate can be written as a function of the abrasive size distribution, down pressure and the hardness H_{w1} of the upper layer based on the material removal model developed in Chapters 3 and 4. Considering that the *active* abrasive size is approximately equal to $x_{avg}+3\sigma$, the following simplified material removal rate formulation can be obtained:

$$MRR = k_1 \frac{C}{H_{w1}^{3/2} x_{avg}^3} \left(x_{avg} + 3\sigma\right)^2 P_0^{1/2} \tag{5.1}$$

where k_1 accounts for the number N of active abrasives (k_1 includes the portion of the *active* abrasives represented by a normal distribution density function Φ), C is the abrasive weight concentration and $(x_{avg}+3\sigma)^2 P_0^{1/2}$ accounts for the material removed by a single *active* abrasive. Note that if $(x_{avg}+3\sigma)/\sigma$ is approximately constant, the portion of *active* abrasive number represented by Φ is almost independent of abrasive size distribution, which is the case for the experimental data from [5.3] to be used in later sections. From Eq. 5.1, the slope of the material removal as a function of abrasive weight concentration is inversely proportional to the hardness H_{w1} of the upper film. Therefore, the material removal increases rapidly with abrasive weight concentration when the upper layer composed of softer hydrated materials is removed.

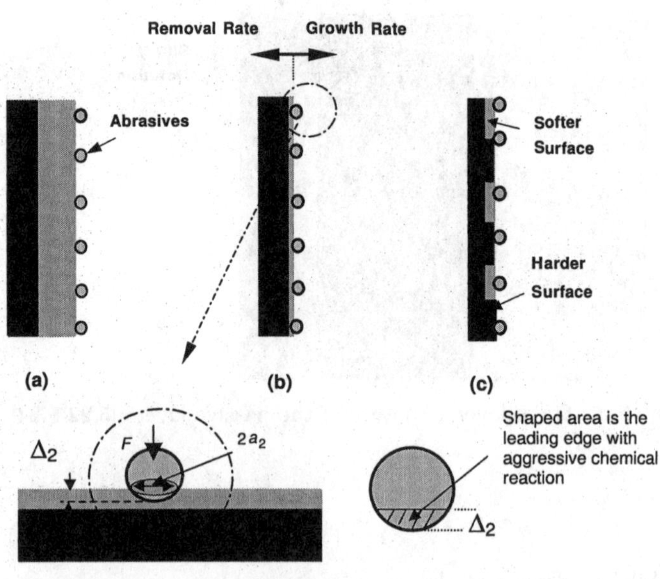

Fig. 5.4. Different layer removed with increase of abrasive weight concentration (or material removal rate) (a) small $MRR < GR$ of upper layer (b) $MRR = GR$ of upper layer (c) $MRR > GR$ of upper layer

With the increase of the weight concentration C of abrasives, the material removal rate increases. At a certain concentration C_t, Fig. 5.1, the material removal rate MRR_1 will be equal to the generation rate GR_1 of the upper softer layer. Then the softer upper layer will be removed as soon as generated, Fig. 5.4 (b). This soft layer is thin and its generation is expected to be reaction controlled. With a continued increase of the abrasive weight concentration, the material removal rate MRR becomes larger than the generation rate GR_1 of the upper layer materials. Under this larger material removal rate, the upper layer material is generated slower than removed and part of the harder bottom layer is exposed. Therefore, instead of a single upper layer of materials, a layer of bi-material composed of both the upper softer materials and the bottom harder materials on the wafer surface (see Fig. 5.4 (c)) will be continuously formed and removed. The ratio of the area of softer materials to the harder materials is determined by the ratio of MRR_1 to MRR-MRR_1. It is proposed that of all the *active* abrasives, part of the abrasives with a number proportional to C_t/x_{avg}^3 is removing the softer upper layer with a in-situ generation rate GR_1, while the other part with a number proportional to $(C-C_t)/x_{avg}^3$ is removing the exposed harder bottom layer. Considering that the material removal rate at concentration C_t is MRR_1 and material removal increases linearly with the *active* abrasive number, the material removal rate with concentration larger than C_1 can be written as

$$\begin{aligned} MRR &= k_1 \frac{C_t}{H_{w1}^{3/2} x_{avg}^3} \left(x_{avg} + 3\sigma\right)^2 P_0^{1/2} + k_1 \frac{C - C_t}{H_{w2}^{3/2} x_{avg}^3} \left(x_{avg} + 3\sigma\right)^2 P_0^{1/2} \\ &= MRR_1 + k_1 \frac{C - C_t}{H_{w2}^{3/2} x_{avg}^3} \left(x_{avg} + 3\sigma\right)^2 P_0^{1/2} \qquad \forall C > C_t \end{aligned} \tag{5.2}$$

This transition of the material removal rate from region 1 to region 2 is shown in Fig. 5.5 schematically. H_{w2} is larger than H_{w1}. Therefore, the linear increase of material removal in region 2 is slower.

The possibility that two different materials, one, the metal substrate, and the other, the metal oxide, are removed simultaneously with the increase of material removal has been noted by Kuo and Tsai [5.26], although they did not realize that a transition may occur on the interface of this bi-layer structure of passive films, instead of on the metal/oxide interface. The transition on the metal/oxide interface, however, is possible with a further increase of the material removal rate, which is not covered in this book.

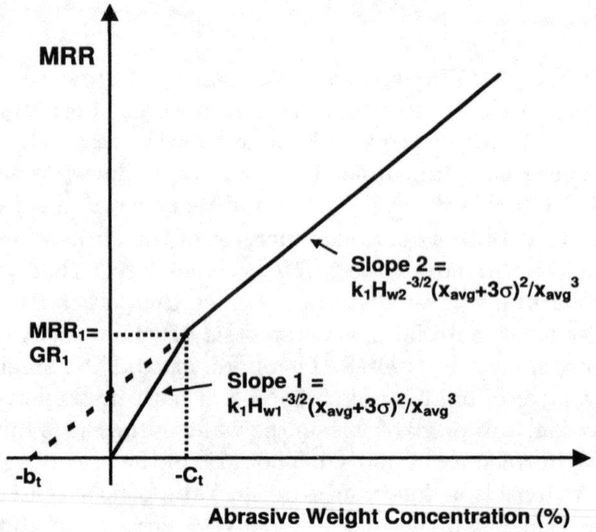

Fig. 5.5. The transition from rapid increase region 1 to slower linear increase region 2

From the above discussion, the chemicals influence the material removal in the region 2 from two aspects: the generation rate of the upper softer layer, and the hardness value of the upper and bottom layers. The addition of oxidation reagents, increasing the growth rate of the upper soft layer, will lead to a higher portion of soft materials on the wafer surface, and therefore, an increase of material removal. That the material removal increases with the addition of oxidization reagents has been observed in experiments for various wafer materials including tungsten [5.4] [5.10] [5.27] and copper [5.5] [5.14] [5.28]. Aluminum can be passivated easily even with exposure to DI water and ambient air. Therefore, the generation rate GR_1 saturates easily. This explains why MRR in aluminum CMP does not change much with the increase of oxidizer concentrations [5.9]. The increases of oxidizer concentrations do not change the basic micrographic structure of passive layers. This implies that their hardness values do not change much with concentrations. Therefore, the lines in the second region should move up and down parallel with each other only with changes of the oxidizer concentration. Fig. 5.6 (a) shows this idea schematically. In later sections, experimental results supporting this will be presented. Changing the type of oxidizers, however, may lead to a change of the hardness. If the metal oxide is too hard, MRR may be small even with a high generation rate. This idea is shown schematically in Fig. 5.6 (b). Therefore, a proper oxidizer for CMP should satisfy two conditions simultaneously: first, high oxidization rate, and second, soft enough metal oxide after the passivation. The hydration of the passive film might be indispensable for CMP in consideration of the softness requirement.

However, it is noteworthy that in the first region the chemicals contribute to the material removal through the hardness value of the upper softer layer only. The passivation rate does not contribute to the *MRR*. This implies that unlike that in the second region, increasing oxidizer concentration will not yield an increase of *MRR* in the first region. Gutmann [5.39] noted that in copper CMP when the *MRR* is large, an increase of oxidizer concentration yields an apparent increase of *MRR*. When the material removal is small, however, it does not cause an apparent change. An explanation of this is that, in the first case, the material removal is in the second region, where the passivation rate contributes to the overall material removal. In the second case, however, the material removal is in the first region, where the passivation rate does not contribute to the overall material removal. This idea can be seen schematically from Fig. 5.6 (a). The material removal saturation with oxidizer concentrations for a fixed abrasive concentration can also be explained by Fig. 5.6 (a) schematically.

Better understanding of the growth mechanism of the passive films is needed to evaluate the transition concentration and passivation rate exactly. An exact evaluation of this value is beyond the scope of this book. Here it is proposed only that the value of transition concentration C_t is independent of the abrasive size distribution and pressure P_0 applied on the wafer top surface. This could be based on the following understanding on the coupling effects of mechanical and chemical elements. At the transition concentration C_t, the material removal rate is equal to the generation rate of the upper porous film. Therefore, from Eq. 5.1, there is

$$GR_1 = k_1 \frac{C_t}{H_{w1}^{3/2} x_{avg}^3} (x_{avg} + 3\sigma)^2 P_0^{1/2}. \tag{5.3}$$

The abrasives are indented into the upper layer surface under force F, Fig. 5.4. The polishing consists of the passage of abrasive particles under load across the wafer surface. The area of the leading edge between a single abrasive and the surface layer is equal to the radius a_2 of the projected indentation circle times the indentation depth Δ_2 and is proportional to $(x_{avg}+3\sigma)^2 P_0^{1/2}$. The *active* abrasive number N is proportional to C_t/x_{avg}^3, Chapters 3 and 4 [5.1-5.2]. Therefore, the total area of leading edge at transition concentration C_t is proportional to $k_1 C_t (x_{avg}+3\sigma)^2 / x_{avg}^3 P_0^{1/2}$. Cook [5.20] proposed that the temperature at the leading edge is much higher due to the bonding breakage. Therefore, the chemical reactions there are more intensive than those on the

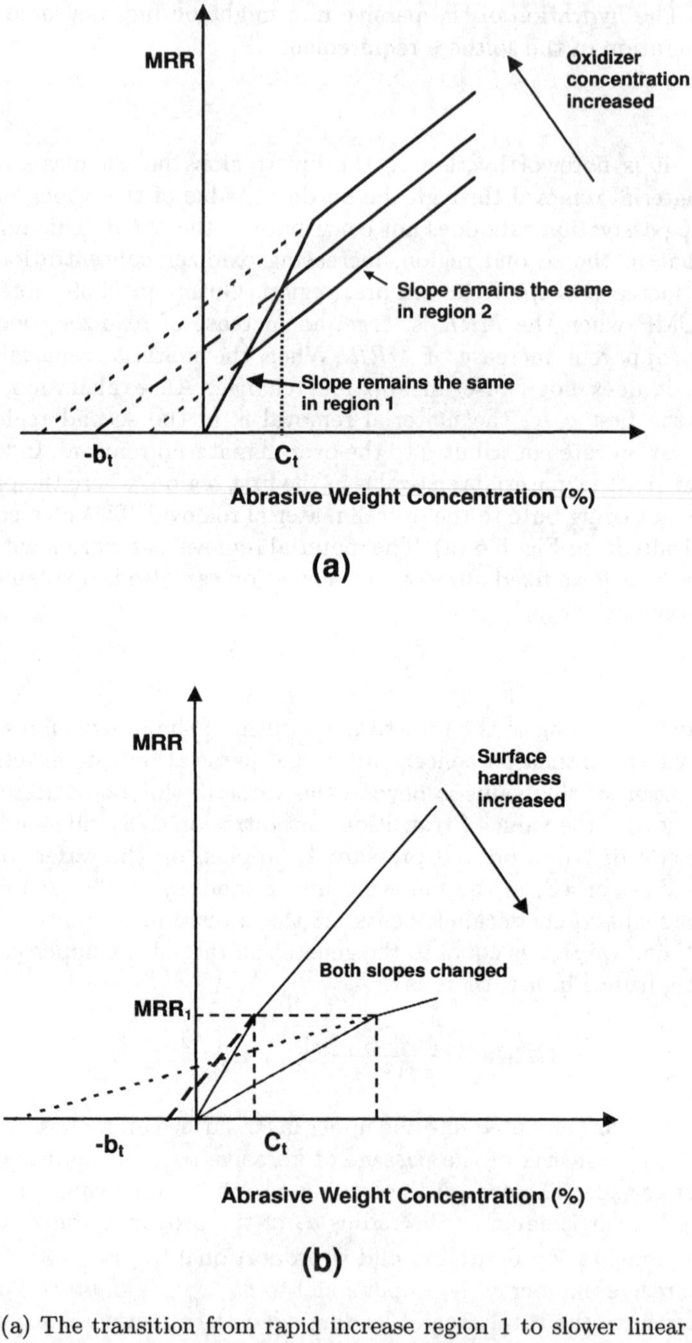

Fig. 5.6. (a) The transition from rapid increase region 1 to slower linear increase region 2 under different upper layer generation rates but the same hardness (b) The transition from rapid increase region 1 to slower linear increase region 2 under the same generation rates but different hardness of upper layer

other areas at the wafer surface. Furthermore, due to the high-energy state of the stressed metal, a higher intensity of broken bonds in the neighborhood of the leading edge may lower the energy barrier to oxidation [5.29]. Therefore, it is reasonable to assume that the generation rate of the upper layer is proportional to this direct contact-leading edge area. This relationship can be simply written as

$$GR_1 = \frac{k_1}{H_{w1}^{3/2} x_{avg}^3} (x_{avg} + 3\sigma)^2 P_0^{1/2} m(C_t, j) \tag{5.4.a}$$

This function decouples the effects of the mechanical elements including down pressure and abrasive size distribution from other elements. The function $m(C_t, j)$ is introduced here to account for the effects of other chemical and mechanical factors j such as the applied potential, chemical concentration, ion diffusion constant, slurry pH values and temperatures. The C_t may influence the generation rate through other elements than the leading edge area and this is accounted for in function $m(C_t, j)$. An assumption in Eq. 5.4.a is that the abrasive size distribution and down pressure do not affect the generation rate through parameters other than the leading edge area. Substitution of Eq. 5.3 into Eq. 5.4.a yields

$$\begin{aligned} GR_1 &= k_1 \frac{C_t}{H_{w1}^{3/2} x_{avg}^3} (x_{avg} + 3\sigma)^2 P_0^{1/2} \\ &= \frac{k_1}{H_{w1}^{3/2} x_{avg}^3} (x_{avg} + 3\sigma)^2 P_0^{1/2} m(C_t, j) \\ &\Rightarrow m(C_t, j) = const \end{aligned} \tag{5.4}$$

which is independent of the abrasive size distribution and down pressure $\Rightarrow$ transition concentration $C_t = const$ independent of abrasive size distribution and down pressure.

For convenience, Eq. 5.2 can be written in another form:

$$MRR = \frac{k_1 (C + b_t)}{H_{w2}^{3/2} x_{avg}^3} (x_{avg} + 3\sigma)^2 P_0^{\frac{1}{2}} \tag{5.5}$$

where $-b_t$ is the intercept of the MRR line with the concentration axis, Fig. 5.5, representing the effects of the passivation rate of the upper softer layer on the total material removal rate. It is easy to know that $b_t = C_t (H_{w2}/H_{w1})^{3/2}$ from Fig. 5.5. Since C_t is independent of the abrasive size distribution and down pressure, b_t should be independent of the abrasive size distribution and down pressure as well. The prediction of the transition under different abrasive size distribution based on this model is shown in Fig. 5.7. Separating the portion of active abrasive from k_1 we obtain the material removal as a function of slurry chemicals, abrasive size distribution and down pressure as follows:

$$MRR = \underbrace{\frac{k(C+b_t)}{H_{w2}^{\frac{3}{2}} x_{avg}^3}}_{part\ 1} \underbrace{\left\{1 - \Phi\left[3 - C_8\left(\frac{x_{avg}+3\sigma}{\sigma}\right) P_0^{\frac{1}{3}}\right]\right\}}_{part\ 2} \underbrace{(x_{avg}+3\sigma)^2}_{part\ 3 = x_{avg-a}^2} P_0^{\frac{1}{2}} \qquad \forall C > C_t \quad (5.6)$$

where $k = k_1/\Phi$. $C + b_t$ andH_{w2} in Eq. 5.6 are taking the place of C and H_{w1} in Eq. 5.1, for the range of $C > C_t$. $C + b_t$ can be taken as an *effective* concentration, and the wafer surface can be taken as covered with a single material with an *effective* hardness H_{w2}. The independence of b_t on the down pressure guarantees the validity of the down pressure dependency of material removal rate in the range of $C > C_t$, which can be written as

$$MRR = C_3\left[1 - \Phi\left(3 - C_2 P_0^{\frac{1}{3}}\right)\right] P_0^{\frac{1}{2}} \qquad (5.7.a)$$

where

$$C_3 = k(C+b_t)\frac{(x_{avg}+3\sigma)^2}{H_{w2}^{3/2} x_{avg}^3} \qquad (5.7.b)$$

is dependent on the chemicals, abrasives and pad, and

$$C_2 = C_8\frac{x_{avg}+3\sigma}{\sigma} \qquad (5.7.c)$$

is dependent on the pad topography (C_8), pad materials (C_8), and abrasive size distribution but independent of chemicals. Note that C_2 reflects the sensitivity of material removal on the pressure distribution and can be used to optimize the non-uniformity from the viewpoint of consumable effects, Chapter 8.

In summary, the rapid increase of material removal with a small increase of the abrasive concentration in the first region is due to the nature of low hardness of the upper porous softer layer. This transition may not be a function of the abrasive weight concentration only. Based on the same reasoning, it may be a function of other mechanical elements such as down pressure and velocity. At small down pressures and velocities, the material removal increase may be much faster since the material removed is the softer upper layer. This is supported by experimental data from Ouma [5.24].

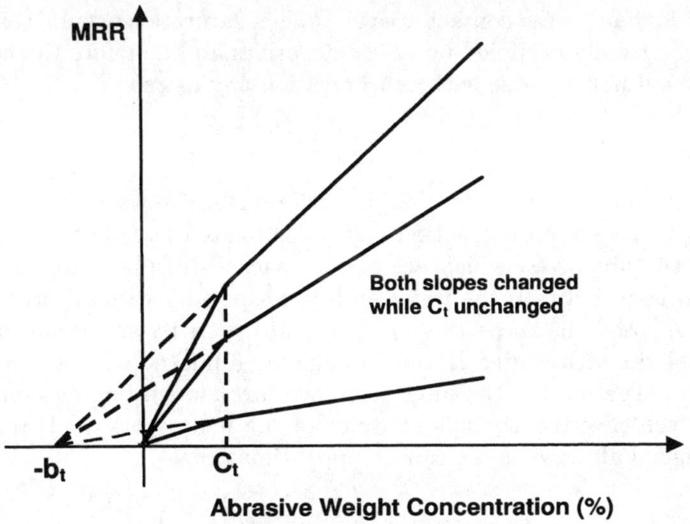

Fig. 5.7. The transition from rapid increase region 1 to slower linear increase region 2 under different abrasive size distributions

5.3 Transition from the Second Region to the Third Region: Effects of Abrasive Size Distribution and Wafer-Pad Contact Area

The second transition is the transition of the material removal rate from the linear slower increase region 2 to a saturation region 3 where material removal stops increasing with the weight concentration of abrasives. It is proposed that this transition is because the contact area between wafer and pad is totally occupied by *active* abrasives.

The material removal formulation as a function of abrasive size distribution in region 2 before the material removal saturates is developed in the last section. The value of $k_1(C+b_t)/x_{avg}^3$ is proportional to the number $N(=N_1+N_2)$ of abrasives on the contact area A [5.2]. When the weight concentration is small, most of the contact is direct contact between the wafer and pad asperities. The contact area A is dependent on the down pressure, pad material and pad topography but independent of the abrasive geometry and abrasive size, Fig. 4.1. When the area is totally occupied by abrasives, however, the abrasives behave as an interfacial layer between the wafer and pad asperities, Fig. 5.8. In this case, the pad asperities are considered to have a higher *effective* Young's modulus. With larger abrasives, the *effective* Young's modulus of the pad is larger. This changes the contact area A'. When saturation occurs, this contact area will be totally occupied by *active* abrasives and therefore the number of abrasives will not increase with concentration any more.

At the material removal saturation, the relationship between the contact area and the abrasive size distribution can be estimated as follows. As shown in Fig. 5.8, the abrasives, which are closely packed together, are taken as an interfacial layer between the wafer and pad asperities with effective Young's modulus E_a^* and thickness x_{avg-a}. The pad asperities are assumed to have a spherical tip with radius R and Young's modulus E_p, as used in Chapter 3. The pad asperity and the interfacial layer are modeled as two springs in a series, so the effective Young's modulus of the pad asperity E_p^*, considering the interfacial abrasive layer, can be approximated as

$$E_p^* = E_a^* E_p \left(\frac{R + x_{avg-a}}{E_a^* R + E_p x_{avg-a}} \right) \tag{5.8}$$

Since not all of the interfacial space between the dashed curve in Fig. 5.8 (b) and the pad asperity is occupied by abrasives, the effective Young's modulus E_a^* of the interfacial layer of abrasives is much smaller than the real Young's modulus E_a of the abrasive materials. This E_a^* can be estimated as $E_a^* = E_a A''/A'$ where A'' is the real contact area between the abrasives and the wafer, and A' the pad contact area occupied by the abrasives, as shown in Fig. 5.8 (b). The A'' is equal to the number of active abrasives times the projected area of the indentation of a single abrasive into the wafer. A' is equal to the number of abrasives times the area occupied by a single abrasive. Therefore, the ratio of A'' and A' can be estimated as $A''/A' = P_c/H_w$ based on Eq. (3.4), where P_c is the contact pressure and H_{w2} the *effective* wafer hardness. Therefore, $E_a^* = E_a P_c/H_{w2}$. Since the contact pressure P_c (around 10^6Pa [5.1]) is much smaller than the wafer hardness (around 10^9Pa for tungsten and 10^{10}Pa for silicon oxide and silicon [5.34]), the effective E_a^* is much smaller than the real E_a. For alumina abrasives, the real E_a is around 500GPa [5.5]. Therefore, the effective E_a^* is around 500GPa/(1000~10,000)= 0.05~0.5GPa.

The effective radius R^* of the asperity considering the interfacial abrasive layer is $R + x_{avg-a}$, Fig. 5.8. Based on contact mechanics ([5.33] & Eq. 3.2 in Chapter 3), an approximate relationship between the abrasive size x_{avg-a} and the contact area A' under force F can be obtained as

$$A' \propto \left(\frac{R^*}{E_p^*} \right)^{2/3}$$

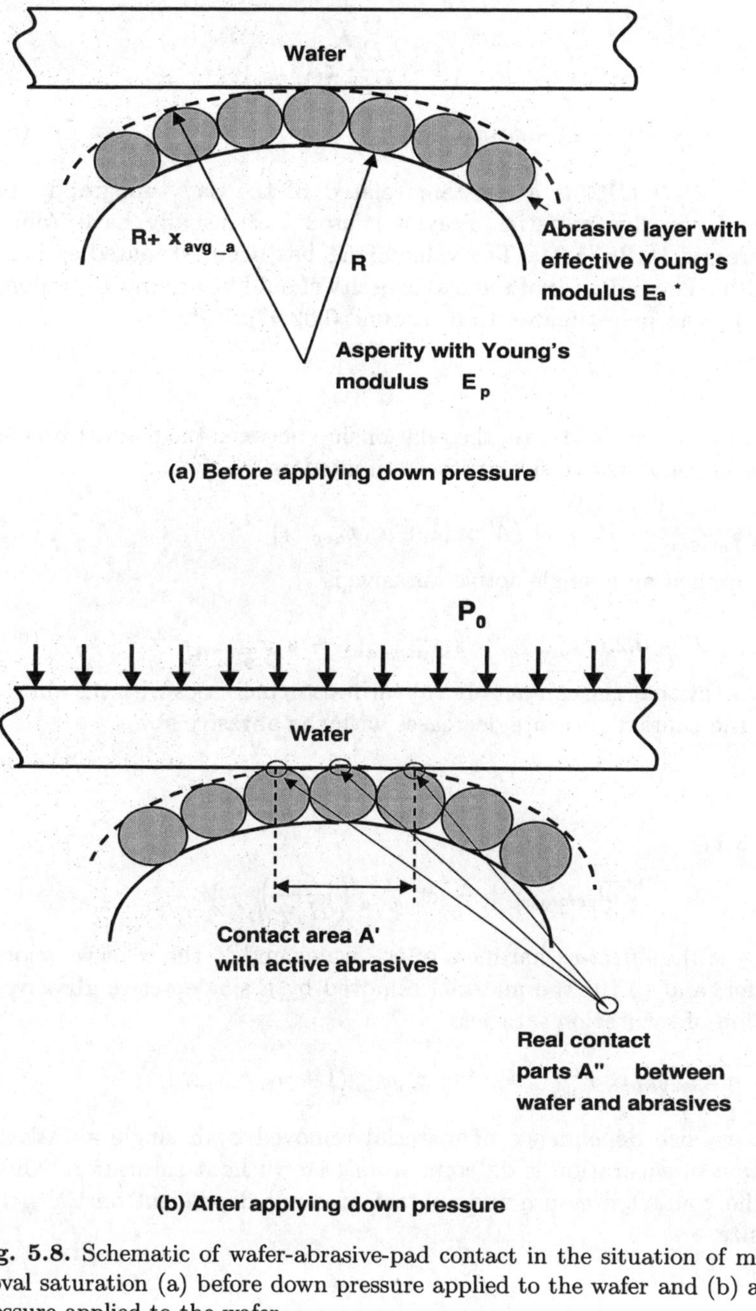

Fig. 5.8. Schematic of wafer-abrasive-pad contact in the situation of material removal saturation (a) before down pressure applied to the wafer and (b) after down pressure applied to the wafer

$$= \left(\frac{E_a^* R + E_p x_{avg-a}}{E_a^* E_p} \right)^{2/3}$$

$$= \left(\frac{R}{E_p} \right)^{2/3} \left(1 + \frac{E_p}{R E_a^*} x_{avg-a} \right)^{2/3}$$

$$\propto (1 + m_1 x_{avg-a})^{2/3} \tag{5.9}$$

where $m_1 = E_p/(RE_a^*)$ is a constant related to the pad topography, pad material and abrasive material. Polymers used here usually have Young's modulus around 1GPa [5.34]. The value of E_a^* has been estimated earlier as 0.05~0.5GPa. The radius R of the pad asperity should be around 10 ~100μm. Therefore m_1 can be estimated to be around 0.02~2μm^{-1}.

Once the contact area is known, the relationship between the contact pressure P_c and the *active* abrasive size can be estimated as:

$$P_c \propto P_0/A' \propto (1 + m_1 x_{avg-a})^{-2/3} \tag{5.10}$$

The force applied on a single active abrasive is

$$F = 0.25\pi x_{avg-a}^2 P_c \propto x_{avg-a}^2 (1 + m_1 x_{avg-a})^{-2/3} \tag{5.11}$$

It is noteworthy that the contact area at saturation increases with the abrasive size while the contact pressure decreases with the abrasive size.

From Eq. 3.11

$$\overline{Vol_{removed}} = \frac{\sqrt{2}}{4} x_{avg-a}^2 \left(\frac{P_c}{H_{w2}} \right)^{3/2} V,$$

where H_{w2} is the *effective* hardness of the wafer and V the relative velocity of the wafer, and (5.10) the material removed by a single active abrasive in the situation of saturation satisfies:

$$\overline{Vol_{removed}} \propto x_{avg-a}^2 P_c^{3/2} \propto x_{avg-a}^2 (1 + m_1 x_{avg-a})^{-1} \tag{5.12}$$

The abrasive size dependence of material removed by a single abrasive in the situation of saturation is different from that without saturation. This is because the contact pressure and contact area are dependent on the *active* abrasive size.

The number N of *active* abrasives in the situation of saturation is:

$$N = A' / \left(0.25\pi x_{avg-a}^2\right) \propto \left(1 + m_1 x_{avg-a}\right)^{2/3} x_{avg-a}^{-2} \tag{5.13}$$

Therefore, the material removal rate MRR_s at saturation should satisfy

$$MRR_s = N\overline{Vol}_{removed} \propto \left(1 + m_1 x_{avg-a}\right)^{-1/3} \propto A'^{-1/2} \tag{5.14.a}$$

The saturation material removal rate decreases with the increase of the abrasive size. It is noted that E_p^* in (5.8) may be approximately equal to E_p when R is much larger than x_{avg-a}. Following similar steps as above by substituting $E_p^* = E_p$ and $R^* = R + x_{avg-a}$, into (5.9), a simplified relationship between MRR_s and the active abrasive size can be obtained as

$$MRR_s = N\overline{Vol}_{removed} \propto \left(1 + x_{avg-a}/R\right)^{-1/3} \propto A'^{-1/2} \tag{5.14.b}$$

Based on the above discussions, the material removal rate can be written as a function of abrasive size distribution and concentration:

$$MRR \propto k_1 \left(C + b_t\right) \left(x_{avg} + 3\sigma\right)^2 / {x_{avg}}^3 \tag{5.15}$$

when $C \leq C_s$,which is a function of weight concentration, and

$$MRR \propto \left(1 + m_1 x_{avg-a}\right)^{-1/3} = \left[1 + m_1 \left(x_{avg} + 3\sigma\right)\right]^{-1/3} \propto A'^{-1/2} \tag{5.16}$$

when $C \geq C_s$,which is independent of weight concentration. If the slope of the linear region is Sl, which satisfies

$$Sl \propto \frac{k_1 \left(x_{avg} + 3\sigma\right)^2}{x_{avg}^3} \tag{5.17}$$

as shown in (5.15), then the relationship between C_s and abrasive size can be written as

$$C_s = MRR_s/Sl - b_t \propto \left[1 + m_1 \left(x_{avg} + 3\sigma\right)\right]^{-\frac{1}{3}} x_{avg}^3 / \left[k_1 \left(x_{avg} + 3\sigma\right)^2\right] - const \tag{5.18}$$

In summary, when $C_t < C < C_s$, the contact area is independent of the abrasive size. A linear relationship between the concentration and the material removal rate exists in this region. When $C > C_s$, the contact area is dependent on the *active* abrasive size. The material removal saturates since the contact area has been totally occupied by the abrasives [5.3], [5.7]. It is noteworthy that this second transition of material removal regions was also modeled by Fu et al. [5.40] and Paul [5.41-5.42] using different approaches.

5.4 Experimental Verification

5.4.1 Two Transitions

Figs. 5.9 and 5.10 show tungsten (W) CMP *MRR* experimental results by Jairath et al. [5.4]. They proposed that the passivation/oxidation of the tungsten surface occurs as: W+ $Ox^{++} \rightarrow WO_3 + Ox^{+}$, where Ox is the oxidant used in the slurry. It is proposed in the model that this passive film is composed of two layers, one a hydrated softer layer, the other a compact harder layer. Both silica and alumina abrasives are used in the polishing experiments. The abrasives and oxidizer concentrations are changed from 1% to 8% and 1× to 5×. It is seen that the material removal increases proportionally with the abrasive weight concentrations. Higher oxidizer concentration yields a higher material removal rate. This is due to the higher growth rate of the upper softer layer. The hardness value of the bottom layer, however, may not change since the micrographic structure does not change. Therefore, for different oxidizer concentrations, the slopes of material removal in Figs. 5.9 and 5.10 keep as constant. This agrees with the model prediction in Fig. 5.6 in the previous section 5.2. Jairath [5.4] did not measure the material removal rate in the absence of abrasives; however, experimental data [5.10] from other sources show that it is minimal. This is also a requirement to avoid isotropic etching for industry-standard slurries. Therefore, the increase of material removal in the first region must be much more rapid than that in the second region. The dashed line in Fig. 5.9 represents a conservative estimation of the material removal increase in the first region. Based on this estimation, the hardness ratio of the bottom layer to the upper layer should be at least 5.4. More experimental data in the first region in the future will be helpful. It is noted that the increase of material removal in the second region is more rapid for alumina abrasives than that for silica abrasives. This indicates that the effects of the abrasive morphology may not be neglected. A more detailed model to account for the effect of abrasive morphology, and probably the chemical durability of the abrasives, can be developed in the future. This transition of material removal regions has been observed in copper and tantalum CMP [5.42] as well, as shown in Fig. 5.11. Note that the material removal in the absence of abrasives is zero.

Jairath et al. [5.4] did not change the abrasive size distribution. Therefore, their experimental results are not sufficient to verify the proposed material removal formulation as a coupling function of abrasive size distribution, slurry chemicals and the wafer-pad contact area. Bielmann et al. [5.3] did tungsten CMP experiments using five different distributions of abrasives and slurry

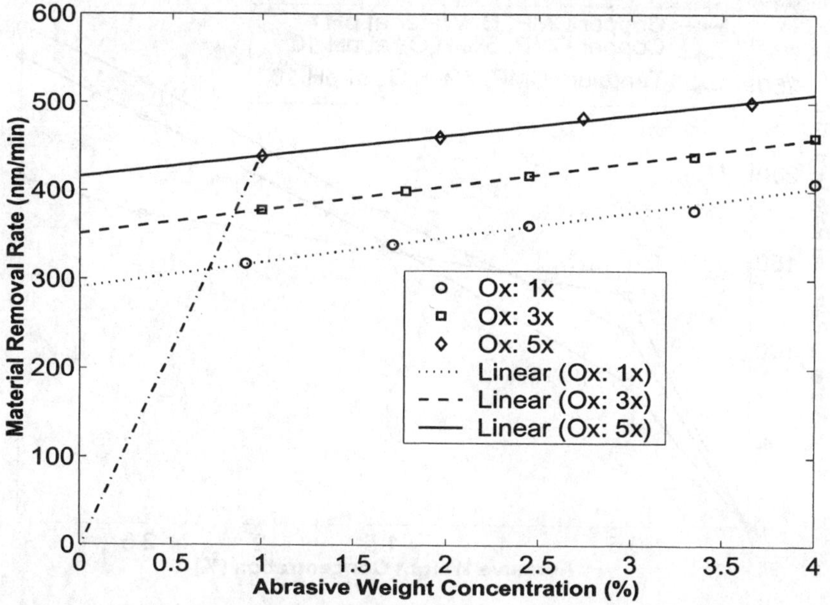

Fig. 5.9. Removal rate of tungsten CMP as a function of oxidizer concentration and silica abrasive weight concentration

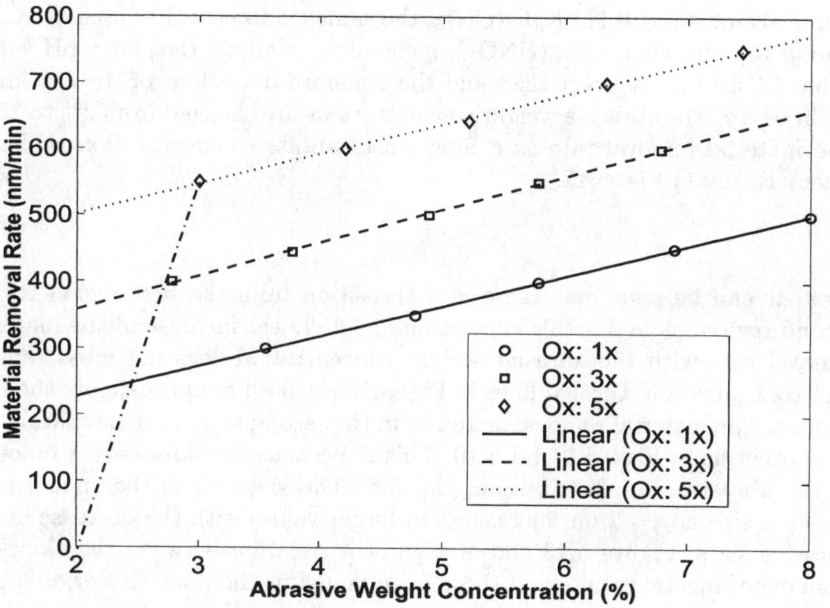

Fig. 5.10. Removal rate of tungsten CMP as a function of oxidizer concentration and alumina abrasive weight concentration

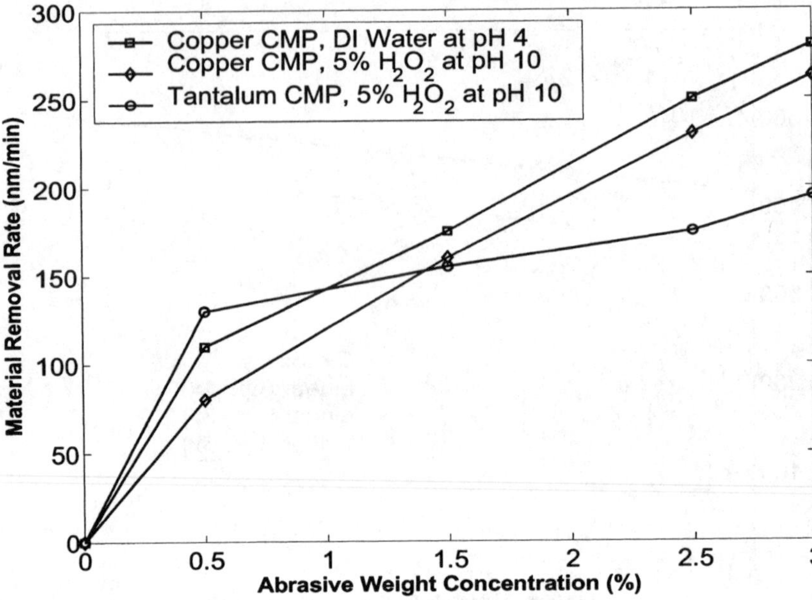

Fig. 5.11. Removal rate of copper and tantalum CMP as a function of alumina abrasive weight concentration [5.38]

chemicals including 0.1M $K_3Fe(CN)_6$, the same oxidizer as that used by Kaufman [5.10], and nitric acid (HNO_3), a chemical to adjust the slurry pH value. Table 4.1 lists the average sizes and the standard deviations of the five kinds of abrasives. The abrasive weight concentrations are changed from 2% to 15%. The material removal rate as a function of abrasive concentration is shown schematically in Fig. 5.12.

First, it can be seen that there is a transition from the first region to the second region, as in Jairth's experiments. The linear increase of the material removal rate with the abrasive weight concentration does not cross zero at zero concentration. Dashed lines in Fig. 5.12 are used to approximate the first regions. The material removal increases in the second region are not parallel to each other as in Figs. 5.9 and 5.10. This is because the slopes are a function of the abrasive size distribution, Eq. 5.6. The slope Sl is the smallest for abrasive size $x_{avg}= 2\mu$m, increasing to larger values with the decrease of the abrasive sizes. Figure 5.13 shows a good correlation between the slopes Sl from experimental results and those from model predictions. In section 5.2, it has been proposed that the transition concentration C_t is independent of the abrasive size distribution. This is demonstrated by the independence of the value of b_t, the intersect of the MRR lines in region 2 with the concentration

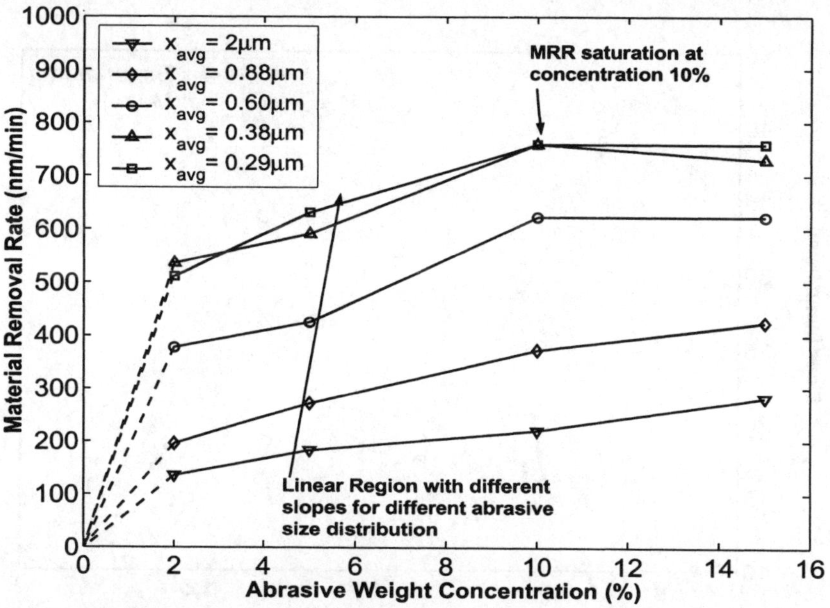

Fig. 5.12. Material removal rate as function of weight concentrations under five different abrasive size distributions [5.3]

axis, on the abrasive size distribution. A constant value of 10 is obtained for b_t by fitting the experimental results, Fig. 5.14.

The second transition can be seen from Bielmann's experimental results as well. The material removal saturation occurs when the concentration is larger than 10% for abrasive sizes 0.29μm, 0.38μm and 0.60μm, while the linear relationship holds for abrasive sizes 0.88μm and 2μm, Fig. 5.12. Based on (5.14), the saturation *MRR* for an abrasive size of 0.29μm should be larger than that for abrasive size 0.6 μm. This can be seen from the data in Fig. 5.12 as well. Based on the model, the ratio of saturation *MRR* for abrasive sizes 0.29μm and 0.6μm should be equal to the ratio 758/622= 1.22 of the contact areas $A'^{-1/2}$ for these two abrasive sizes, Fig. 5.12. The x_{avg-a} for x_{avg}=0.29 and 0.6μm can be obtained approximately as 0.495 and 1.22μm, respectively. Using (5.14), m_1 is obtained as 3. In the earlier section, m_1 is estimated to be 0.02~2. Considering the estimation and experimental errors, the fitted value m_1=3 is reasonable. Using (5.14-b) with the assumption that $E_p^* = E_p$, the fitted asperity tip radius R is equal to 3μm. This range of R value is quite reasonable. Moreover, it is noted that the material removal saturates earlier for smaller abrasives. This agrees with (5.18), which indicates that the saturation concentration is smaller for smaller abrasive sizes.

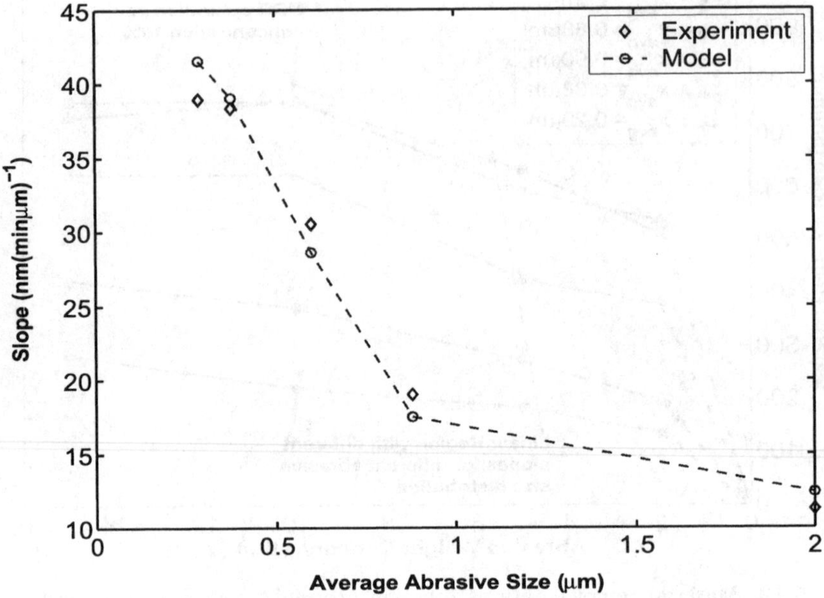

Fig. 5.13. Experimental slopes vs. predictions

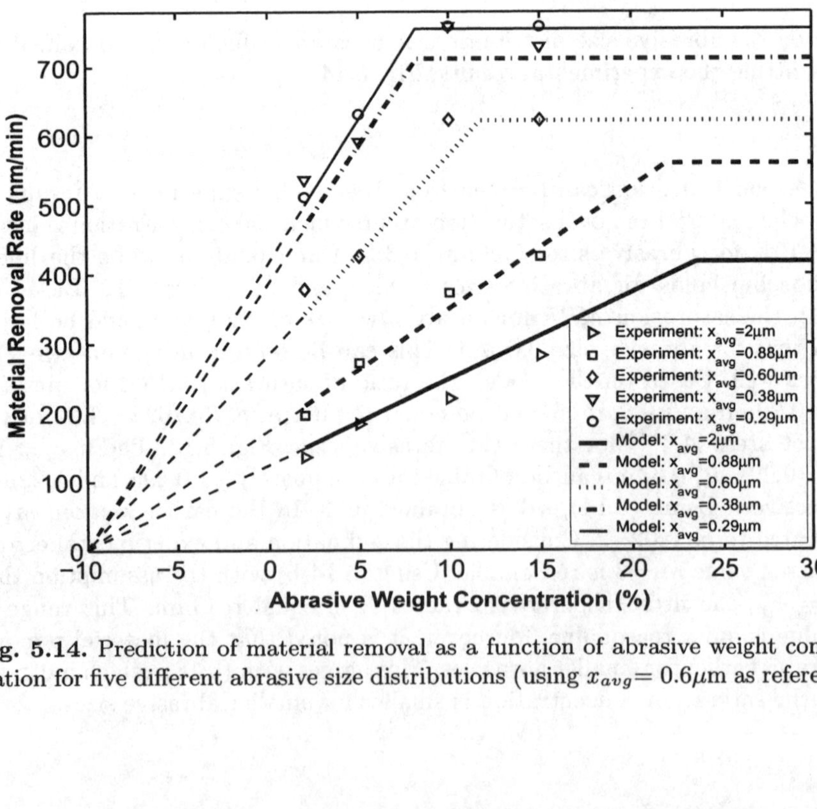

Fig. 5.14. Prediction of material removal as a function of abrasive weight concentration for five different abrasive size distributions (using x_{avg}= 0.6μm as reference)

Substituting m_1 into (5.16), we can predict the ratio of the saturation MRR_s for x_{avg}= 0.29 and 0.38μm as 1.077, indicating that they are close to each other. This correlates with the experimental results in Fig. 5.12. Using Eq. 5.18, the saturation concentrations, C_s, for x_{avg}= 0.29, 0.38 and 0.6μm can be calculated. The predicted slopes in the linear region are 41.573, 39.035 and 28.614 respectively for the above three abrasive sizes. The value of b_t is equal to 10. The MRR_s= 758, 758/1.069=709 and 622nm/min. So the values of C_s are predicted as 8.23%, 8.16% and 11.73%, respectively.

Based on the above discussion, the saturation material removal rate and saturation concentration for x_{avg}= 0.88 and 2μm can be estimated approximately. Using (5.16), the ratio of saturation material removal rate for abrasive size x_{avg}=0.6 and 0.88μm can be obtained as 1.11. Similarly, the ratio for abrasive size x_{avg}=0.6 and 2μm is approximately 1.5. Therefore, the saturation MRR_s for x_{avg}= 0.88 and 2μm are predicted as 560nm/min and 416nm/min, respectively. Using (5.18) and the slope data, the saturation concentration for x_{avg}= 0.88 and 2μm can be predicted as 21.9% and 23.5%, respectively. However, more experimental data is needed to verify this.

In summary, using the material removal curve at x_{avg}= 0.6μm as reference, the material removal as a function of concentration for other abrasive sizes can be predicted and these are plotted schematically in Fig. 5.14.

5.4.2 Coupling Effects of Abrasive Sizes, Polishing Pad and Chemicals on Down Pressure Dependency of *MRR*

Most of the time CMP works in the second region. It is worth looking at experimental results supporting the down pressure dependence of *MRR* predicted by Eq. (5.7-a). This can be seen from the following three aspects: i. the correlation between the Eq. (5.7-a) and the experimental down pressure dependence; ii. the dependence of C_2 in the Eq. (5.7-a) on polishing pads and abrasives, as indicated by Eq. (5.7-c); and iii. the independence of C_2 on slurry chemicals. Jairath et al. [5.4] did silica CMP experiments using three different abrasives, while other parameters including slurry chemicals were kept the same. Fig. 5.15 shows a good correlation between the model predictions and Jairath's experimental results. C_2 changes apparently with the abrasives. The down pressure dependence of material removal using two different polishing pads (one grooved IC1000 pad, and the other perforated IC1000 pad) and two different slurry abrasives (one colloidal abrasives, and the other fumed abrasives) has been investigated by Clark et al. [5.35]. Again, the model prediction correlates with the experimental results, Figs. 5.16 (a) and (b). It is also found that C_2 changes with the polishing pads and abrasives. This is reasonable considering that C_2 is a function of abrasive size distribution, pad materials and pad topography. It is found in Fig. 5.16 (a) that when using colloidal abrasives, the ratio of C_2 for the two different polishing pads is 2. This ratio does not change when using the same two polishing pads for fumed abrasives, Fig. 5.16 (b). This implies the correctness that C_2 was a product of the C_8, a parameter related to pad topography and pad material, and $(x_{avg}+3\sigma)/\sigma$, the abrasive size distribution, as described by Eq. (5.7-c). Passivation dominant copper CMP was done by Ramarajan et al. [5.36] using different chemicals. It is seen from Fig. 5.17 that the model correlates with the experimental results quite well. All $C_2 = 0.24$ do not change with the slurry chemicals, agreeing with the conclusion that C_2 is independent of the chemicals.

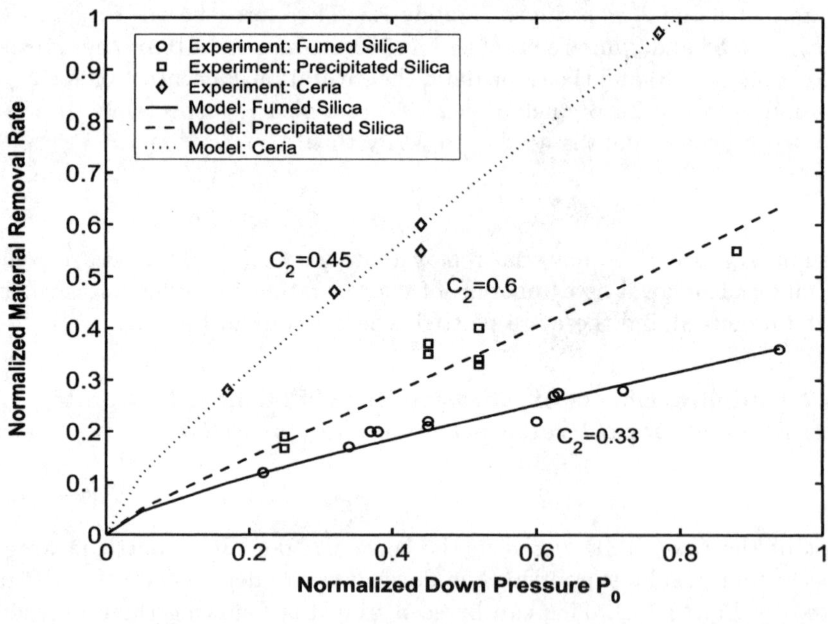

Fig. 5.15. Model prediction vs. experimental data for silicon oxide CMP using three kinds of abrasives

Moreover, a possible transition of down pressure dependency from region 1

$$MRR = \frac{k_2 C}{H_{w1}^{3/2}} \left[1 - \Phi\left(3 - C_2 P_0^{1/3}\right)\right] P_0^{1/2} \tag{5.19}$$

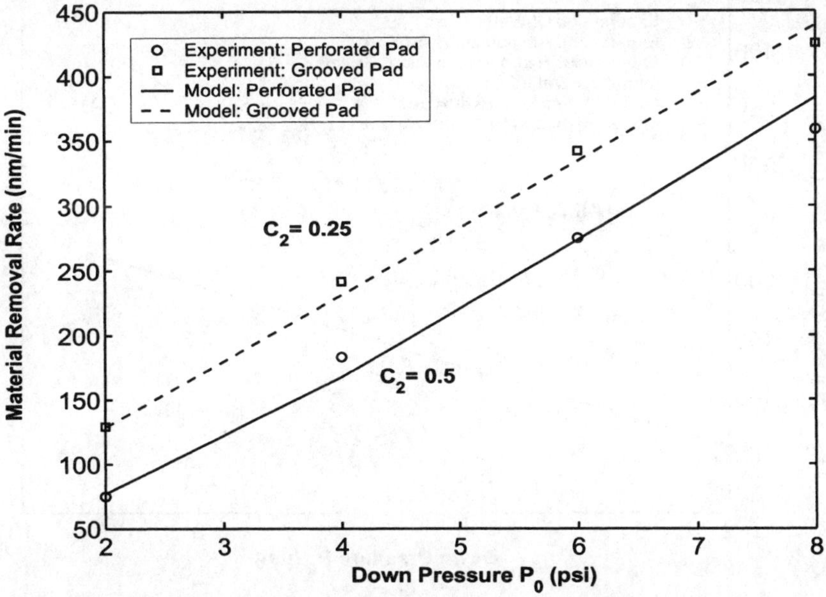

Fig. 5.16.a. Model prediction vs. experimental data for silicon oxide CMP using collodial abrasives and different pads

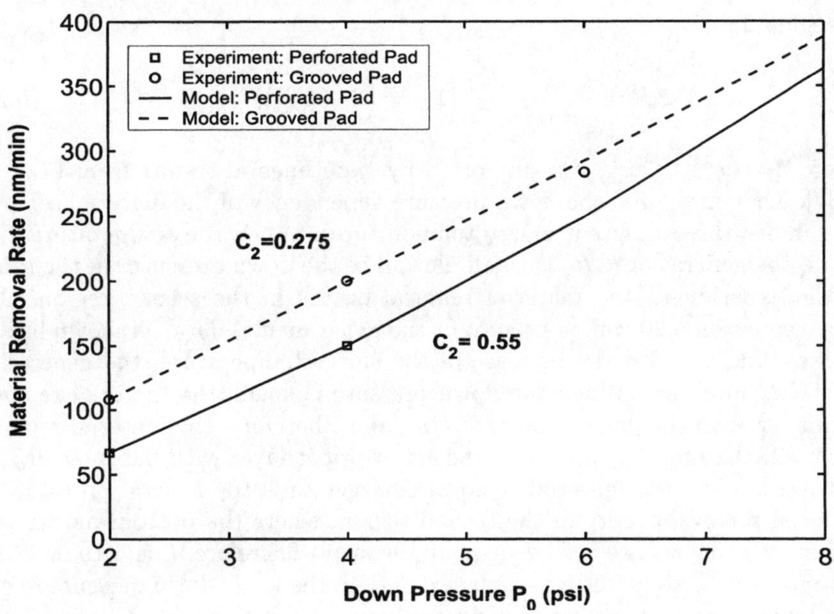

Fig. 5.16.b. Model prediction vs. experimental data for silicon oxide CMP using fumed abrasives and different pads

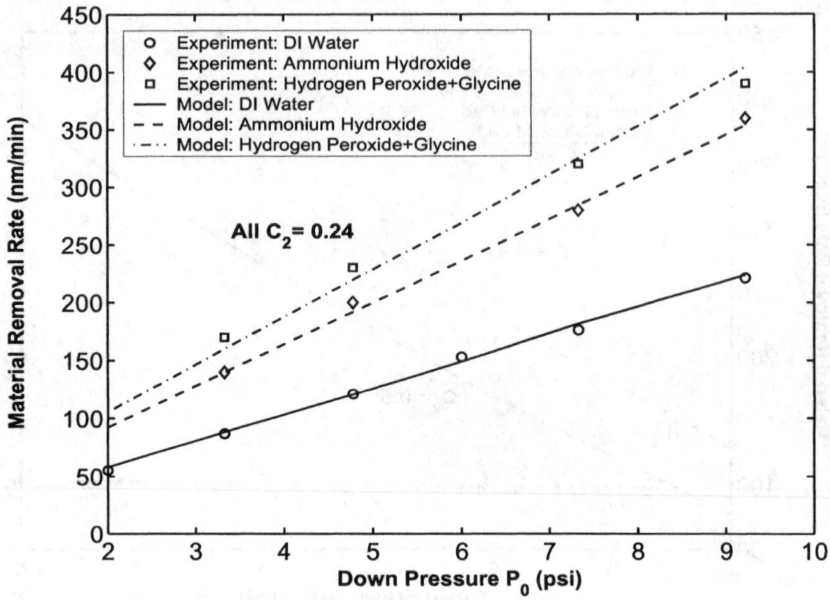

Fig. 5.17. Model prediction vs. experimental data for copper CMP using different passivation chemicals

to region 2

$$MRR = \frac{k_2\,(C+b_t)}{H_{w2}^{3/2}}\left[1-\Phi\left(3-C_2P_0^{1/3}\right)\right]{P_0}^{1/2} \qquad (5.20)$$

where $k_2{=}kx^2_{avg-a}/x^3_{avg}$, is supported by experimental results from Li et al. [5.37]. They measured the down pressure dependency of the material removal rate under three different slurry temperatures. When the temperature T is small, the generation GR_1 is small. For all of the down pressures in the range of the experiment, the material removal occurs in the second region. One single equation 5.20 can be used to fit the experimental data, as shown in Fig. 5.18 by line 5. With the increase of the slurry temperature, the generation rate GR_1 increases. When the down pressure is small, the material removal is smaller than the generation rate GR_1 and, therefore, the material removal occurs in the first region, where the softer upper layer with hardness H_{w1} is removed. When the material removal is larger than the generation rate, the material removal occurs in the second region, where the bottom harder and upper softer layers are removed simultaneously. Therefore, Eqs 5.19 and 5.20 have to be used to fit the experimental data in the lower down pressure region and higher down pressure region respectively, as shown by line 1 to line 3 and line 2 to line 4 in Fig. 5.18.

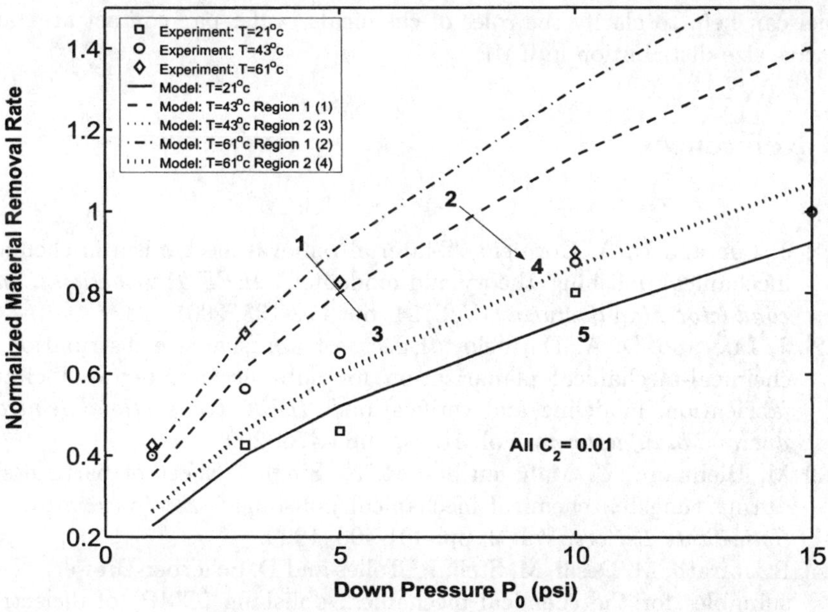

Fig. 5.18. Down pressure dependency of material removal rate under different slurry temperatures

5.5 Conclusion

A material removal rate model as a function of abrasive weight concentration has been developed by extending a material removal model developed earlier. With an increase of the weight concentration of abrasives, three regions of material removal exist: first, a chemically dominant and rapidly increasing region, whose range is determined by the passivation rate and hardness of the surface passivation layer; second, a mechanically dominant linear region, where the material removal is proportional to the weight concentration, and third, a mechanical dominant saturation region, where the material removal saturates because the total contact area is fully occupied by the abrasives. The passive layer of the wafer surface is proposed to be a bi-layer structure. A detailed model is developed to explain that the transition from the first to the second region is due to a transition from a wafer surface covered with a single soft material to a surface covered with both soft and hard materials. The chemicals contribute to the material removal through the generation rate of the upper softer layer of the passive films. The slope of the linear region is a function of abrasive size distribution, and the saturation removal rate is a function of abrasive size distribution and the wafer-pad contact area. These model predictions and relationships are verified by experimental results. The model can help to clarify the roles of chemicals, wafer-pad contact area and abrasive size distribution in CMP.

5.6 References

[5.1] J. Luo and D. A. Dornfeld, "Material removal mechanism in chemical mechanical polishing: theory and modeling," *IEEE Transaction: Semiconductor Manufacturing,* Vol. **14**, pp. 112-133, 2001.

[5.2] J. Luo and D. A. Dornfeld, "Effects of abrasive size distribution in chemical-mechanical planarization for sub-micron integrated circuit fabrication: modeling and verification," *IEEE Transaction: Semiconductor Manufacturing*, Vol. **16**, pp. 469-476, 2003.

[5.3] M. Bielmann, U. Mahajan and R. K. Singh, "Effect of particle size during tungsten chemical mechanical polishing," *Electrochemical and Solid-State Letters*, Vol. **2**, pp. 401-403, 1999.

[5.4] R. Jairath, M. Desai, M. Stell, R. Tolles and D. Scherber-Brewer, "Consumables for the chemical mechanical polishing (CMP) of dielectrics and conductors," *Materials Research Society Symposium Proceedings*, Vol. **337**, pp. 121-131, 1994.

[5.5] Q. Luo, S.Ramarajan and S. V. Babu, "Modification of the preston equation for the chemical-mechanical polishing of copper," *Thin Solid Films*, Vol. **335**, pp. 160-167, 1998.

[5.6] D. Devlieger, M. Desai and C. Fruitman, "Sol gel abrasives in chemical mechanical polish," *Advanced Metallization for ULSI Applications in 1994, Conference Proceedings ULSI-X, 1995 Material Research Society*, pp. 201-205, 1995.

[5.7] D. J. Stein, D. L. Hetherington and J. L. Cecchi, "Investigation of the kinetics of tungsten chemical mechanical polishing in potassium iodated-based slurries—I. Role of alumina and potassium iodate," *Journal of the Electrochemical Society*, Vol. **146**, pp. 376-381, 1999.

[5.8] F. M. Doyle and S. Aksu, "Copper CMP: electrochemistry in slurries containing organic complexing agents" *Small Feature Reproducibility (SFR) Weekly Meeting, University of California at Berkeley,* Berkeley, CA, U. S. A., Feb. 2001.

[5.9] J. Hernandez, P. Wrschka, Y. Hsu, T.-S. Kuan, G. S. Oehrlein, H. J. Sun, D. A. Hansen, J. King and M. A. Fury, "Chemical mechanical polishing of Al and SiO_2 thin films: the role of consumables," *Journal of the Electrochemical Society*, Vol. **146**, pp. 4647-4653, 1999.

[5.10] F. B. Kaufman, D. B. Thompson, R. E. Broadie, M. A. Jaso, W. L. Guthrie, D. J. Guthrie, D. J. Pearson and M. B. Small, "Chemical-mechanical polishing for fabricating patterned W metal features as chip interconnects," *Journal of the Electrochemical Society,* Vol. **138**, pp. 3460-3465, 1991.

[5.11] Y. Moon, "Mechanical aspects of the material removal mechanism in chemical mechanical polishing (CMP)," *Ph.D. Dissertation, Department of Mechanical Engineering, University of California at Berkeley*, Berkeley, CA, U. S. A., 1999.

[5.12] S. Mischler, S. Debaud and D. Landolt, "Wear-accelerated corrosion of passive metals in tribocorrosion systems," *Journal of the Electrochemical Society*, Vol. **145**, pp. 750-758, 1998.

[5.13] I. Garcia, D. Drees and J. P. Celis, "Corrosion-wear of passivating materials in sliding contacts based on a concept of active wear track area," *Wear*, Vol. **249**, pp. 452-460, 2001.

[5.14] J. M. Steigerwald, S. P. Murarka, D. J. Duquette and R. J. Gutmann, "Surface layer formation during the chemical mechanical polishing of copper thin films," *Materials Research Society Symposium Proceedings*, Vol. **337**, pp. 133-138, 1994.

[5.15] J. Larsen-Basse and H. Liang, "Probable role of abrasion in chemo-mechanical polishing of tungsten," *Wear*, Vol. **233-235**, pp. 647-654, 1999.

[5.16] D. Zeidler, Z. Stavreva, M. Plötner and K. Drescher, "Characterization of Cu chemical-mechanical polishing by electrochemical investigations" *Microelectronic Engineering*, Vol. **33**, pp. 259-265, 1997.

[5.17] M. A. Fury, D. L. Scherber and M. A. Stell, "Chemical mechanical planarization of aluminum based alloy for multilevel metallization," *Material Research Society Bulletin*, Vol. **20**, pp. 61-64, 1995.

[5.18] J. A. Trogolo and K. Rajan, "Near surface modification of silica structure induced by chemical/mechanical polishing," *Journal of Materials Science*, Vol. **29**, pp. 4554-4558, 1994.

[5.19] E. I. Hwang and D. A. Dornfeld, "Scratch testing of silicon wafers for surface characterization," *Small Feature Reproducibility (SFR) Annual Workshop, University of California at Berkeley*, Berkeley, CA, U. S. A., Nov. 2000.

[5.20] L. M. Cook, "Chemical processes in glass polishing," *Journal of Non-Crystalline Solids*, Vol. **120**, pp. 152-171, 1990.

[5.21] R. Carpio, J. Farkas and R. Jairath, "Initial study on copper CMP slurry chemistries," *Thin Solid Films*, Vol. **266**, pp. 238-244, 1995.

[5.22] D. D. Macdonald and M. Urquidi-Macdonald, "Theory of steady-state passive films," *Journal of the Electrochemical Society*, Vol. **137**, pp. 2395-2402, 1990.

[5.23] D. D. Macdonald, S. R. Biaggio and H. Song, "Steady-state passive films–Interfacial kinetic effects and diagnostic criteria," *Journal of the Electrochemical Society*, Vol. **139**, pp. 171-177, 1992.

[5.24] N. Sato, K. Kudo and R. Nishimura, "Depth analysis of passive films on iron in neutral borate solution," *Journal of the Electrochemical Society*, Vol. **123**, pp. 1419-1423, 1976.

[5.25] J. Hernandez, P. Wrschka and G. S. Oehrlein, "Surface chemistry studies of copper chemical mechanical planarization," *Journal of the Electrochemical Society*, Vol. **148**, pp. G389-G397, 2001.

[5.26] H-S. Kuo and W-T. Tsai, "Effect of applied potential on the chemical mechanical polishing of aluminum in phosphoric acid base slurry," *Journal of the Electrochemical Society*, Vol. **147**, pp. 2136-2142, 2000.

[5.27] J. Farkas, R. Carpio, R. Bajaj, C. Galanakis, R. Jairath, B. P. Jones and S-M. Tzeng, "Oxidation and etching of tungsten in CMP slurries," *Advanced Metallization for ULSI Applications in 1994, Conference Proceedings ULSI-X, 1995 Material Research Society*, pp. 25-32, 1995.

[5.28] Q. Luo, D. R. Campbell and S. V. Babu, "Chemical-mechanical polishing of copper in alkaline media," *Thin Solid Films*, Vol. **311**, pp. 177-182, 1997.

[5.29] F. C. M. J. M. Van Delft, P. asiski and G. S. A. M. Theunisse, "Plasma etch mechanistic framework used for tribochemical reactions," *Journal of Materials Science Letter*, Vol. **13**, pp. 1370-1374, 1994.

[5.30] J. Luo and D. A. Dornfeld, "Improvement of non-uniformity in chemical mechanical polishing from the viewpoint of consumable effects based on a developed material removal model: a research proposal," *ESRC Reports (01-2001), University of California at Berkeley*, Berkeley, CA, U. S. A, Jan. 2001.

[5.31] J. Luo and D. A. Dornfeld, "Optimization of chemical mechanical planarization from the viewpoint of consumable effects," *Journal of the Electrochemical Society*, Vol. **150**, pp. G807-G815, 2003.

[5.32] D. O. Ouma, "Modeling of chemical mechanical polishing for dielectric planarization," *Ph.D. Dissertation, Department of Electrical Engineering and Computer Science, MIT*, Cambridge, MA, U. S. A., 1999.

[5.33] K. L. Johnson, *Contact Mechanics*, Cambridge University Press, Cambridge, U. K., 1985.

[5.34] J. F. Shackelford, W. Alexander and J. S. Park, *CRC Material Science and Engineering Handbook,* CRC Press, Boca Raton, Florida, U. S. A., 1994.

[5.35] A. J. Clark, K. B. Witt and R. L. Rhoades, "Oxide removal rate interactions between slurry, pad, down force, and conditioning," *1999 CMP-MIC Conference,* pp. 401-404, Santa Clara, CA, U. S. A., Feb., 1999.

[5.36] S. Ramarajan and S. V. Babu, "Modified Preston equation-revisited," *Materials Research Society Symposium Proceedings: Chemical- Mechanical Polishing—Fundamentals and Challenges*, Vol. **556**, pp. 149-154, 2000.

[5.37] W. Li, D. W. Shin, M. Tomozawa and S. P. Murarka, "The effect of the polishing pad treatments on the chemical-mechanical polishing of SiO_2 films," *Thin Solid Films*, Vol. **270**, pp. 601-606, 1995.

[5.38] A. Jindal, S. Hegde and S. V. Babu, "Evaluation of alumina/silica mixed abrasive slurries for chemical mechanical polishing of copper and tantalum," *The Eighteenth International IEEE VLSI Multilevel Interconnection Conference*, pp. 297-306, Santa Clara, CA, U. S. A., Sept. 2001.

[5.39] R. J. Gutmann, "Copper damascene patterning: challenges for chemical mechanical planarization," *Seminar at Dept. of Mechanical Engineering, University of California at Berkeley*, Berkeley, CA, U. S. A., Feb. 2002.

[5.40] Y-L. Wang and W-T. Tseng, "Chemical-mechanical polish of aluminum alloy thin films: slurry chemistries and polish mechanism," *Thin Solid Films*, submitted for publication, 2003.

[5.41] H-S. Kuo and W-T. Tsai, "Effects of alumina and hydrogen peroxide on the chemical-mechanical polishing of aluminum in phosphoric acid base slurry," *Materials Chemistry and Physics*, Vol. **69**, pp. 53-61, 2001.

[5.42] W. Cho, Y. Ahn, C-W. Baek and Y-K. Kim, "Effect of mechanical process parameters on chemical mechanical polishing of Al thin films," *Microelectronic Engineering*, Vol. **65**, pp. 13-23, 2003.

6

ONE AND SEMI-TWO DIMENSIONAL FEATURE- AND DIE-SCALE MODELING FOR THE DAMASCENE PROCESS

6.1 Introduction

The resistance-capacitance (RC) delay due to interconnect metal layers needs to be reduced as the semiconductor industry drives toward faster circuits. Due to its low resistivity and high electromigration resistance, copper has been applied in sub-micron and deep sub-micron integrated circuits as an interconnection substitute for aluminum [6.1]. However, the problem for copper is that it is hard to be patterned by conventional dry and wet etchings. One method to resolve this problem is the copper damascene process.

The whole process flow of the damascene process is shown schematically in Fig. 1.5. First, trenches and vias are formed in the inter metal dielectric (IMD) layer using the conventional etching methods. A barrier layer and copper are subsequently filled into the trenches and vias. CMP is used afterwards to remove the bulk/overburden copper and barrier layer from the top of the oxide surface, with the copper left in the trenches/vias as interconnect lines/plugs.

As it is the only method that can provide global planar surfaces and pattern copper with a large process window, CMP is indispensable for the copper damascene process. However, there are two problems with copper CMP, one, copper dishing, and the other, oxide erosion [6.2-6.12]. Ideally, CMP should be stopped right over the dielectrics surface, Fig. 6.1 (a); unfortunately, to prevent the circuit short, an overpolishing after the exposure of the dielectric surface is usually needed to ensure that all the coppers are cleared off the oxide. The material removal of copper is non-uniform at both wafer and die scales (usually referred to as with-in wafer non-uniformity (WIWNU) and with-in die non-uniformity (WIDNU)). Therefore, the overpolishing will cause oxide or IMD thickness reduction in regions where the copper is removed faster. This phenomenon is called oxide erosion [1]. Besides the oxide erosion, as the material removal rate of the copper is usually larger than that of the oxide, a step will be formed between the oxide dielectrics and copper lines during the overpolishing, leading to a copper thickness reduction relative to the oxide. This is referred to as copper dishing. The total copper thickness loss, equaling the oxide erosion plus the copper dishing, is defined as copper thinning. Fig. 6.1 (b) shows the copper dishing and oxide erosion schematically [2]. Excepting copper CMP, erosion and dishing are also problems associated with other damascene processes such as shallow trench isolation.

The oxide erosion and copper thinning are undesirable for several reasons. First, the thickness reduction of dielectrics leads to a larger parasite capacitance between interconnection metal layers. Second, the copper thinning increases the copper line resistance. Both of them cause a larger RC delay and a slower circuit speed. Moreover, the copper dishing generates a non-planar surface, propagating the non-planarization to the IMD layers to be deposited above the copper surface.

Topography evolution models are needed to explain the formation of copper dishing and oxide erosion, at both feature and die scales. The key topography evolution model is a step height evolution model. One such model was first developed by Stine et al. [6.18] for oxide or ILD/IMD CMP. They assumed that the pad is hard and touches the raised part of the oxide features only. Based on the model, the step height decreases linearly with time. When the pad is soft or the initial step height is small, the pad may touch both the upper and bottom parts of the steps simultaneously. In this case, Tseng et al. [6.9] and Lai et al. [6.8] proposed that the material removal rates at the upper and bottom parts of the step converge exponentially to the blanket MRR, leading to an exponential reduction of the step height. These two models were combined by Smith et al. [6.20] into a two-stage model. They proposed

[1]In the past, the thickness reduction relative to the field oxide was taken as oxide erosion. This is probably due to its easiness of measurement using commercial metrology tools. However, it is inaccurate. The field oxide also has a thickness reduction, which is referred as the field oxide erosion. The oxide erosion should be equal to the oxide thickness relative to the field oxide plus the field oxide erosion, Fig. 6.1 (b).

[2]Besides the dishing and erosion, the surface scratches caused by the abrasive abrasion is often observed with copper CMP. The particle-scale material removal model in Chapters 3, 4 and 5 [6.13-6.17] will help understand the formation mechanism of the scratch and its reduction.

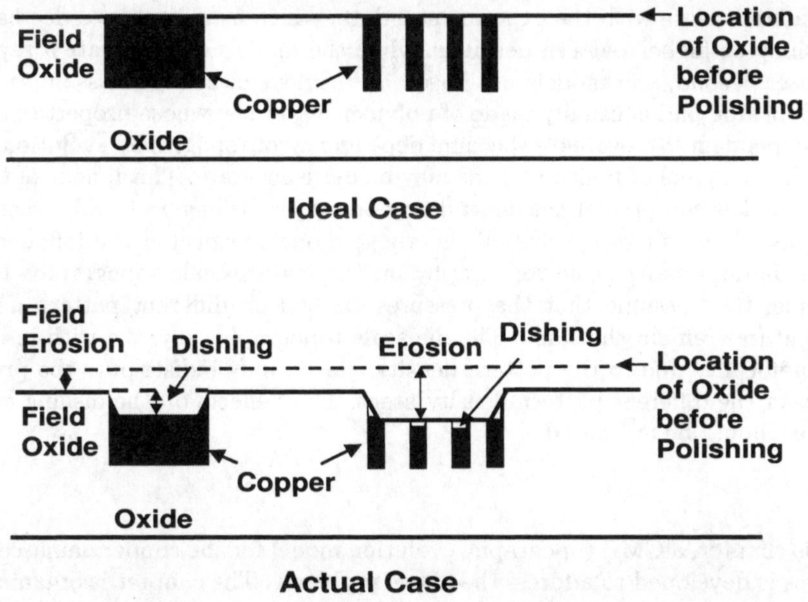

Fig. 6.1. Ideal and actual topography after CMP in copper damascene process

that at the beginning of the planarization, the pad touches the raised part of the step only and with the reduction of the step, the pad touches both the upper and bottom parts of the step at a certain time point, depending on the local pad compression and oxide removal properties. Tseng's model [6.9] is for the planarization of one single material. It is extended by Grillaert et al. [6.21] and Tseng et al. [6.22] into a step height evolution model on the planarization of two different materials. Depending on the material removal rates of the two materials and the stiffness of the pad, the step is proposed to increase exponentially with time. The above models of Stine [6.18], Tseng [6.19], and Smith [6.20] can be used to model the bulk copper removal in the copper damascene process before the oxide dielectric is exposed, while the models of Grillaert [6.21] and Tseng [6.22] can be applied to model the overpolishing stage after the oxide dielectric is exposed. A combination of these models yields a copper damascene model by Tugbawa et al. [6.23-6.24], covering both the bulk and overpolishing stages.

Although the copper damascene process model by Grillaert [6.21], Tseng [6.22] and Tugbawa [6.23-6.24] are successful in certain respects, say, the fact that the prediction of the oxide erosion as a function of pattern densities correlates qualitatively well with the experimental data, there have several shortcomings. First, the prediction of the copper dishing as a function of pattern densities does not correlate with the experimental data, which usually show a decrease of dishing at higher pattern densities, while the models demonstrate a rapid increase. Second, the models are based on a linear elastic pad assumption. The polishing pad is usually made of polymer materials, whose properties are time dependent. To evaluate the time dependency of topography evolution, a viscoelastic model of polishing pads may be more accurate. Third, none of the above models can predict the dependency of copper dishing and oxide erosion on copper line width. Fourth, all the above models neglected the influences of the die and wafer scale topography on the feature-scale topography. For example, they assume that the pressures applied on different pattern density features remain the same. The die scale topography, say, the with-in die non-uniformity due to the pattern density effects, may redistribute the pressure over the different pattern density areas. Their effects on the dishing and erosion should be evaluated.

In this chapter, a CMP topography evolution model for the copper damascene process is developed to address the above problems. The chapter is organized as follows:

i. First, a one dimensional topography evolution model is developed using spring and dashpot models of the polishing pad. The dishing and erosion are modeled as functions of input parameters including pattern density, pad stiffness and material removal properties of copper and oxide. Different pad models are used at the feature and die scales, one considering the pad asperity layer only, and the other considering the bulk pad only.
ii. Then, to account for the line width dependency of the dishing and erosion, a two-dimensional boundary element model and a contact mechanics model are applied to evaluate the vertical pad deformation over the features. An *effective* stiffness of pad asperity layer is obtained from the two-dimensional deformation. It is subsequently input into the one-dimensional model to evaluate the dishing and erosion as a function of the line width. This model is not a pure two dimensional model and is referred to as a semi-two dimensional model.
iii. At the die-scale, the effects of vertical directional die-scale topography on feature scale topography evolution are evaluated.
iv. In the current version of the model, the die is assumed to be composed of pattern density cells, each with a fixed pattern density. The size of the cells is assumed to be sufficiently large that they can be considered isolated from each other. In this case, the influences of the lateral directional die-scale topography on the feature is negligible. In reality, the distribution of the feature pattern density may be random and in this case the lateral die-scale topography will affect the topography evolution at the feature scale. Based on a physical understanding of the semi-empirical concept of the *effective* pattern density proposed by Ouma [6.26], a scheme of

lateral directional die-feature scale integration for topography evolution modeling is discussed in the last section of this chapter.

6.2 Framework of Topography Evolution Model

The topography evolution in CMP is a contact-wear problem, as formulated by Chekina et al. [6.25]. The pressure distribution over the wafer-pad interface is uneven due to the non-uniform topography. With the planarization, this pressure non-uniformity leads to an update of the topography due to different material removal rates at different locations. The updated topography then will redistribute the pressure distribution over the interface. Based on this understanding, the topography evolution in CMP is an iteration of two processes: one, a pressure re-distribution process and the other, a topography update/non-uniform material removal process. This iteration of pressure distribution, non-uniform removal, topography update, and then pressure re-distribution continues until the planarization is finished.

Thus, two sub-models can be formulated for the topography evolution: first, a mechanics model to redistribute the pressure and second, a material removal model (function of pressure) to update the topography. The pressure distribution prediction by the mechanics model is used as input by the material removal model to update the topography, and visa verse. If dividing the total polishing time into N polishing time blocks Δt, the relationship between the two models at time t_i is

$$P(t_i) = x[Y(t_{i-1})] \tag{6.1}$$

$$Y(t_i) = y[Y(t_{i-1}), P(t_{i-1})] \tag{6.2}$$

where index i denotes the ith time block, t_i is equal to $i\Delta t$, functions x and y represent the mechanics model and the material removal model respectively, $P(t_i)$, $P(t_{i-1})$, $Y(t_i)$ and $Y(t_i)$ are the pressure distribution and topography at time t_i and t_{i-1} respectively. Knowing the initial topography, the topography at time t_i can be obtained by repeating the above process from $t=$ 0 to $t=$ t_i for i times. The framework of the above process is shown schematically in Fig. 6.2.

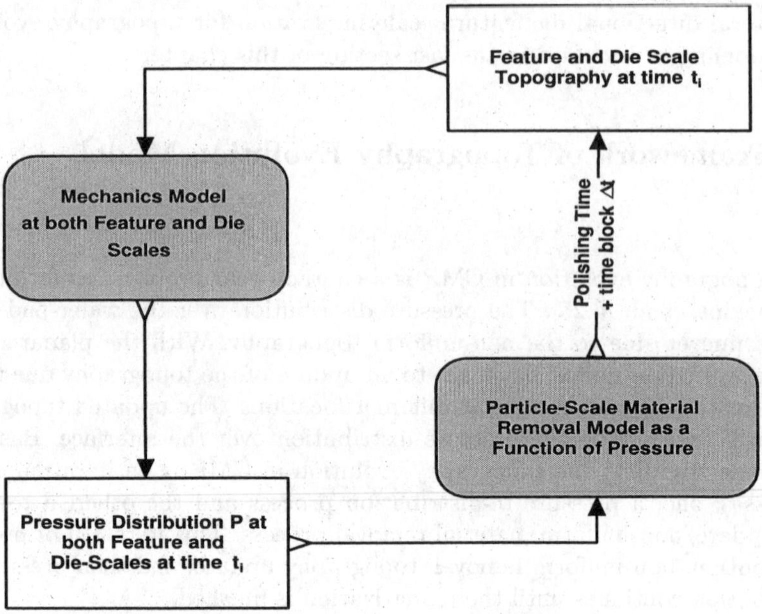

Fig. 6.2. Framework of topography evolution model

6.3 Definition of Model Parameters

6.3.1 Feature-Scale Wafer Topography

At the feature scale, the wafer topography can be represented by a series of steps S. Fig. 6.3 (a) and (b) show schematically a step before and after dielectrics are exposed. These steps are initially introduced from processes before CMP such as chemical vapor deposition and electroplating. Usually, the initial step height is around several hundred nanometers. During the planarization process, the step height S continuously changes. Before the oxide is exposed, it is primarily a 'copper' step—both upper and bottom parts of the step are copper materials. After the oxide is exposed, the step is composed of two different materials—the upper part is the oxide material while the bottom part is the copper material. The final step height between the oxide and copper, as shown in Fig. 6.3 (b), is the copper dishing d.

Besides the step height S, several other parameters are needed to describe the topography: first, the height H of the upper part of the step to the silicon substrate (equal to the copper thickness H_{cu} plus the oxide thickness H_{ox}); second, the copper line width W_{cu}; and third, the oxide space width W_{ox}.

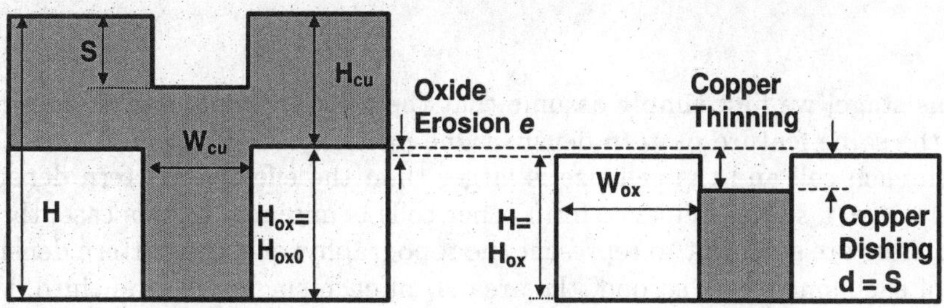

Fig. 6.3. (a) Feature scale topography before dielectric material is exposed and (b) feature scale topography after dielectric material is exposed

These are shown schematically in Fig. 6.3 (a) and (b). Depending on applications, the copper line width varies from hundreds of nanometers to hundreds of micrometers. The reduction of oxide thickness H_{ox} from the point that the oxide is exposed to the end of planarization is the oxide erosion *e*. The copper thinning is equal to the oxide erosion plus the copper dishing, Fig. 6.3 (a). Except for the above parameters, an important parameter at the feature scale is the feature pattern density *PD*. It is equal to the copper line width (area) divided by the pitch width (area) [3], and can be expressed as $PD = W_{cu}/(W_{cu} + W_{ox})$. It is noteworthy that the definition of *PD* in the copper damascene process is different from that in ILD/IMD planarization. In ILD/IMD, *PD* is defined as the ratio of the width (area) of the upper part of the step to the overall pitch width (area). Therefore, 1- *PD* in copper CMP = *PD* in ILD/IMD CMP.

6.3.2 Die-Scale Wafer Topography

Fig. 6.4 (a) shows schematically the top view of a die. The die area is usually tens of millimeters times tens of millimeters. It can be separated into smaller cells composed of copper line features with different pattern densities, Fig. 6.4 (a). Not all features have the same feature pattern density (calculated in a window size close to the feature size, usually tens of micrometers × tens of micrometers, Fig. 2.6.). Ouma [6.26] developed a concept of *effective* pattern density to account for the random distribution of local pattern density across the die. A calibrated parameter, called planarization length (or *effective* pattern density window size) is needed to calculate the *effective* pattern density. All the features in the *effective* pattern density window surrounding a certain feature will contribute to its effective pattern density. In the later section 6.9.2, we will discuss how to integrate the die and feature scale models based on a physical understanding of the effective pattern density concept.

[3] Pitch width (area)= copper line width (area) plush oxide space width (area)

At this stage, we first simply assume that the cells are composed of features with the same feature pattern density. One pattern density (from 0.1 to 0.9) is with each cell and the cell size is larger than the effective pattern density window size [4], so the influence from other cells is minimal. In this case, three parameters are sufficient to represent the topography: one, the pattern density PD_i of each single cell; second, the area A_i of each single cell; and third, the effective height H_{effi} of each single cells. The effective height H_{effi} of the cells are defined as $PD_i H+(1\text{-}PD_i)\ (H-S)$, where H is the height of the upper part of the steps and $(H-S)$ the height of the bottom part of the steps. Among the three parameters, the cell size and pattern density PD_i represent the die-scale topography in the lateral direction and the *effective* height represents the topography in the vertical direction.

6.3.3 Pad Material Properties

Two different material models are used in this chapter to model the polishing pad; one is a linear elastic model, and the other a linear viscoelastic model.

The Linear Elasticity Model

The polishing pads can be modeled as linear elastic materials. The equation describing the relationship between pressure P and strain γ in linear elasticity is

$$P = E\gamma \tag{6.3}$$

where E is the Young's modulus of the linear elastic materials. A series of springs with stiffness constant K can be used to model the linear elastic polishing pad, as shown in Fig. 6.5. The stiffness constant is formulated as $K = E/L$, where E is the Young's modulus of the polishing pad material and L the length of the spring. The relationship between the pressure supported by the spring and the spring deformation Df can be written as

$$P = E\gamma = E\frac{Df}{L} = K \times Df \tag{6.4}$$

[4] A size in the range of several minimeters to characterize the planarization ability at die-scale. Sometimes it is referred to as planarization length. It was discussed in Chapter 2 and to be discussed in more detail in section 6.9.2.

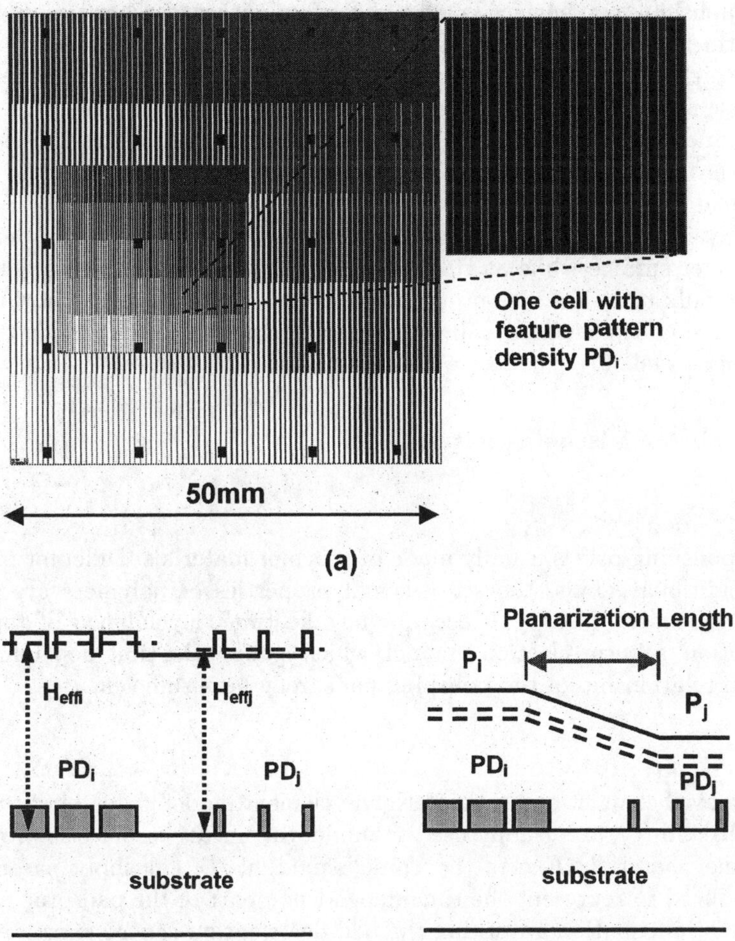

Fig. 6.4. Die scale wafer topography: (a) Top view of a die with cells with different pattern densities and (b) Side view of two different density areas before and after CMP

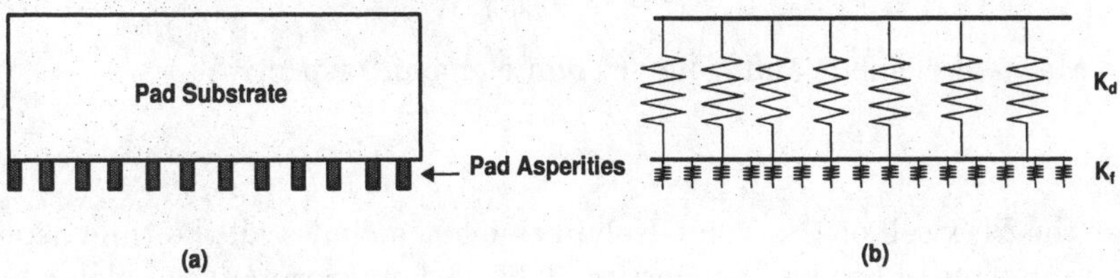

Fig. 6.5. (a) Polishing pad with asperities and (b) separated spring models of pad substrate and pad asperities

The polishing pad surface is a rough surface with pad asperities, Fig. 6.5 (a). The thickness of this asperity layer is different from that of the bulk pad. The SEM pictures of pad topography in Chapter 3 shows that the thickness of the pad asperities is usually in the range of 10 ~100μm, while the bulk thickness is around 1~2mm. Therefore, two different springs, one with stiffness constant K_d, the other with stiffness constant K_f, are needed to model the bulk pad and pad asperity layer, respectively, Fig. 6.5 (b). The pressure distribution at the feature scale is mainly due to the contact between the pad asperities and the wafer surface while at the die scale, it is mainly due to the deformation of the bulk pad. The idea of using two different springs to model the asperity layer and pad substrate has been applied by Runnels et al. [6.27] in their CMP model as well.

The Linear Viscoelasticity Model

The polishing pad is usually made of polymer materials. Different from those of traditional solids, the stress-strain properties of polymers are markedly rate dependent [6.28]. To account for the time dependency of topography evolution, a linear elasticity model, which ignores the time dependent stress-strain relationship of the polishing pads, may be insufficient.

A Maxwell element and a Voigt-Kelvin element model [6.30], as shown in Fig. 6.6 (b) and (c), are usually used to model the linear viscoelastic properties of polymer materials. Except the Young's modulus E, a dashpot parameter η is introduced to represent the time dependency part of the pad properties. The differential equation governing the pad deformation can be expressed as

$$P = \eta\dot{\gamma} - (\eta/E)\,\dot{P} \tag{6.5}$$

$$P = \eta\dot{\gamma} + E\gamma \tag{6.6}$$

for the Maxwell element and Voigt-Kelvin element, respectively.

Neither the Maxwell or the Voigt-Kelvin element includes all the time dependent components of the pad properties [6.30]. A four-parameter model, which is a series combination of a Maxwell element and a Voigt-Kelvin element is sometimes used to model the viscoelastic materials, Fig. 6.6 (d). In this case,

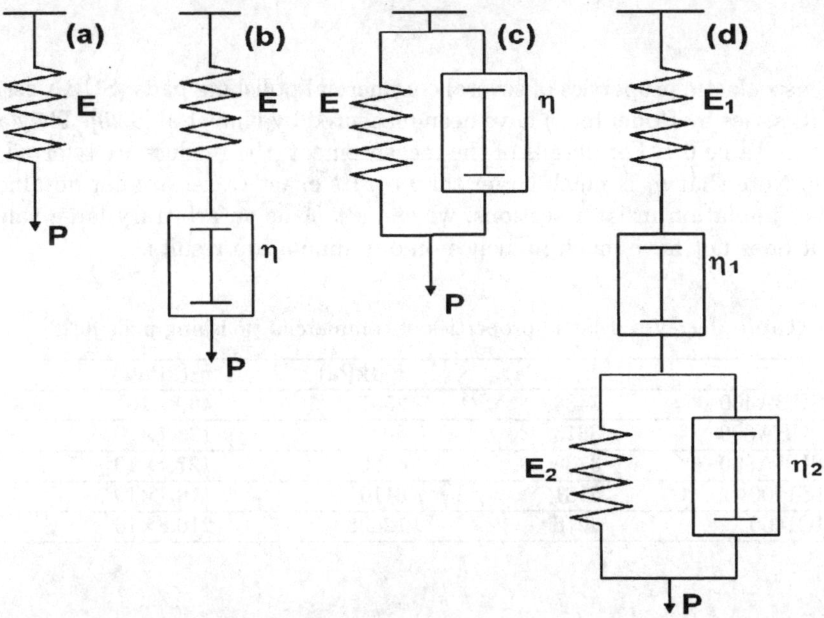

Fig. 6.6. (a) Spring Model (b) Maxwell Element Model (c) Voigt-Kelvin Element Model (d) Four-Parameter Model for Polishing Pad

the differential equation governing the relationship between the pressure and strain is

$$\ddot{P}+\left(\frac{E_1}{\eta_2}+\frac{E_1}{\eta_1}+\frac{E_2}{\eta_2}\right)\dot{P}+\frac{E_1E_2}{\eta_1\eta_2}P=E_1\,\ddot{\gamma}+\frac{E_1E_2}{\eta_2}\,\dot{\gamma} \tag{6.7}$$

where E_1 and η_1 are parameters related to the Maxwell element and E_2 and η_2 are parameters related to the Voigt-Kelvin element. Similar to the stiffness $K = E/L$ in the linear elasticity case, terms $K_1 = E_1/L, K_2 = E_2/L, \chi_1=\eta_1/L$ and $\chi_2=\eta_1/L$ can be introduced into Eq. 6.7. Then the relationship between the pressure and deformation is obtained:

$$\ddot{P}+\left(\frac{K_1}{\chi_2}+\frac{K_1}{\chi_1}+\frac{K_2}{\chi_2}\right)\dot{P}+\frac{K_1K_2}{\chi_1\chi_2}P=K_1\left(\ddot{Df}\right)+\frac{K_1K_2}{\chi_2}\left(\dot{Df}\right) \tag{6.8}$$

Different K_1, K_2, χ_1 and χ_2 have to be applied at feature and die scales as those in the linear elasticity case. They are assumed to be K_{1f}, K_{2f}, χ_{1f} and χ_{2f} and K_{1d}, K_{2d}, χ_{1d} and χ_{2d} respectively, with the index f referring to parameters at feature scale and the index d referring to parameters at the die scale.

The viscoelastic properties of several commercial polishing pads (SUBA series and IC series by Rodel Inc.) have been measured by Kim et al [6.29]. They are listed in Table 6.1. For details of the measurement, the readers are referred to [6.29]. Note that η_1 is much larger than η_2. Its exact values are not obtained. In the simulation in later sections, we can set η_1 as an arbitrary large value, and it does not have much influence on our simulation results.

Table 6.1. Viscoelastic properties of commercial polishing pads [6.28]

	E_1 (kPa)	E_2(kPa)	η_2(kPa·s)
SUBA400	1438	2557	46.8×10^3
SUBA600	3813	6021	123.7×10^3
SUBA800	3738	4486	127.4×10^3
IC1000	5181	6116	118.1×10^3
IC1400	4276	12828	216.8×10^3

6.3.4 Material Removal Properties

The material removal model in particle scale has been discussed in Chapters 3, 4 and 5 in detail. The topography change at each single point x can be obtained by substituting the pressure distribution $P(x,t)$ into the material removal rate formulation 3.22

$$\begin{aligned} \Delta H(x,t) &= \int_0^t MRR(x,t)\,dt \\ &= \int_0^t \left\{ C_3 \left[1 - \Phi\left(3 - C_2 P(x,t)^{\frac{1}{3}}\right)\right] \sqrt{P(x,t)V} \right\} dt \end{aligned} \quad (6.9)$$

where C_3 and C_2 are two parameters representing the consumable effects, $\Delta H(x,\, t)$ the height change of point x at time t, and Φ a probability density function to account for the abrasive size distribution. This material removal rate formulation correlates well with the experimental revised Preston's equation $MRR = K_e P + MRR_0$ (Chapters 3 and 5). Note that the velocity can be assumed to be independent of the location at the die scale and therefore it is included in the K_e term in this chapter. The two revised Preston's Equations used to describe the material removal rate of copper and oxide are:

$$MRR_{cu} = K_{ecu}(C_3, C_2)P + MRR_{cu0}(C_3, C_2) \tag{6.10}$$

$$MRR_{ox} = K_{eox}(C_3, C_2)P + MRR_{ox0}(C_3, C_2) \tag{6.11}$$

where K_{ecu}, K_{eox}, MRR_{cu0} and MRR_{ox0} are the Preston's coefficients for copper and oxide, respectively, and they are function of consumable parameters C_3 and C_2.

With the existence of the particle scale material removal model, the major issue in topography evolution modeling will be the mechanics model to evaluate the pressure evolution at both feature and die scales.

6.4 Modeling of Wafer-Pad Contact at the Feature-Scale

The removal of a copper line feature in the ith cell with pattern density PD_i can be separated into three stages: copper bulk removal stage I, copper bulk removal stage II, and the over-polishing stage, as shown schematically in Fig. 6.7 (a)-(c). The idea of the three-stage removal has been proposed by Grillera [6.21], Tseng [6.22], Tumgma [6.23] and other researchers. In the first removal stage ((a:1)-(a:2) in Fig. 6.7 (a)), only the upper/raised parts of the copper steps are in contact with the polishing pad. In the second removal stage ((b:1)-(b:2) in Fig. 6.7 (b)), both upper and lower parts of the copper steps are in contact with the polishing pad and are removed until the dielectric materials are exposed. In the third stages ((c:1)-(c:2) in Fig. 6.7 (c)), dielectric materials on the upper part and coppers in the bottom part of the steps are removed simultaneously.

In the following, the feature-scale contact model is first formulated based on the assumption of a linear elastic pad, then the model based on the assumption of a linear viscoelastic pad is formulated. The wafer features are assumed to be a rigid body. The pressure distribution is therefore a function of pad properties only. The same assumption is to be applied in the die-scale modeling.

6.4.1 Linear Elastic Pad Assumption

Stage 1

In the first stage, the following relationships between pressure distribution and feature-scale topography exist

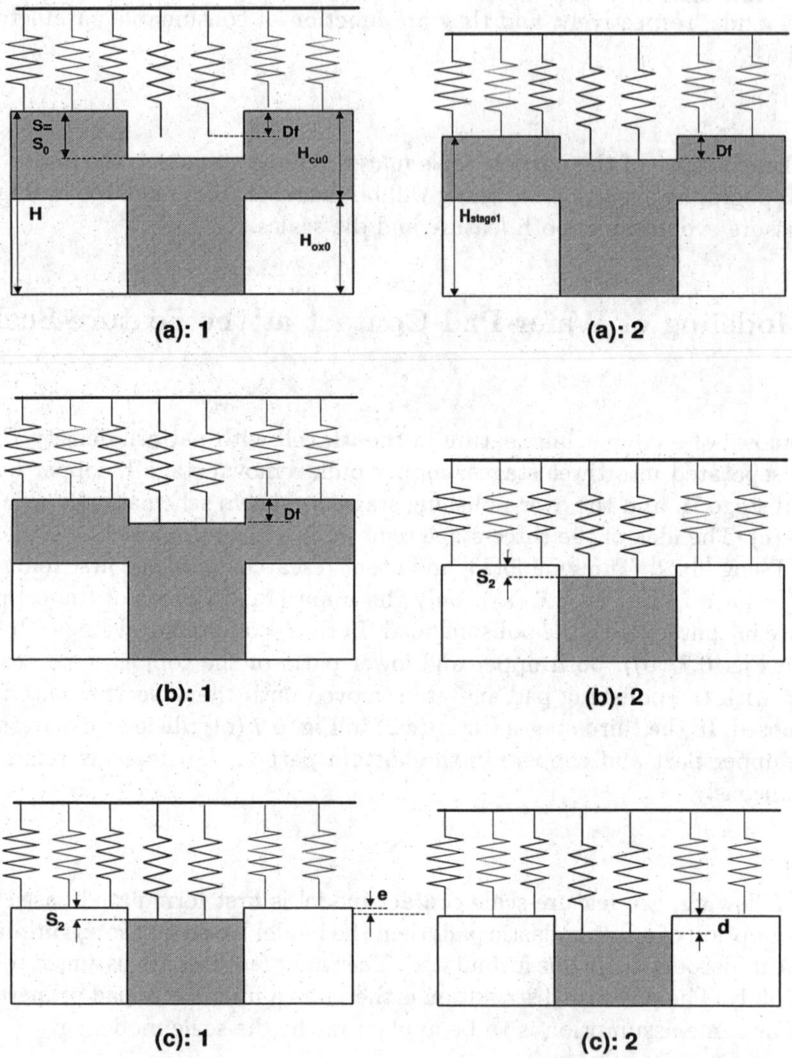

Fig. 6.7. Three stages of feature-scale topography evolution in copper damascene process: (a) bulk copper removal stage I (b) bulk copper removal stage II and (c) overpolishing stage III

$$P_{upperi}(t) = P_i(t)/(1 - PD_i) \tag{6.12}$$

$$P_{bottomi}(t) = 0 \tag{6.13}$$

$$H_{upperi}(t) = H_{cu0} + H_{ox0} - \int_0^t \left[K_{ecu} P_i(t)/\left(1 - PD_i\right) + MRR_{cu0}\right] dt \tag{6.14}$$

$$H_{bottomi}(t) = H_{cu0} + H_{ox0} - S_0 \tag{6.15}$$

$$S(t) = S_0 - \int_0^t \left[K_{ecu} P_i(t)/\left(1 - PD_i\right) + MRR_{cu0}\right] dt \tag{6.16}$$

when

$$S(t) > Df(t) = \frac{P_i(t)}{K_f\left(1 - PD_i\right)} \tag{6.17}$$

where $P_i(t)$ is the average pressure supported by the ith cell at time t, $P_{upperi}(t)$ the pressure supported by the upper part of the step, $P_{bottomi}(t)$ the pressure supported by the bottom part of the step, $H_{upperi}(t)$ the height of the upper part of the step, $H_{bottomi}(t)$ the height of the bottom part of the step, H_{cu0} and H_{ox0} the initial thickness of copper and oxide, respectively, S_0 the initial step height, and $Df(t)$ the deformation of polishing pad under pressure $P_i(t)/(1\text{-}PD_i)$. The pressure P_i is a function of time, and its evaluation will be addressed in detail in section 6.5. Eqs. 6.12 and 6.13 indicate that in stage 1 the pressure P_i is totally supported by the upper part of the step. Therefore, only the upper part is removed in this stage, Eqs. 6.14 and 6.15. Note that Eqs 6.12 and 6.13 correspond to the mechanics/pressure redistribution model x in Eq. 6.1 and Eqs. 6.14 - 6.16 correspond to the topography update/non-uniform material removal model y in Eq. 6.2.

The integrals in Eqs. 6.14 and 6.16 can be discretized by separating the polishing time t into N time block Δt :

$$\int_0^t \left[K_{ecu} P_i(t)/\left(1 - PD_i\right) + MRR_{cu0}\right] dt$$
$$= \sum_{j=0}^{N} \left\{K_{ecu}\left[P_i(j\Delta t)/\left(1 - PD_i\right)\right] + MRR_{cu0}\right\} \Delta t \tag{6.18}$$

This can facilitate the computer implementation of the integrals.

Formulation 6.17 sets the boundaries for stage 1. Based on the linear elastic model, the polishing pad over the upper part of the step has a deformation $K_f P_i(t)/(1 - PD_i)$ at time t. When the step height $S(t)$ is equal to this deformation, both upper and bottom parts of the step will be in contact with the wafer, indicating the start of stage 2.

Moreover, it is noteworthy that the above pad deformation $Df = \frac{P_0}{(1-PD_i)K_f}$ separating stage 1 from stage 2 is obtained based on the one-dimensional spring model. If considering the lateral pad deformation, the deformation will be much smaller. An effective stiffness $K_{f_{eff}}$ based on a two dimensional model will be introduced in section 6.7 to replace K_f, which can address the line width dependency of the topography evolution.

Stage 2

In stage 2, both upper and bottom parts of the step are in contact with the wafer. In this case, the following relationships between pressure distribution and feature-scale topography exist:

$$P_{upperi}(t)(1 - PD_i) + P_{bottomi}(t)PD_i = P_i(t) \tag{6.19}$$

$$\Delta P(t) = P_{upperi}(t) - P_{bottomi}(t) = K_f S(t) \tag{6.20}$$

$$H_{upperi}(t) = H_{stage1} - \int_{t_1}^{t} [K_{ecu} P_{upperi}(t) + MRR_{cu0}]\, dt \tag{6.21}$$

$$H_{bottomi}(t) = H_0 - S_0 - \int_{t_1}^{t} [K_{ecu} P_{bottomi}(t) + MRR_{cu0}]\, dt \tag{6.22}$$

$$S(t) = H_{upperi}(t) - H_{bottomi}(t) \tag{6.23}$$

$$\dot{S}(t) = -K_{ecu} \Delta P \tag{6.24}$$

when

$$H_{upperi}(t) > H_{ox0} \tag{6.25}$$

where $\Delta P(t)$ is the pressure difference between the upper and lower parts of the step $S(t)$, H_{stage1} the height of the upper part of the step at the end of stage 1, H_{ox0}, the original oxide thickness as shown in Fig. 6.3 (a), and t_1 is the end time of stage 1. Eq. 6.19 is to satisfy the force equilibrium condition. The pad deformation on the upper and lower parts of the step is different, leading to a non-uniform pressure distribution. Eq. 6.20 indicates that the pressure difference $\Delta P(t)$ over the upper and bottom parts is equal to the spring stiffness K_f times the step height $S(t)$. From Eqs. 6.19 and 6.20 the pressures applied on the step can be calculated as follows:

$$P_{upperi}(t) = P_i(t) + PD_i \Delta P = P_i(t) + PD_i K_f S(t) \tag{6.26}$$

$$P_{bottomi}(t) = P_i(t) - (1 - PD_i)\,\Delta P = P_i(t) - (1 - PD_i)\,K_f S(t) \tag{6.27}$$

This pressure difference leads to a difference in material removal rates over the upper and bottom areas, Eqs. 6.21 and 6.22, and a nonlinear time dependent step height reduction, Eqs. 6.23 and 6.24. Formulation 6.25 sets the boundary for stage 2. Similar to those in stage 1, Eqs. 6.19, 6.20, 6.26 and 6.27 correspond to the pressure re-distribution function x and Eqs. 6.21~6.24 correspond to the topography update/non-uniform material removal function y. Note that the end time t_2 of stage 2 and t_1 of stage 1 are both functions of pattern density PD_i, pad materials, and material removal properties. The start time of overpolishing stage 3 is different for features with different pattern densities.

Stage 3

Stage 3 is the overpolishing stage. The copper clearing time t_2 of stage 2 is different for different pattern density cells. This indicates that when some of the oxide is cleared of copper, the rest may still be covered with coppers. To ensure that all coppers over the oxide surface are cleared, a long enough polishing time is required [5]. This causes oxide thickness reduction in regions where the oxide is exposed earlier. This has been called as the oxide erosion.

The pressure distribution mechanism in stage 3 is the same as that in stage 2. The difference is that the upper part is oxide instead of copper. Therefore, Eqs. 6.21 and 6.24 are replaced by the following Eqs. 6.28 and 6.29 while the other equations describing the relationships between pressure distribution and feature-scale topography remain the same.

[5] Sometimes a time point at which all copper over the wafer is cleared from the oxide surface is called the nominal polishing time. The polishing time longer than the fixed nominal polishing time is referred to as the overpolishing time. The ratio of the overpolishing time to the nominal polishing time is called the overpolishing percentage. Note that this definition of overpolishing is different from the definition in this work. In our work, the overpolishing definition is feature/pattern density dependent while the overpolishing percentage is a wafer-scale concept and independent of features.

$$H_{upperi}(t) = H_{ox0} - \int_{t_2}^{t} \left[K_{eox} P_{upperi}(t) + MRR_{ox0}\right] dt \tag{6.28}$$

$$\begin{aligned}\dot{S}(t) &= (K_{ecu} P_{bottomi} + MRR_{cu0}) - (K_{eox} P_{upperi} + MRR_{ox0}) \\ &= \{K_{ecu} \left[P_i(t) - (1 - PD_i)\Delta P\right] + MRR_{cu0}\} \\ &\quad - \{K_{eox} \left[P_i(t) + PD_i \Delta P\right] + MRR_{ox0}\} \\ &= -[K_{ecu}(1 - PD_i) + K_{eox} PD_i]\Delta P \\ &\quad + K_{ecu} P_i(t) + MRR_{cu0} - K_{eox} P_i(t) - MRR_{ox0}\end{aligned} \tag{6.29}$$

Note that Eqs. 6.26 and 6.27 are used in formulating Eq. 6.29.

The boundary for stage 3 will be $t = T$ where T is the overall planarization time. The final dishing is the step between the upper oxide and bottom copper, or

$$d = S(T) = H_{upperi}(T) - H_{bottomi}(T). \tag{6.30}$$

The erosion e is equal to the thickness of oxide removed, or

$$e = \int_{t_2}^{T} \left[K_{eox} P_{upperi}(t) + MRR_{ox0}\right] dt \tag{6.31}$$

Similar to those in stage 1, the integrals in Eqs. 6.22, 6.28 and 6.31 can be discretized for computer implementation.

6.4.2 Linear Viscoelastic Pad Assumption

When the pad is modeled as the four parameter model shown in Fig. 6.6 (d), the time dependency of step height reduction will be different from that in the linear elastic model. In this section, the formulation of the model based on the linear viscoelasticity assumption is discussed. As in the linear elasticity case, three stages of wafer-pad contact exist.

Stage 1

In the first stage, only the upper part of the step is in contact with the pad. There are the following relationships between pressure distribution, pad deformation and feature-scale topography:

$$P_{upperi}(t) = P_i(t)/(1 - PD_i) \tag{6.32}$$

$$P_{bottomi}(t) = 0 \tag{6.33}$$

$$H_{upperi}(t) = H_0 - \int_0^t [K_{ecu} P_i(t)/(1 - PD_i) + MRR_{cu0}]\, dt \tag{6.34}$$

$$H_{bottomi}(t) = H_0 - S_0 \tag{6.35}$$

$$S(t) = S_0 - \int_0^t K_{cu} [P_i(t)/(1 - PD_i) + MRR_{cu0}]\, dt \tag{6.36}$$

$$\ddot{P}_{upperi}(t) + \left(\frac{K_{1f}}{\chi_{2f}} + \frac{K_{1f}}{\chi_{1f}} + \frac{K_{2f}}{\chi_{2f}}\right) \dot{P}_{upperi}(t) + \frac{K_{1f}K_{2f}}{\chi_{1f}\chi_{2f}} P_{upperi}(t) = K_{1f}\, \ddot{D}f(t) + \frac{K_{1f}K_{2f}}{\chi_{2f}} \dot{D}f(t) \tag{6.37}$$

when

$$S(t) > Df(t) \tag{6.38}$$

The only difference between the above formulations and those based on the linear elasticity assumption is that Eq. 6.37 which governs the deformation of the polishing pad takes the place of Eq 6.17. The differential Eq. 6.37 can be discretized as follows to facilitate the numerical implementation:

$$\begin{aligned} &\frac{P_{upperi}[(j+2)\Delta t] - 2P_{upperi}[(j+1)\Delta t] + P_{upperi}(j\Delta t)}{\Delta t^2} \\ &+ \left(\frac{K_{1f}}{\chi_{2f}} + \frac{K_{1f}}{\chi_{1f}} + \frac{K_{2f}}{\chi_{2f}}\right) \frac{P_{upperi}[(j+2)\Delta t] - P_{upperi}[(j+1)\Delta t]}{\Delta t} \\ &+ \frac{K_{1f}K_{2f}}{\chi_{1f}\chi_{2f}} P_{upperi}[(j+2)\Delta t] \\ &= K_{1f} \left\{ \frac{Df[(j+2)\Delta t] - 2Df[(j+1)\Delta t] + Df(j\Delta t)}{\Delta t^2} \right\} \\ &+ \frac{K_{1f}K_{2f}}{\chi_{2f}} \left\{ \frac{Df[(j+2)\Delta t] - Df[(j+1)\Delta t]}{\Delta t} \right\} \end{aligned} \tag{6.39}$$

where the index j indicates the jth polishing block and Δt the time step.

As in the linear elasticity case, when considering the lateral directional deformation, the vertical pad deformation before touching the bottom of the step may be much smaller than that predicted using the above one-dimensional model. A two-dimensional model addressing the line width dependency of the pad deformation will be discussed in later sections.

Stage 2 and Stage 3

As that in stage 1, in stages 2 and 3, only Eq. 6.20 needs to be replaced by the following Eqs. 6.40 and 6.41 to accommodate the time dependency of the pressure distribution:

$$P_{upperi}(t) - P_{bottomi}(t) = \Delta P(t) \tag{6.40}$$

$$\Delta \ddot{P}(t) + \left(\frac{K_{1f}}{\chi_{2f}} + \frac{K_{1f}}{\chi_{1f}} + \frac{K_{2f}}{\chi_{2f}}\right) \Delta \dot{P}(t) + \frac{K_{1f}K_{2f}}{\chi_{1f}\chi_{2f}} \Delta P(t) = K_{1f}\,\ddot{S}(t) + \frac{K_{1f}K_{2f}}{\chi_{2f}}\,\dot{S}(t) \tag{6.41}$$

Boundary Conditions at Each Stage

We assume that pressure and pad deformation are continuous during the topography evolution. The pressure derivative is assumed to be zero at the beginning of each stage.

6.5 Modeling of Wafer-Pad Contact at the Die-Scale

The model of wafer-pad contact at the die scale is to predict the average pressure distribution $P_i(t)$ among different pattern density cells. The effective height as discussed in section 6.3.2 is used to represent the wafer topography at the die scale. The height difference may be either from the electroplating process [6.30] before CMP, or from the different material removal rates in the different pattern density cells.

The pressure $P_i(t)$ supported by each pattern density cell is different due to the effective height difference. If the die is separated into n different pattern density cells and the corresponding pattern densities are PD_i and PD_j for the ith and jth cells, Fig. 6.4, assuming that the areas of ith and jth pattern density cells are A_i and A_j, then, based on the linear elastic model of the polishing pads, the relationship among the average down pressure P_0, P_i and P_j can be expressed as

$$P_0 \sum_{i=1}^{n} A_i = \sum_{i=1}^{n} [P_i(t) A_i] \tag{6.42}$$

$$\Delta P_{ij}(t) = P_i(t) - P_j(t) \tag{6.43}$$

$$\Delta H_{ij}(t) = H_{effi}(t) - H_{effj}(t) \tag{6.44}$$

$$\Delta P_{ij}(t) = K_d \Delta H_{ij}(t) \tag{6.45}$$

where $\Delta H_{ij}(t)$ is the effective height difference between cells i and j, Eq. 6.42 is to satisfy the force equilibrium condition and Eq. 6.45 is from the linear elastic property of the polishing pads. As in section 6.3.2, the effective H_{effi} in the area of pattern density PD_i is defined as

$$H_{effi}(t) = H_{upperi}(t)(1 - PD_i) + H_{bottomi}(t) PD_i \tag{6.46}$$

Note here that all cells are assumed to keep contact with the pad so that Eq. (6.45) is correct.

From Eqs. 6.41 to 6.45, n linear independent equations are obtained and solving them yields pressure $P_i(t)$. For the linear viscoelastic polishing pads, replacing Eq. 6.45 governing the relationship between the pressure distribution and topography with the following equation

$$\begin{aligned} \Delta \ddot{P}_{ij}(t) + \left(\frac{K_{1d}}{\chi_{2d}} + \frac{K_{1d}}{\chi_{1d}} + \frac{K_{2d}}{\chi_{2d}} \right) \Delta \dot{P}_{ij}(t) + \frac{K_{1d} K_{2d}}{\chi_{1d} \chi_{2d}} \Delta P_{ij}(t) \\ = K_{1d} \Delta \ddot{H}_{ij}(t) + \frac{K_{1d} K_{2d}}{\chi_{2d}} \Delta H_{ij}(t) \end{aligned} \tag{6.47}$$

yields n linear independent equations. Solving them yields P_i.

6.6 Analytical Solution Assuming Die-Scale Pressure Independent of Polishing Time

The pressure P_i at the die scale is a function of time and changes with the die-scale topography. However, when the spring stiffness K_d is smaller than the K_f, its change may be small. In this section, we will formulate the solution of feature scale topography evolution by assuming that the cell pressure P_i is constant and equal to the down pressure P_0. Analytical solutions exist in this case for both linear elastic and viscoelastic pads. They may offer more physical insights on the model.

6.6.1 Linear Elastic Pad Assumption

Analytical solutions of feature-scale topography evolution based on the linear elasticity assumption have been obtained by different researchers [6.21-6.23]. Most of them considered only one or two stages proposed in above sections. For example, models by Stine [6.18] and Smith [6.20] covered the first and second stages while models by Tseng [6.19][6.21] and Grillaert [6.22] covered the second and third stages. As a natural extension of their models, Tugbawa et al. [6.23] developed mathematical formulations to cover all the above three stages. However, they do not consider the MRR_{ox0} and MRR_{cu0} terms in the MRR equation. In this chapter, these terms are included for more accurate solutions.

Stage 1

Since the pressure $P_i = P_0$ applied on the cell is independent of time, from Eqs. 6.16 and 6.17 there are

$$S(t) = S_0 - \left[K_{ecu} \frac{P_0}{1 - PD_i} + MRR_0 \right] t \tag{6.48}$$

$$S_1 = \frac{P_0}{(1 - PD_i) K_f} \tag{6.49}$$

Therefore,

$$\begin{aligned} t_1 &= \frac{S_0 - S_1}{K_{ecu} \frac{P_0}{1 - PD_i} + MRR_0} \\ &= \frac{S_0 - \frac{P_0}{(1 - PD_i) K_f}}{K_{ecu} \frac{P_0}{1 - PD_i} + MRR_0} \end{aligned} \tag{6.50}$$

$$H_{stage1} = H_0 - S_1 = H_0 - \frac{P_0}{(1 - PD_i) K_f} \tag{6.51}$$

where t_1 is the time at which stage 1 ends, S_1 the step height at time t_1 and H_{stage1} the height of upper part of the step at time t_1.

It is seen that when the pattern density PD_i is large, the pad deformation over the upper part of the step is larger and therefore the step S_1 (= pad deformation Df) at the end of the stage 1 is larger. This indicates that the pad will contact the bottom part of the step earlier.

Stage 2

Based on the linear elasticity (Eq. 6.20) and Preston's Eq. 6.24, there is a relationship between the step height S and pressure difference ΔP:

$$S(t)=\Delta P\,(t)\,/K_f \tag{6.52}$$

$$-K_{ecu}\Delta P = \dot{S}\,(t) \tag{6.53}$$

Considering that the ΔP is equal to $P_0/(1-PD_i)$ at $t=t_1$,

$$\Delta P(t) = \frac{P_0}{1-PD_i}e^{-(K_{ecu}K_f)(t-t_1)} \tag{6.54}$$

Therefore,

$$P_{upperi}(t) = P_0 + \Delta P(t)PD_i = P_0\left[1+\frac{PD_i}{1-PD_i}e^{-(K_{ecu}K_f)(t-t_1)}\right] \tag{6.55}$$

$$P_{bottomi}(t) = P_0 - \Delta P(t)(1-PD_i) = P_0\left[1-e^{-(K_{ecu}K_f)(t-t_1)}\right] \tag{6.56}$$

$$S(t) = S_1 e^{-K_{ecu}K_f(t-t_1)} = \frac{P_0}{(1-PD_i)\,K_f}e^{-K_{ecu}K_f(t-t_1)} \tag{6.57}$$

The end time t_2 of stage 2 can be obtained by solving the following equation:

$$\begin{aligned} H_{ox0} &= H_{stage1} - \int_{t_1}^{t_2}\left[K_{ecu}P_{upperi}(t) + MRR_{cu0}\right]dt \\ &= H_0 - \frac{P_0}{(1-PD_i)K_f} - \left[K_{ecu}P_0 + MRR_{cu0}\right](t_2-t_1) \\ &\quad + P_0\frac{PD_i}{(1-PD_i)K_f}\left[1-e^{-K_{ecu}K_f(t_2-t_1)}\right] \end{aligned} \tag{6.58}$$

From this it is seen that the step height reduction and pressure evolution in stage 2 are exponential functions of time. The step height decreases exponentially. When the time t is long enough, the static step height $S(\infty)$ will go to zero, indicating a full planarization. It is noted that the $S(t)$ is always smaller for a larger pattern density. This is because t_1 is smaller and stage 2 starts earlier. The pressure on the upper part of the step decreases exponentially while the pressure on the lower part increases. The time constant $1/(K_f K_{ecu})$ of the exponential step reduction is independent of the pattern density PD_i. The planarization rate/efficiency dS/dt can be defined as the ratio of the step height reduction to the polishing time. From Eqs. 6.53 and 6.54, larger feature pattern density, down pressure, copper Preston's coefficient K_{ecu} and pad stiffness K_f yield a higher planarization rate.

The end time t_2 of step 2 and final step height S_2 will depend on the copper thickness, pattern density and other parameters. The overpolishing time $T-t_2$ of stage 3 therefore depends on the pattern density and copper thickness. This will affect the erosion, as will be addressed in later sections.

Stage 3

Based on linear elasticity, the relationship between the step height and pressure difference over the step in stage 3 is

$$\dot{S}(t) = \Delta\dot{P}/K_f.$$

Substituting Eq. 6.29 into the above equation yields

$$\begin{aligned} &\Delta\dot{P}+K_f[K_{ecu}(1-PD_i)+K_{eox}PD_i]\Delta P \\ &-K_f\left(K_{ecu}P_0+MRR_{cu0}-K_{eox}P_0-MRR_{ox0}\right)=0 \end{aligned} \tag{6.59}$$

$\Rightarrow$

$$\begin{aligned} \Delta P(t) = &\frac{K_{ecu}P_0+MRR_{cu0}-K_{eox}P_0-MRR_{ox0}}{K_{ecu}\left(1-PD_i\right)+K_{eox}PD_i} \\ &-\left[\frac{K_{ecu}P_0+MRR_{cu0}-K_{eox}P_0-MRR_{ox0}}{K_{ecu}\left(1-PD_i\right)+K_{eox}PD_i}-S_2K_f\right] \\ &\times\text{EXP}\left\{-k_f\left[K_{ecu}(1-PD_i)+K_{eox}PD_i\right](t-t_2)\right\} \end{aligned} \tag{6.60}$$

$$\begin{aligned} S(t) = &\frac{K_{ecu}P_0+MRR_{cu0}-K_{eox}P_0-MRR_{ox0}}{K_f\left[K_{ecu}\left(1-PD_i\right)+K_{eox}PD_i\right]} \\ &-\left\{\frac{K_{ecu}P_0+MRR_{cu0}-K_{eox}P_0-MRR_{ox0}}{K_f\left[K_{ecu}\left(1-PD_i\right)+K_{eox}PD_i\right]}-S_2\right\} \\ &\times\text{EXP}\left\{-K_f\left[K_{ecu}(1-PD_i)+K_{eox}PD_i\right](t-t_2)\right\} \end{aligned} \tag{6.61}$$

where S_2 is the step height at the end of stage 2. From the above equations, it is seen that both pressure and step height increases exponentially with time. However, unlike that in stage 2, the time constant

$$\left\{K_f\left[K_{ecu}(1-PD_i)+K_{eox}PD_i\right]\right\}^{-1}$$

depends on pattern density PD_i. Because K_{eox} is smaller than K_{ecu}, the time constant is also larger than that in the second stage. The dishing d is equal to the $S(T)$ at the end of planarization. From Eq. 6.61, a larger pattern density yields larger dishing. This can be shown by the simulations in later sections as well.

Substituting P_{upperi} into Eq. 6.31 yields erosion e

$$
\begin{aligned}
e &= \int_{t_2}^{T} \left(K_{eox} P_{upperi} + MRR_{ox0}\right) dt \\
&= \int_{t_2}^{T} \left[K_{eox} P_0 + MRR_{ox0} + K_{eox} PD_i \Delta P(t)\right] dt \\
&= \left(K_{eox} P_0 + MRR_{ox0}\right)(T - t_2) + K_{eox} PD_i K_f \int_{t_2}^{T} S(t) dt
\end{aligned}
\tag{6.62}
$$

It is seen that higher pattern densities generate larger erosion.

6.6.2 Linear Viscoelastic Pad Assumption

Stage 1

In the case of linear viscoelastic pad, the pressure P_{upperi} applied on the raised area of the step is still constant and equal to $P_0/(1\text{-}PD_i)$ in stage 1. Unlike that in the linear elastic pad case, however, the deformation Df of pad is not a constant but a function of time t. From Eq. 6.37 the deformation satisfies the following differential equation:

$$
\frac{K_{1f} K_{2f}}{\chi_{1f} \chi_{2f}} P_{upperi} = K_{1f}\, \ddot{Df}(t) + \frac{K_{1f} K_{2f}}{\chi_{2f}}\, \dot{Df}(t) \tag{6.63}
$$

$\Rightarrow$

$$
\begin{aligned}
Df(t) &= \frac{P_{upperi}}{K_{1f}} + \frac{P_{upperi}}{\chi_{1f}} t + \frac{1}{K_{2f}} P_{upperi} \left(1 - e^{-\frac{K_{2f}}{\chi_{2f}} t}\right) \\
&= \frac{P_0}{(1 - PD_i)\, K_{1f}} + \frac{P_0}{(1 - PD_i)\, \chi_{1f}} t \\
&\quad + \frac{1}{K_{2f}} \frac{P_0}{1 - PD_i} \left(1 - e^{-\frac{K_{2f}}{\chi_{2f}} t}\right)
\end{aligned}
\tag{6.64}
$$

where boundary conditions $Df(0) = \frac{P_0}{(1-PD_i)K_{f1}}$ is used to obtain the solution. Eq. 6.64 indicates that the pad deformation over the raised area increases exponentially with time. The time constant is equal to χ_{2f}/K_{2f}.

At the end of stage 1, the deformation $Df(t_1)$ is equal to the step height S_1, or,

$$\begin{aligned} S_1 &= S_0 - [K_{ecu}P_0/(1-PD_i)+MRR_{cu0}]\, t_1 \\ &= Df(t_1) \\ &= \frac{P_0}{(1-PD_i)\,K_{1f}} + \frac{P_0}{(1-PD_i)\,\chi_{1f}} t_1 \\ &\quad - \frac{1}{K_{2f}} \frac{P_0}{1-PD_i} \left(1 - e^{-\frac{K_{2f}}{\chi_{2f}} t_1}\right) \end{aligned} \tag{6.65}$$

Solving this equation yields t_1 and the final step S_1 (or pad deformation $Df(t_1)$). Both of them are functions of pattern density. Since the pressure on the upper part of the step is constant, the step height reduction is still a linear function of time as in the linear elasticity case.

Stage 2

In stage 2, from the linear viscoelastic properties of the polishing pad (Eq. 6.41) and Preston's Eq. 6.24, there is

$$\begin{aligned} \Delta \ddot{P}(t) + \left(\frac{K_{1f}}{\chi_{2f}} + \frac{K_{1f}}{\chi_{1f}} + \frac{K_{2f}}{\chi_{2f}}\right) \Delta \dot{P}(t) + \frac{K_{1f}K_{2f}}{\chi_{1f}\chi_{2f}} \Delta P\,(t) \\ = K_{1f}\,\ddot{S}\,(t) + \frac{K_{1f}K_{2f}}{\chi_{2f}} \dot{S}\,(t) \end{aligned}$$

$$\begin{aligned} \dot{S}(t) &= -K_{ecu}\Delta P(t) \\ \ddot{S}(t) &= -K_{ecu}\Delta \dot{P}(t) \end{aligned} \tag{6.66}$$

$\Rightarrow$

$$\begin{aligned} \Delta \ddot{P}(t) \left(\frac{K_{1f}}{\chi_{2f}} + \frac{K_{1f}}{\chi_{1f}} + \frac{K_{2f}}{\chi_{2f}} + K_{ecu}K_{1f}\right) \Delta \dot{P}(t) + \\ \left(\frac{K_{1f}K_{2f}}{\chi_{1f}\chi_{2f}} + K_{ecu}\frac{K_{1f}K_{2f}}{\chi_{2f}}\right) \Delta P\,(t) = 0 \end{aligned} \tag{6.67}$$

$\Rightarrow$

$$\Delta P = Ae^{\lambda_1(t-t_1)} + Be^{\lambda_2(t-t_1)} \tag{6.68}$$

where λ_1 and λ_2 are the roots of the characteristic equation of 6.67 and

$$A = \lambda_2 \Delta P(t_1)/(\lambda_2 - \lambda_1) = \lambda_2 P_0/(\lambda_2 - \lambda_1)/(1 - PD_i) \tag{6.69}$$

$$B = \lambda_1 \Delta P(t_1)/(\lambda_1 - \lambda_2) = \lambda_1 P_0/\lambda_1 - \lambda_2)/(1 - PD_i) \tag{6.70}$$

Boundary conditions $\Delta P(t_1) = P_0/(1\text{-}PD_i)$ and $\Delta \dot{P}(t_1) = 0$ are used to obtain the above solutions. Using Preston's Equation, the step height $S(t)$ is solved as

$$\begin{aligned} S(t) &= -K_{ecu} \int_{t_1}^{t} \Delta P dt \\ &= S_1 + \frac{AK_{ecu}}{\lambda_1}\left[1 - e^{\lambda_1(t-t_1)}\right] + \frac{BK_{ecu}}{\lambda_2}\left[1 - e^{\lambda_2(t-t_1)}\right] \end{aligned} \tag{6.71}$$

It is seen that the pressure distribution and step height is still an exponent function of time. The time constant is independent of the pattern density. However, unlike in the linear elasticity case, two time constants exist. The static step height $S(\infty)$ is

$$\begin{aligned} S(\infty) &= S_1 + \frac{AK_{ecu}}{\lambda_1} + \frac{BK_{ecu}}{\lambda_2} \\ &= S_1 + \frac{\lambda_2 + \lambda_1}{\lambda_1 \lambda_2} K_{ecu} \frac{P_0}{1 - PD_i} \\ &= S_1 - \frac{\frac{K_{1f}}{\chi_{2f}} + \frac{K_{1f}}{\chi_{1f}} + \frac{K_{2f}}{\chi_{2f}} + K_{ecu} K_{1f}}{\left(\frac{K_{1f}K_{2f}}{\chi_{1f}\chi_{2f}} + K_{ecu}\frac{K_{1f}K_{2f}}{\chi_{2f}}\right)} K_{ecu} \frac{P_0}{1 - PD_i} \end{aligned} \tag{6.72}$$

Eq. 6.72 implies that the static step height $S(\infty)$ may not be equal to zero. Therefore, a full planarization cannot be realized based on the linear viscoelastic pad assumption. The static step height is a function of pattern density and pad material properties. A large K_{ecu} or down pressure P_0 will generate a smaller $S(\infty)$ and therefore better planarization. Also note that A and B is larger for larger pattern densities. Thus, the step height decreases faster for larger pattern densities, similar to that in the linear elasticity case. This indicates a better planarization ability.

Note that the static step height $S(\infty)$ obtained above may be smaller than zero when the pattern density is high or pressure and K_{ecu} are large. This indicates that a 'reverse' step may be formed. However, this 'reverse' step is not observed in practice. A different mechanism can be used to explain the material removal after the steps go to zero. When the step height is larger than zero, the pad deformation on the upper and bottom parts of the steps is different. The pressure supported by the upper part is larger and that by the bottom part is smaller. When the step height becomes zero, this pressure difference over the wafer-pad interface will be kept due to the dashpot part of the pad property. However, the whole pattern density cell will have the chance to support the higher and lower pressures. Therefore, the 'average' material removal rate on each point will remain the same value, $K_{ecu}P_0+MRR_{cu0}$, and, therefore, the step height will never be smaller than zero. The boundary condition after the step height goes to zero, thus, should be $dS/dt=0$ and $S=0$. The differential equation 6.41 governing the pad deformation now becomes:

$$\Delta \ddot{P}(t) + \left(\frac{K_{1f}}{\chi_{2f}} + \frac{K_{1f}}{\chi_{1f}} + \frac{K_{2f}}{\chi_{2f}}\right) \Delta \dot{P}(t) + \frac{K_{1f}K_{2f}}{\chi_{1f}\chi_{2f}} \Delta P(t) = 0 \qquad (6.73)$$

Solving Eq. 6.73 yields

$$\Delta P = A_1 e^{\lambda_{11}(t-t_{12})} + B_1 e^{\lambda_{22}(t-t_{12})} \qquad (6.74)$$

where λ_{11} and λ_{22} are the solutions of the characteristic equation of 6.73, t_{12}, the time when the step height becomes zero, $\Delta P(t_{12})$ the pressure difference over the wafer-pad interface at time t_{12} and

$$A_1 = \lambda_{22}\Delta P(t_{12})/(\lambda_{22}-\lambda_{11}) \qquad (6.75)$$

$$B_1 = \lambda_{11}\Delta P(t_{12})/(\lambda_{11}-\lambda_{22}) \qquad (6.76)$$

Based on above understanding, the thickness of copper at the end of stage 2 can be written as

$$H(t_2) = H_{ox0} + H_{cu0} - \int_{t_1}^{t_{12}} [K_{ecu}(P_0 + PD_i\Delta P) + MRR_{cu0}]dt - \int_{t_{12}}^{t_2} [K_{ecu}P_0 + MRR_{cu0}]dt \qquad (6.77)$$

This is for the case where full planarization is realized. When full planarization cannot be realized, (or $S(t)$ is always larger than zero), the second integral on the right side of above equation should be neglected and t_{12} in the first integrals should be replaced by t_2. The end time t_2 of stage 2 can be obtained by equaling Eq. 6.77 to H_{ox0}.

Stage 3

In the third stage, due to the different material removal rate of copper and oxide, a step will be formed over the wafer surface once again. The larger pressure over the wafer-pad interface will be supported by the upper/raised part of the step while the smaller pressure will be supported by the bottom part. From the linear viscoelastic property of the polishing pad and Eq. (6.35), the relationship between the pressure difference and the step height can be written as

$$\Delta \ddot{P}(t) + \left(\frac{K_{1f}}{\chi_{2f}} + \frac{K_{1f}}{\chi_{1f}} + \frac{K_{2f}}{\chi_{2f}}\right) \Delta \dot{P}(t) + \frac{K_{1f}K_{2f}}{\chi_{1f}\chi_{2f}} \Delta P(t) = K_{1f}\,\ddot{S}(t) + \frac{K_{1f}K_{2f}}{\chi_{2f}} \dot{S}(t) \quad (6.78)$$

$$\dot{S} = -\left[K_{ecu}(1\text{-}PD_i) + K_{eox}PD_i\right] \Delta P + \left(K_{ecu}P_0 + MRR_{cu0} - K_{eox}P_0 - MRR_{ox0}\right) \quad (6.79)$$

and

$$\ddot{S} = -\left[K_{ecu}(1-PD_i) + K_{eox}PD_i\right] \Delta \dot{P}(t) \quad (6.80)$$

Therefore,

$$\Delta \ddot{P}(t) + \left\{\frac{K_{1f}}{\chi_{2f}} + \frac{K_{1f}}{\chi_{1f}} + \frac{K_{2f}}{\chi_{2f}} + K_{1f}\left[K_{ecu}(1-PD_i) + K_{eox}PD_i\right]\right\} \Delta \dot{P}(t) + \left\{\frac{K_{1f}K_{2f}}{\chi_{1f}\chi_{2f}} + \frac{K_{1f}K_{2f}}{\chi_{2f}}\left[K_{ecu}(1-PD_i) + K_{eox}PD_i\right]\right\} \Delta P(t) + \frac{K_{1f}K_{2f}}{\chi_{2f}}\left(K_{eox}P_0 + MRR_{ox0} - K_{ecu}P_0 - MRR_{cu0}\right) = 0 \quad (6.81)$$

Solving this differential equation, ΔP is obtained as

$$\Delta P = A_2 e^{\lambda_3(t-t_2)} + B_2 e^{\lambda_4(t-t_2)} + D_2 \quad (6.82)$$

where

$$D_2 = \frac{\left(K_{ecu}P_0 + MRR_{cu0} - K_{eox}P_0 - MRR_{ox0}\right)\chi_{1f}}{1 + \chi_{1f}\left[K_{ecu}(1-PD_i) + K_{eox}PD_i\right]} \quad (6.83)$$

$$A_2 = \frac{D_2\lambda_4 - \Delta P_2\lambda_4}{\lambda_3 - \lambda_4} \quad (6.84)$$

$$B_2 = \frac{D_2\lambda_3 - \Delta P_2\lambda_3}{\lambda_4 - \lambda_3} \quad (6.85)$$

ΔP_2 the pressure difference at the end of stage 2 and λ_3 and λ_4 the roots of the characteristic equation of 6.81. Substitution of Eq. 6.82 into Eq. 6.79 and solving the differential equation yields

$$
\begin{aligned}
S(t) &= \int_{t_2}^{t} \left\{ \begin{array}{l} -\left[K_{ecu}(1-PD_i)+K_{eox}PD_i\right]\Delta P \\ +K_{ecu}P_0+MRR_{cu0}-K_{eox}P_0-MRR_{ox0} \end{array} \right\} dt \\
&= S_2 + \frac{A_2\left[K_{ecu}(1-PD_i)+K_{eox}PD_i\right]}{\lambda_3}\left[1-e^{\lambda_3(t-t_2)}\right] \\
&\quad + \frac{B_2\left[K_{ecu}(1-PD_i)+K_{eox}PD_i\right]}{\lambda_4}\left[1-e^{\lambda_4(t-t_2)}\right] \\
&\quad + \left\{ \begin{array}{l} -D_2\left[K_{ecu}(1-PD_i)+K_{eox}PD_i\right] \\ +K_{ecu}P_0+MRR_{cu0}-K_{eox}P_0-MRR_{ox0} \end{array} \right\}(t-t_2) \\
&= S_2 + A_2\frac{K_{ecu}(1-PD_i)+K_{eox}PD_i}{\lambda_3}\left[1-e^{\lambda_3(t-t_2)}\right] \\
&\quad + B_2\frac{K_{ecu}(1-PD_i)+K_{eox}PD_i}{\lambda_4}\left[1-e^{\lambda_4(t-t_2)}\right] \\
&\quad + \frac{K_{ecu}P_0+MRR_{cu0}-K_{eox}P_0-MRR_{ox0}}{1+\left[K_{ecu}(1-PD_i)+K_{eox}PD_i\right]\chi_{1f}}(t-t_2) \qquad (6.86)
\end{aligned}
$$

Both ΔP and $S(t)$ change exponentially with time. The time constants $1/\lambda_3$ and $1/\lambda_4$ are a function of pattern density PD_i. Besides the exponential terms, a linear term (third term on the right side of Eq. 6.86) exists in the step height function $S(t)$. The coefficients before the exponential and linear term are determined by boundary conditions A_2, B_2 and ΔP_2 and may be either negative or positive. Therefore, unlike that in the linear elasticity case, the step height may not increase monotonically with the pattern density.

6.7 Line Width Dependency of Topography Evolution

In the earlier sections, the pad is modeled as one-dimensional springs and dash-pots, and the deformation of the pad asperity layer in the lateral direction is not considered. Therefore, the topography evolution is independent of feature size. However, it was observed experimentally that step height reduction, dishing and erosion strongly depend on the line width. In this section, the earlier one-dimensional topography model is extended to a semi-two dimensional model with the line width as an input parameter. The basic idea is first to use a two-dimensional model to evaluate the pad deformation as a function of line width and pattern density. Then *effective* stiffness values of the pad are estimated from the deformation and substituted into the previous one-dimensional model to replace the one-dimensional stiffness value and evaluate the line width and pattern density dependency of the topography evolution. The model following this idea is not exactly a two-dimensional model. We called it a semi-two dimensional model.

6.7.1 Two-Dimensional Boundary Element Model of Pad Asperity Layer Deformation

First, the deformation of the polishing pad asperity layer over a feature with line width W_{cu} and pattern density PD_i is modeled using a linear elasticity based boundary element method (BEM) [6.31] [6], as shown in Fig. 6.8 (a). The solid line in Fig. 6.8 represents the pad asperity layer and the dash line represents a feature step. A down pressure $P_0 = 15$kPa is assumed and uniform pressures equal to $P_0/(1 - PD_i)$ are applied on the pad-copper contact region. The influence of other features on the deformation is neglected. The copper line width is assumed to vary from 0.32μm to 200μm (0.32, 1, 3.2, 10, 20, 32, 50, 100, 200μm) and the pattern density PD_i varies from 0.1 to 0.9. The pad is assumed to be IC 1000 with E_1= 5181kPa, Table 8.1, and Poisson's ratio $\upsilon_p = 0.5$ for polymer materials, as that in Chapter 3. Note that for the linear viscoelasticity case, the initial deformation does not depends on E_2, η_1 and η_2 but depends on E_1 only. The pad asperity layer thickness l is usually in the range of 10$\sim$100μm [6.32]. In the model, it is taken as 58μm, as that in Chapter 3. The values of 20, 32 and 100μm will also be tried in the simulation to see their effects. Eight quadratic elements are used to mesh the boundary of the pad asperity layer, six on the bottom and upper parts and two on left and right sides. This rough mesh is used for several considerations. First, the computation time is minimal. A number of pattern density and line width combinations (in our simulation, 81 combinations with 9 pattern densities and 9 line widths) will be tried in the simulation. A minimal computation time for each simulation will be preferred. Second, even with such a rough mesh, sufficient accuracy can be obtained for a boundary element method based model [6.31]. Third, at this stage we are more interested in how the deformation changes with line width and pattern density instead of its accurate value. A rough mesh would be sufficient for this purpose.

Fig. 6.9 shows the BEM results of the relative pad deformation over features with line widths in the range of 0.32$\sim$320μm. The pattern density is set as 0.5. The line widths are normalized, with point B (edge of the copper line) in Fig. 6.8 (a) corresponding to -1 and point A (center of the copper line) corresponding to 0. It is seen that when the line width is 0.32μm, the deformation is minimal. It is increased from 0.7nm at W_{cu}= 0.32μm to 120nm at W_{cu}= 320μm. Apparently, the deformation is much less than that predicted by the one dimensional model, even for a large line width, Fig. 6.9. Before touching the bottom of the feature step, the deformation will remain the same

[6]BEM is to be used in Chapter 8 to evaluate the wafer-scale pressure distribution too. A brief introduction of BEM is given in the appendix.

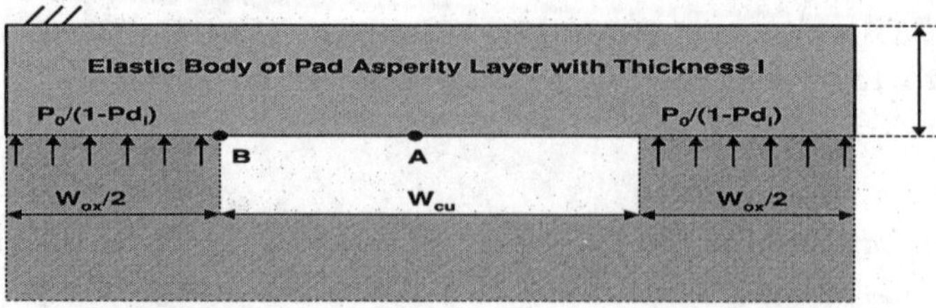

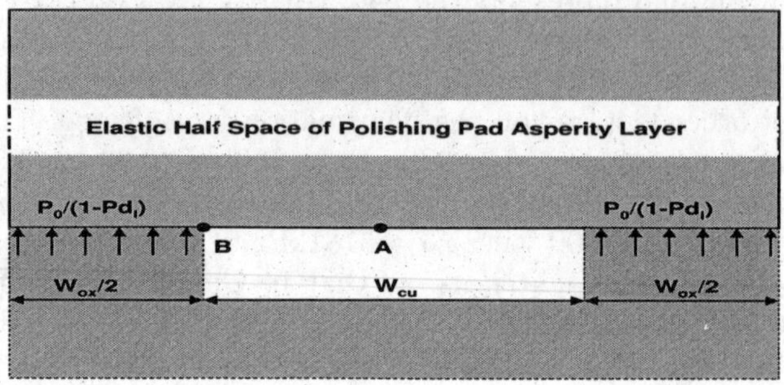

Fig. 6.8. Two-dimensional (a) boundary element model and (b) contact mechanics model of pad deformation over a feature

for a linear elastic case. For the linear viscoelasticity case, this is the initial deformation and it will vary with time. When the feature size is larger, the pad asperity layer will touch the bottom of the step earlier.

To demonstrate more clearly how the pad deforms with the copper line widths and pattern density, more boundary element simulations were done. The deformation of point B relative to point A is plotted in Fig. 6.10 as a function of copper line widths and pattern densities. It is seen that two pad deformation regions exist. When the line width is small, say, $0.32\sim50\mu m$, the deformation increases rapidly with the line width following approximately a linear relationship. When the line width becomes larger, the increase becomes slower. It will be shown later that the increase of dishing with line width follows the same trend.

6.7.2 Two-Dimensional Contact Mechanics Model of Pad Asperity Layer Deformation

The boundary element model can be used to evaluate the pad asperity deformation numerically for an arbitrary asperity layer thickness l. When the

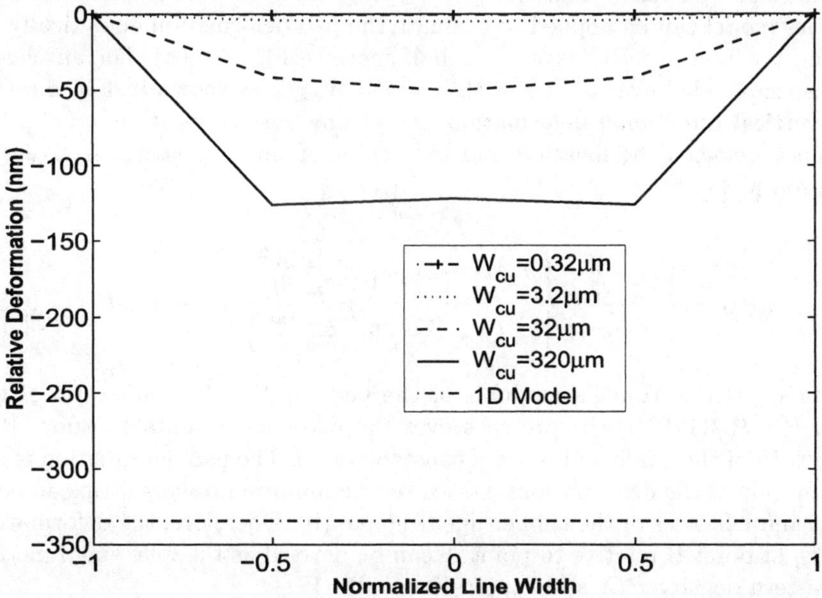

Fig. 6.9. Relative pad (IC1000) deformation over features with different line width (pattern density= 0.5)

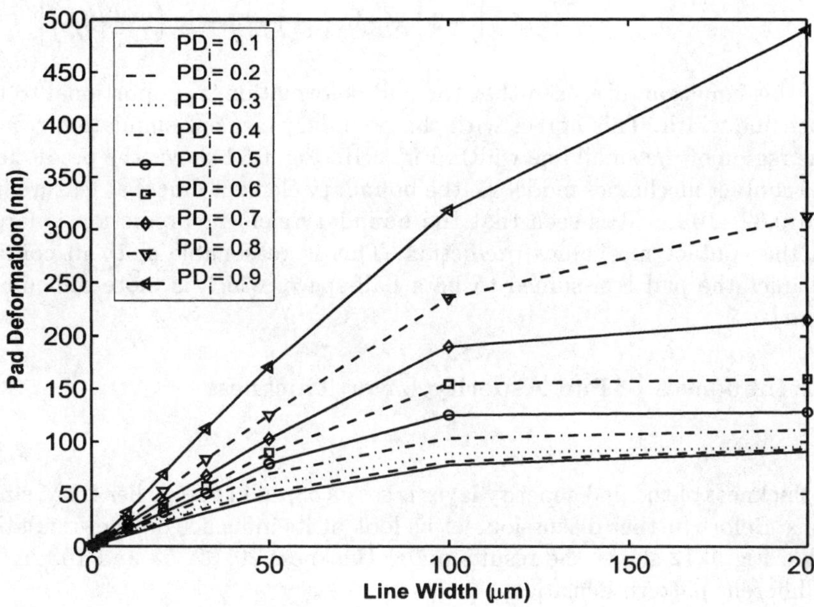

Fig. 6.10. Pad (IC1000) deformation Df_{AB} over features as a function of line width and pattern density

feature size is smaller than the pad asperity layer thickness, a contact mechanics model can be applied to evaluate the pad deformation analytically by taking the pad asperity layer as a half-space, [7] Fig. 6.8 (b). For an elastic half space loaded over a strip with a width $W_{ox}/2$ as shown in Fig. 6.8 (b), the vertical directional deformation Df at any surface location x (x is the distance between the location and the center of the load strip.) is given by Johnson [6.33] as

$$Df(x) = -\frac{1-v_p^2}{\pi E_1}P\left\{\begin{array}{l}(x+\frac{W_{ox}}{4})\ln\left(\frac{x+\frac{W_{ox}}{4}}{\frac{W_{ox}}{4}}\right)^2 \\ -(x-\frac{W_{ox}}{4})\ln\left(\frac{x-\frac{W_{ox}}{4}}{\frac{W_{ox}}{4}}\right)^2\end{array}\right\} + const \tag{6.87}$$

where E_1 is the Young's modulus of the pad, v_p the Poisson's ratio of the pad, $P= P_0/(1\text{-}PD_i)$ the pressure over the pad-copper contact region, W_{ox} the width of the oxide and *const* a constant value. The pad deformation is the summation of the deformations caused by the uniform pressure strips on both right and left sides of the copper line, Fig. 6.8 (b). Therefore, the deformation Df_{AB} of point B relative to point A can be derived as the following function of pattern density PD_i and copper line width W_{cu}:

$$Df_{AB} = -\frac{1-v_p^2}{\pi E_1}\frac{P_0}{1-PD_i}W_{cu}\left\{\begin{array}{l}\frac{(1-PD_i)\ln 2}{PD_i} \\ +\left[\frac{1+PD_i}{PD_i}\ln\left(2\frac{1+PD_i}{1-PD_i}\right)-2\ln\left(\frac{4PD_i}{1-PD_i}\right)\right] \\ -2\left[\frac{1}{PD_i}\ln\left(\frac{2}{1-PD_i}\right)-\ln\left(2\frac{PD_i}{1-PD_i}\right)\right]\end{array}\right\} \tag{6.88}$$

From this equation, it is seen that the pad deformation is proportional to the copper line width. This agrees with the boundary element simulation results at the region of the small line width, Fig. 6.10. Fig. 6.11 shows the predictions of the contact mechanics model vs. the boundary element model at feature line width 0.32~10μm. It is seen that the boundary element prediction is larger than the contact mechanics prediction. This is reasonable since in contact mechanics the pad is assumed to be a half-space, which is more difficult to deform.

6.7.3 Influences of Pad Asperity Layer Thickness

The thickness of the pad asperity layer is set as 58μm in the earlier BEM simulations. Before further discussion, let us look at its influence on the simulation results. Fig. 6.12 shows the results under thickness 20, 32, 58 and 100μm for two different pattern densities.

[7] A half-space of the contact body is a fundamental assumption in contact mechanics [6.33]. This requires the loading area is smaller than the thickness of the contact body.

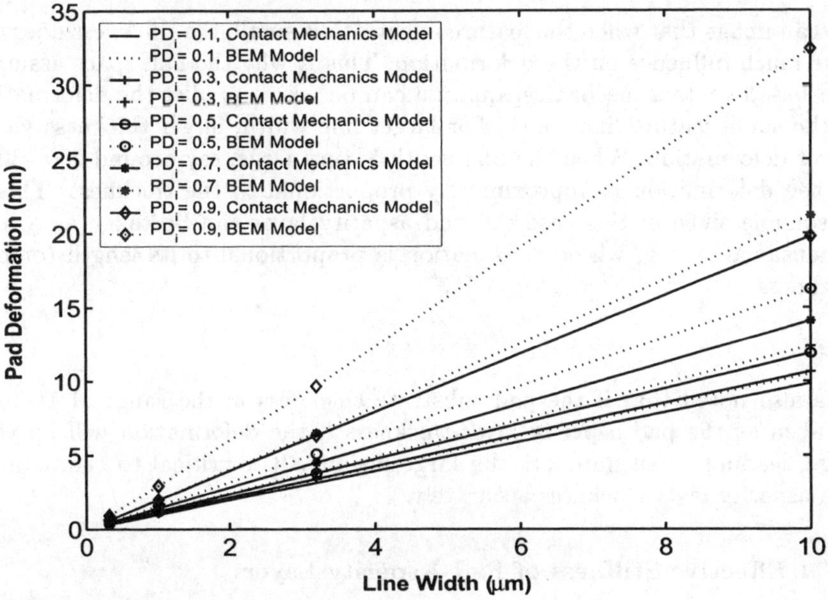

Fig. 6.11. Predictions of deformations at small line widths using BEM and contact mechanics models

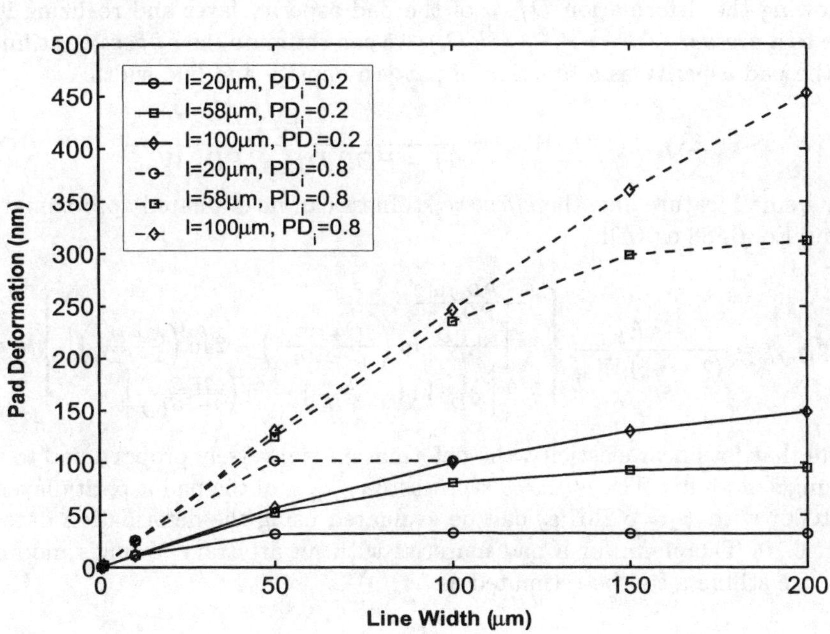

Fig. 6.12. Prediction of deformation under different asperity layer thickness

We can image that when the feature line width is small, the thickness does not have much influence on the deformation. This is why the half-space assumption based contact mechanics equation can be use to predict the deformation at the small feature line width. For larger line width, larger thickness yields larger deformation. When the line width is very large, say around 200~300μ m, the deformation is approximately proportional to the thickness. This is reasonable since in this case the pad asperity layer can be taken as a one-dimensional spring, whose deformation is proportional to its length (or pad thickness).

It is also noted that if the pad substrate thickness in the range of 1~2mm is taken as the pad asperity layer thickness l, the deformation will be very large, leading to an impractically large dishing. It is critical to characterize this asperity layer thickness accurately.

6.7.4 Effective Stiffness of Pad Asperity Layer

Knowing the deformation Df_{AB} of the pad asperity layer and realizing it is due to a pressure $\Delta P = P_0/(1-PD_i)$, we can estimate the $effective$ stiffness of the pad asperity as a function of pattern density and line width:

$$K_{1f_{eff}}\left(PD_i, W_{cu}\right) = \frac{P_0}{\left(1 - PD_i\right) Df_{AB}\left(PD_i, W_{cu}\right)} \tag{6.89}$$

For a small feature line, the *effective* stiffness can be evaluated approximately using Eq. 6. 88 as well:

$$K_{1f_{eff}} = \frac{-\pi E_1}{\left(1 - v_p^2\right) W_{cu}} \left\{ \begin{array}{l} \frac{(1-PD_i)\ln 2}{PD_i} \\ + \left[\frac{1+PD_i}{PD_i} \ln\left(\frac{2(1+PD_i)}{1-PD_i}\right) - 2\ln\left(\frac{4PD_i}{1-PD_i}\right)\right] \\ -2\left[\frac{1}{PD_i} \ln\left(\frac{2}{1-PD_i}\right) - \ln\left(\frac{2PD_i}{1-PD_i}\right)\right] \end{array} \right\}^{-1} \tag{6.90}$$

Note that for linear elasticity, the deformation is inversely proportional to the Young's modulus. The *effective* stiffness $K_{1f_{IC1000}}$ of the pad asperity layer of IC1000 (with E_1= 5.2MPa) can be evaluated using the deformation data in Fig. 6.10. Therefore, for a pad material with an arbitrary Young's modulus E_1, the stiffness can be estimated as

$$K_{1f_{eff}} = \frac{K_{1f_{IC1000}}}{5.2} E_1. \tag{6.91}$$

Similarly, the *effective* values of K_{2f}, χ_{1f}, χ_{2f} can be evaluated by

$$K_{2f_{eff}} = \frac{K_{1f_{IC1000}}}{5.2} E_2 \tag{6.92}$$

$$\chi_{1f_{eff}} = \frac{K_{1f_{IC1000}}}{5.2} \eta_1 \tag{6.93}$$

$$\chi_{2f_{eff}} = \frac{K_{2f_{IC1000}}}{5.2} \eta_2 \tag{6.94}$$

.

We call $K_{1f_{IC1000}}/5.2$ with unit $kPa/nm \cdot MPa$ the normalized stiffness. It is independent of the pad material but a function of pad asperity layer thickness. Its values are tabulated in tables 6.2, 6.3, and 6.4 for different pattern densities PD_i, line widths W_{cu} and two different asperity layer thicknesses $l = 58\mu$m and 32μm. When line width is smaller than 10μm, the normalized stiffness values are approximately independent of the layer thickness l and they are tabulated in table 6.2. The values for a layer thickness of 58 μm and a line width larger than 10μm are tabulated in table 6.3. The values for a layer thickness of 32 μm and a line width larger than 10μm are tabulated in table 6.4. The values of E_1, E_2, η_1 and η_2 in Table 6.1 times the normalized stiffness values equal to the *effective* stiffness values $K_{2f_{eff}}$, $\chi_{1f_{eff}}$ and $\chi_{2f_{eff}}$ of the commercial pads. Similar tables can be made for other layer thickness l using boundary element simulation.

Table 6.2. Normalized stiffness values (with unit kPa/(nm·Mpa)) for small line width

	W_{cu}=0.32μm	W_{cu}=1μm	W_{cu}=3.2μm	W_{cu}=10μm
PD_i=0.1	9.541827	3.051718	0.953575	0.305431
PD_i=0.2	9.698977	3.10067	0.966086	0.310214
PD_i=0.3	10.62797	3.391614	1.051387	0.333312
PD_i=0.4	10.62372	3.390259	1.050965	0.333178
PD_i=0.5	11.48885	3.661458	1.132245	0.355344
PD_i=0.6	12.83299	4.081411	1.258246	0.390861
PD_i=0.7	15.14759	4.808125	1.476798	0.453649
PD_i=0.8	19.44002	6.155209	1.881606	0.571009
PD_i=0.9	31.1	9.82	2.98	0.891

As a summary of the above discussions, we plot the flow chart of the semi-two dimensional model in Fig. 6.13.

Table 6.3. Normalized stiffness values (with unit kPa/(nm·MPa)) for large line widths and pad asperity layer thickness l= 58μm

	W_{cu} =10μm	W_{cu} =20μm	W_{cu} =32μm	W_{cu} =50μm	W_{cu} =100μm
PD_i=0.1	0.305431	0.155507	0.1009	0.067874	0.041227
PD_i=0.2	0.310214	0.155317	0.099816	0.069288	0.044494
PD_i=0.3	0.333312	0.168972	0.10581	0.069437	0.046676
PD_i=0.4	0.333178	0.168904	0.105767	0.069409	0.046657
PD_i=0.5	0.355344	0.179871	0.11391	0.07329	0.046198
PD_i=0.6	0.390861	0.195744	0.125317	0.08093	0.046731
PD_i=0.7	0.453649	0.223207	0.143031	0.093955	0.050711
PD_i=0.8	0.571009	0.274958	0.174039	0.115725	0.061261
PD_i=0.9	0.891	0.417	0.257	0.170	0.0907

6.8 Simulation vs. Experimental Results

A Matlab program was written to implement the model developed in the earlier sections under the framework of the iteration in Fig. 6.2 and the semi-two dimensional model in Fig. 6.13. Two simulators, one based on the analytical solution in section 6.5, the other based on the numerical implementation of the integrals and differential equations in section 6.4, are developed for both linear elastic and viscoelastic models of the polishing pad.

Table 6.4. Normalized stiffness values (with unit kPa/(nm·MPa)) for large line widths and pad asperity layer thickness l= 32μm

	W_{cu} =10μm	W_{cu} =20μm	W_{cu} =50μm	W_{cu} =100μm
PD_i=0.1	0.30942	0.163481	0.078739	0.064417
PD_i=0.2	0.310302	0.162432	0.084749	0.068961
PD_i=0.3	0.337256	0.169461	0.087467	0.078337
PD_i=0.4	0.33712	0.169394	0.087432	0.078305
PD_i=0.5	0.358412	0.182325	0.086751	0.080519
PD_i=0.6	0.390243	0.201425	0.088991	0.080467
PD_i=0.7	0.446259	0.230977	0.098477	0.079636
PD_i=0.8	0.552229	0.281378	0.120623	0.084046
PD_i=0.9	0.843864	0.414153	0.179435	0.110512

The accurate characterization of parts of the inputs, specially, the pad properties, is a challenge. The values of inputs in this work are based on the literature

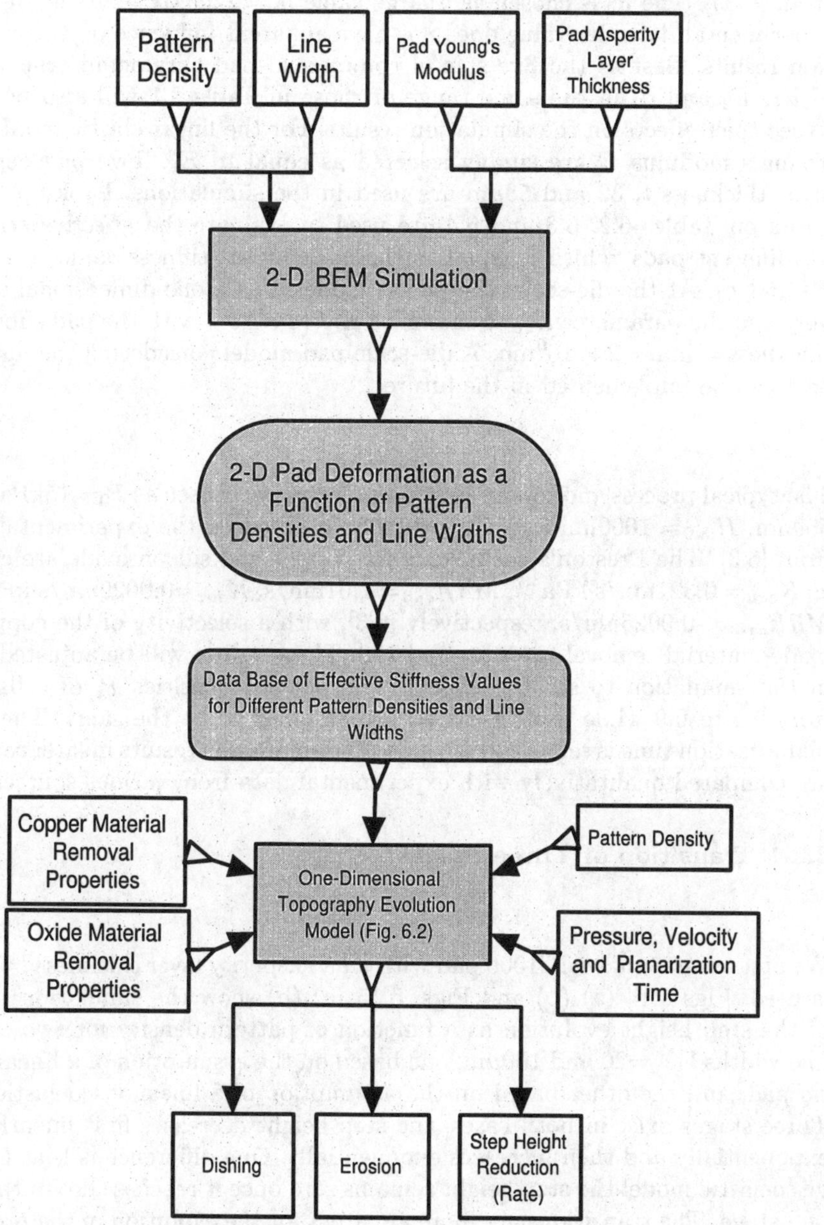

Fig. 6.13. Flow chart of the semi-two dimensional topography evolution model

or estimations. For the linear viscoelastic model, the material parameters of the polishing pads are chosen to be those in Table 8.1, as reported by Kim et al. [6.29]. The η_1 is chosen as a large value η_1= 2200MPa·s based on their experimental data. Its value does not have apparent influence on the simulation results. Besides the five sets of commercial pad parameters, the values of E_1, E_2 and η_2 in the same range of those in Table 8.1 will also be tried to see their effects on the simulation results. For the linear elastic model, the Young's modulus E are simply selected as equal to E_1. Two pad asperity layer thickness l, 32 and 58μm are used in the simulations. Look-up tables based on Tables 6.2, 6.3 and 6.4 are used to evaluate the *effective* stiffness for different pads, which is equal to the normalized stiffness values time E_1, E_2 and η_2. At the die-scale, the pad is modeled as a one-dimensional spring only and the parameter K_{1d} is chosen as $E_1/\left(2 \times 10^6\right)$ with the pad substrate thickness=2mm= 2×10^6nm. A die-scale pad model considering the dashpot part can be implemented in the future.

The typical process and topography parameters are chosen as P_0= 15kPa, S_0= 500nm, H_{cu0}= 1000nm and H_{ox0}= 1000nm based on the experimental data from [6.3]. The Preston's coefficients for copper and silicon oxide are chosen as K_{ecu}= 0.321nm/s·kPa^{-1}, MRR_{cu0}= 1.01nm/s, K_{eox}=0.0029nm/s·kPa and MRR_{ox0}= -0.0025nm/s, respectively [6.3], with a selectivity of the copper to oxide material removal rates around 130. These values will be adjusted later in the simulation to see the effects. The pattern densities P_i of cells vary from 0.1 to 0.9 while their areas A_i are assumed to be the same. The total planarization time is set as 200 seconds. The simulation results in later sections are compared qualitatively with experimental data from various sources.

6.8.1 Transition of Three Stages

We first assume that a IC1000 pad with a pad asperity layer thickness l=58μm is used. Figs 6.14 (a) (b) and Figs. 6.15 (a) (b) show the simulation results of the step height evolution as a function of pattern density for two copper line widths W_{cu}=20 and 100μm, one based on the assumption of a linear elastic pad, and the other based on the assumption of a linear viscoelastic pad. Three stages exist in both cases: the step height decreases first linearly and exponentially and then increases exponentially. One difference is that for the viscoelastic model the step height remains zero once it reaches zero in the second stage. The transitions are more apparent for the evolution of the pressure difference ΔP over the feature step, Figs. 6.16 (a), (b) and 6.17 (a), (b). The ΔP in the first stage keeps as constant $P_0/(1\text{-}PD_i)$. In the second stage, it decreases exponentially. In the linear elasticity case it decreases to zero until a full planarization (zero step height) is realized. In the viscoelastic case, the full planarization is realized earlier. Due to the time dependency of the polishing pad material property, the pressure difference, however, continues decreasing exponentially after full planarization with a larger time constant, Fig. 6.17. Once the oxide below the copper is exposed, steps are formed again and the pressure difference starts to increase exponentially. Three stages with time boundaries t_1, t_{12}, t_2 are plotted in Fig. 6.17 for a pattern density PD_i= 0.9. Comparing simulation results for the two different copper line widths, we can also see that stage 1 is longer for a small line width. This is because the deformation of the pad asperity layer is smaller and therefore it touches the bottom of the step later.

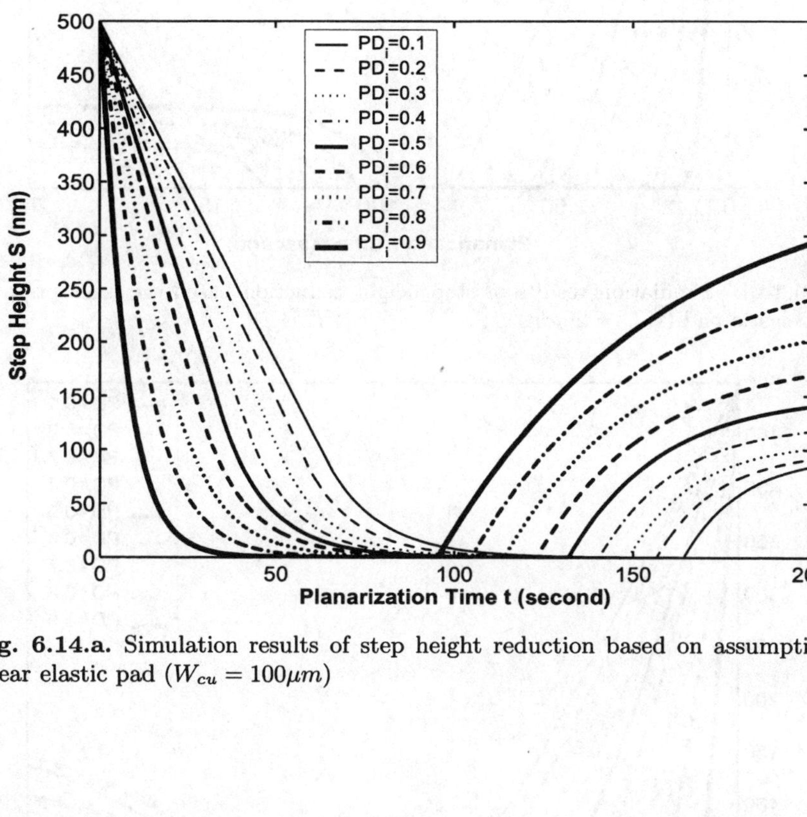

Fig. 6.14.a. Simulation results of step height reduction based on assumption of linear elastic pad ($W_{cu} = 100\mu m$)

Experimental data supporting the transition of three stages is shown schematically in Figs. 6.18 [6.5], 6.19 [6.3] and 6.20 [6.7], respectively. Fig. 6.18 shows the data through the three stages, while Fig. 6.19 shows the step height reduction data in the first and second stages and Fig. 6.20 shows the step height increase data in the third stage.

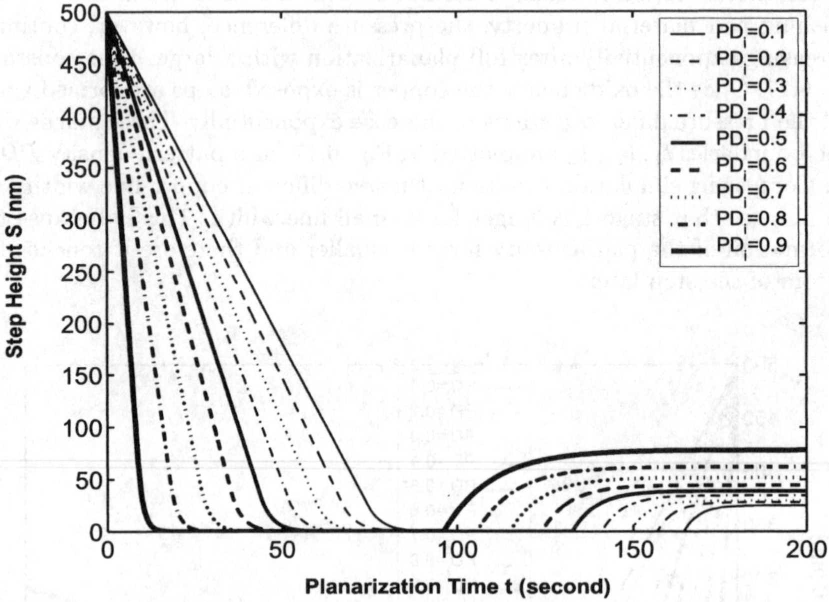

Fig. 6.14.b. Simulation results of step height reduction based on assumption of linear elastic pad ($W_{cu} = 20\mu m$)

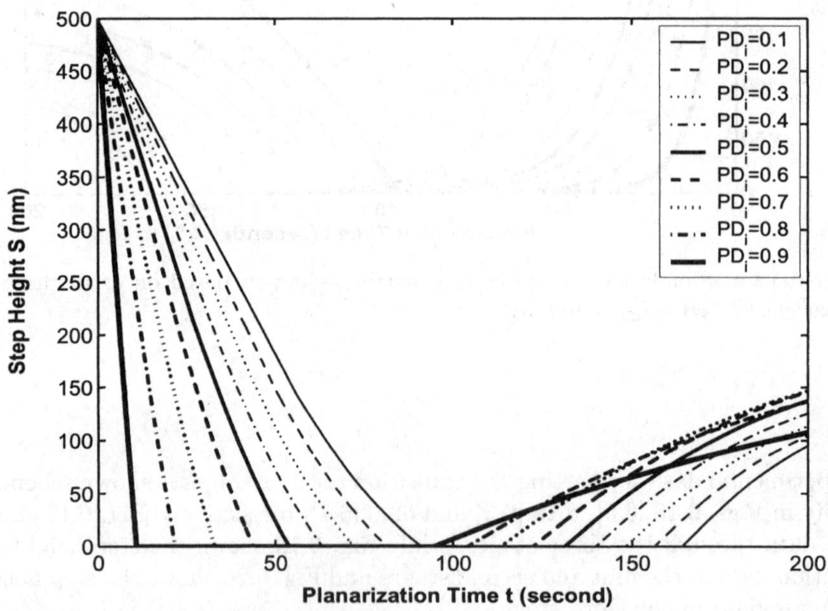

Fig. 6.15.a. Simulation results of step height reduction based on assumption of linear viscoelastic pad ($W_{cu} = 100\mu m$)

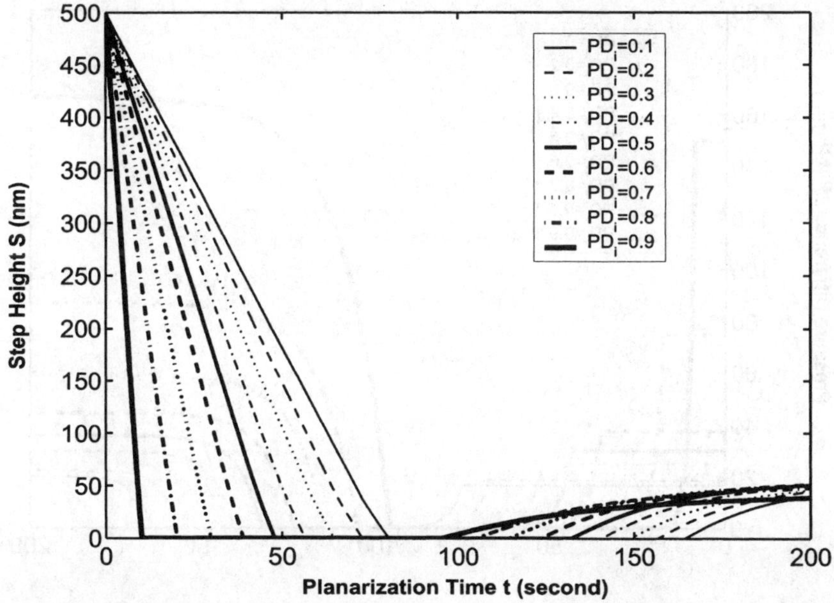

Fig. 6.15.b. Simulation results of step height reduction based on assumption of linear viscoelastic pad ($W_{cu} = 20\mu m$)

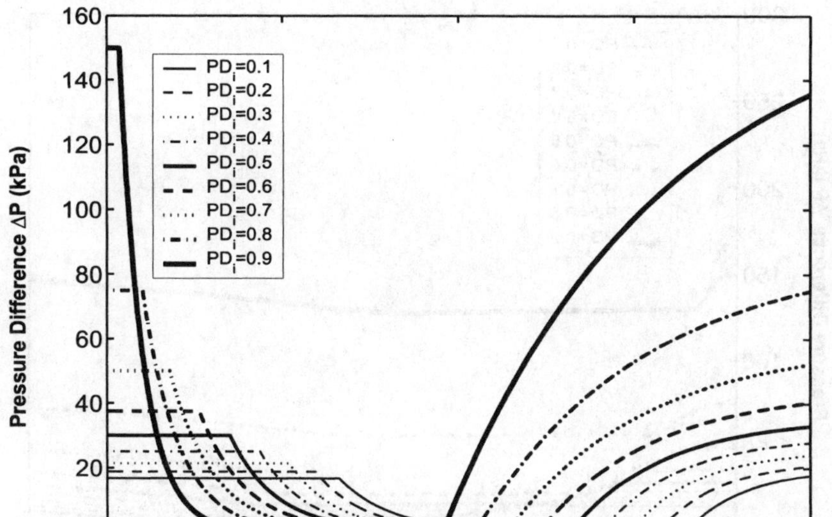

Fig. 6.16.a. Simulation results of pressure difference evolution based on assumption of linear elastic pad ($W_{cu} = 100\mu m$)

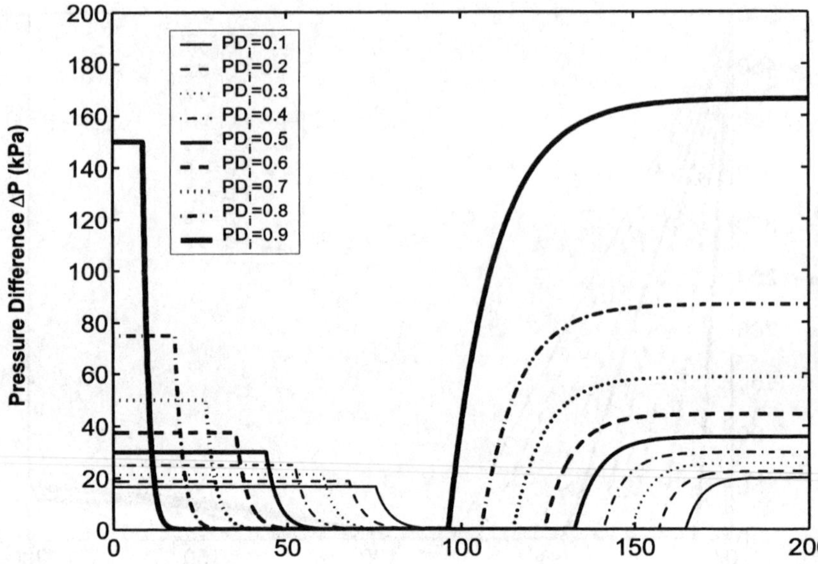

Fig. 6.16.b. Simulation results of pressure difference evolution based on assumption of linear elastic pad ($W_{cu} = 20\mu m$)

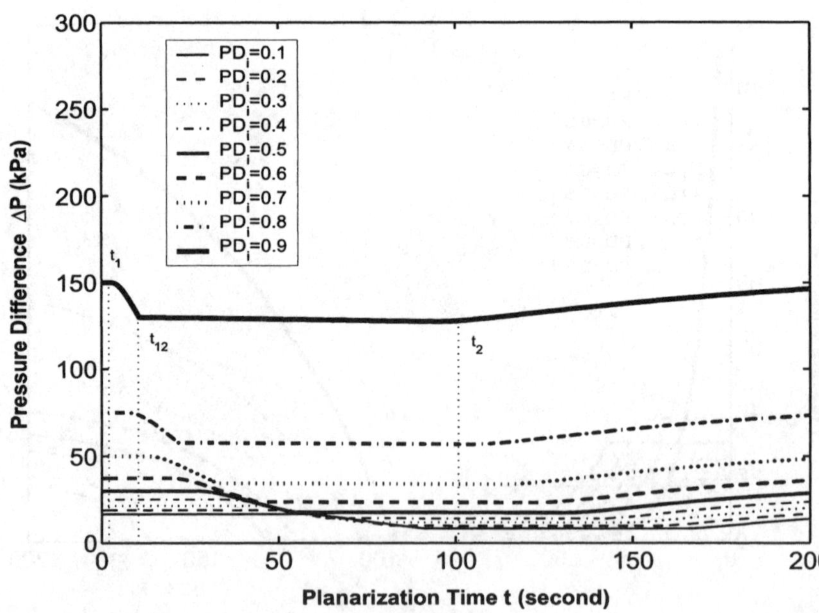

Fig. 6.17.a. Simulation results of pressure difference evolution based on assumption of linear viscoelastic pad ($W_{cu} = 100\mu m$)

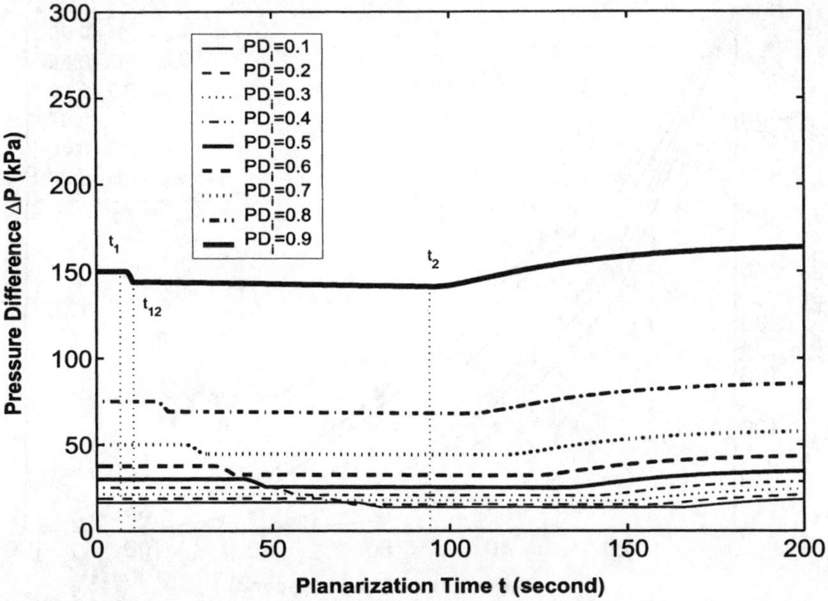

Fig. 6.17.b. Simulation results of pressure difference evolution based on assumption of linear viscoelastic pad ($W_{cu} = 20\mu m$)

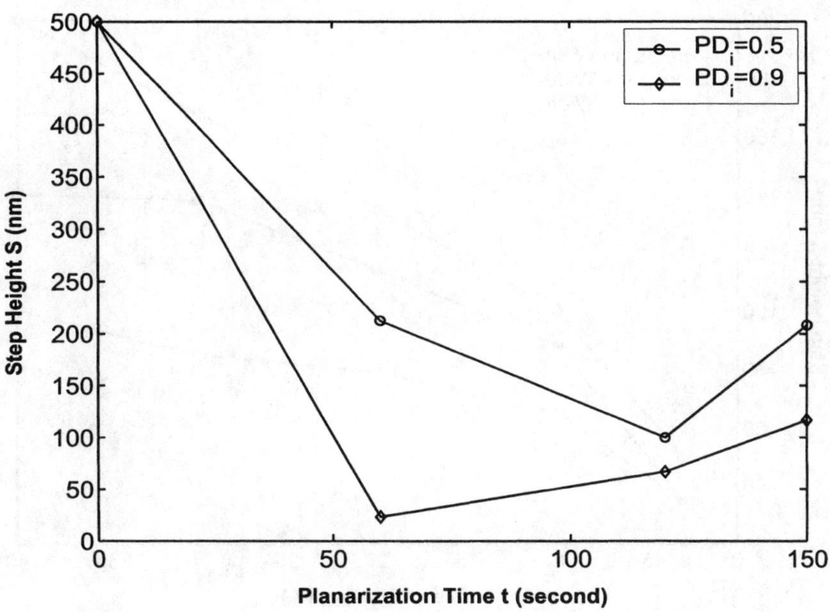

Fig. 6.18. Experimental step height evolution as a function of planarization time [6.5]

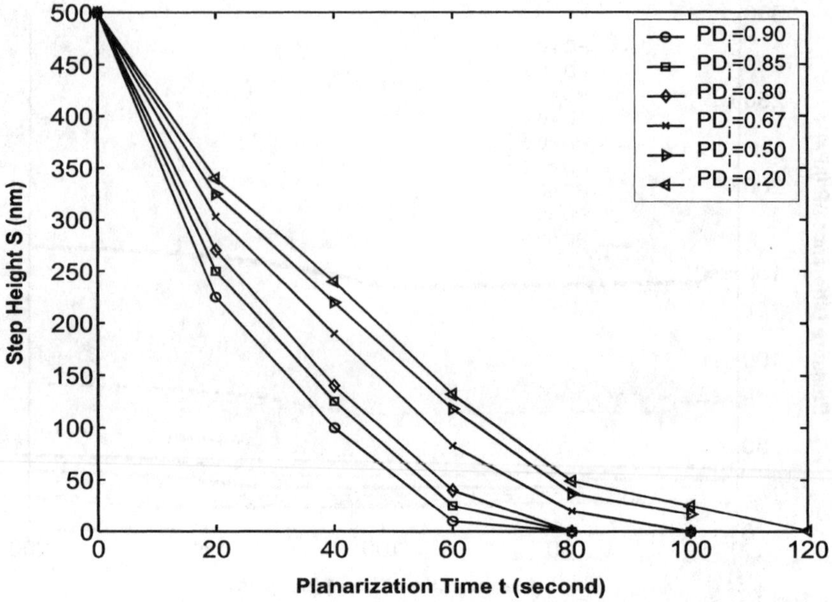

Fig. 6.19. Experimental step height evolution as a function of planarization time in stage 2 [6.3]

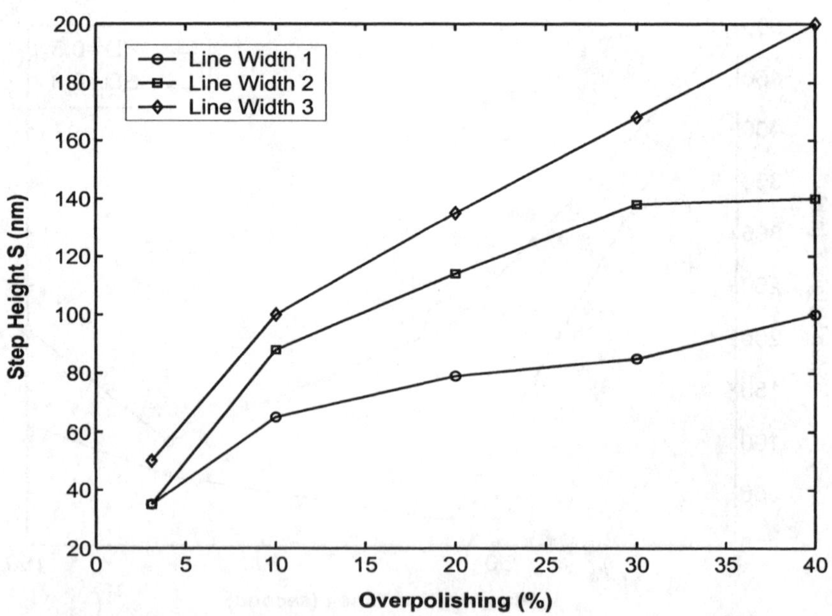

Fig. 6.20. Experimental step height evolution as a function of planarization time in stage 3 [6.7]

6.8.2 Oxide Erosion as a Function of Pattern Density and Line Width

Oxide erosion has long been observed to be a function of pattern density PD_i. Figs 6.21 and 6.22 show the simulation results of erosion vs. pattern density based on assumptions of a linear elastic pad and a linear viscoelastic pad, respectively. The oxide removal rate is small. Therefore, the down pressure P_0 is set as 25kPa instead of 15kPa to have a sufficiently large erosion. Other parameters remain the same as that in the earlier simulations. The simulation results based on assumptions of linear elastic and linear viscoelastic pads are similar to each other: the oxide erosion increases rapidly with the pattern density. This is supported by erosion data for four different copper line widths [6.3], Fig. 6.23. Data from other sources [6.2, 6.5] support this as well. In addition, it is noted that oxide erosion decreases with the increase of the line width. This is supported by the experimental data in Fig. 6.23 as well. However, the sensitivity of the simulation results on line widths is smaller than that in the experimental data. A more accurate model may be needed in the future to explain this sensitivity.

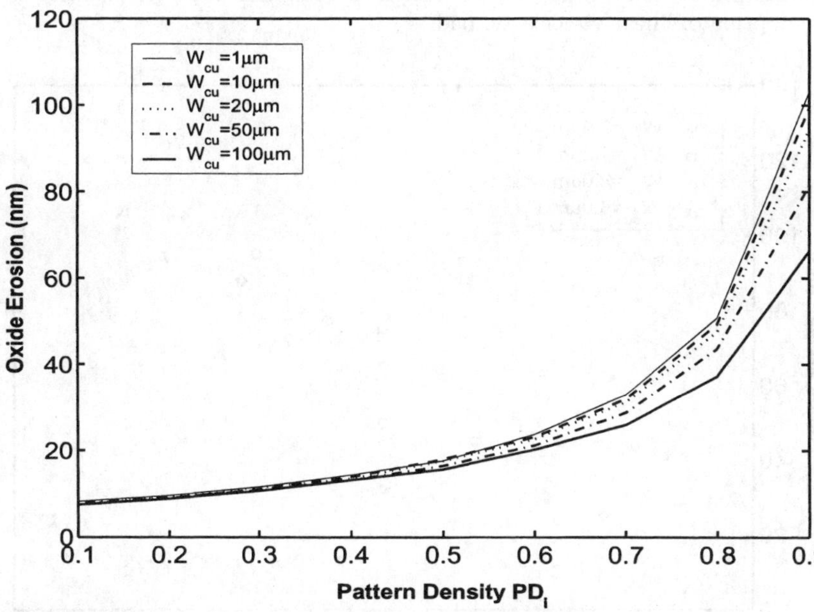

Fig. 6.21. Simulation results of oxide erosion as a function of pattern density based on assumption of linear elastic pad

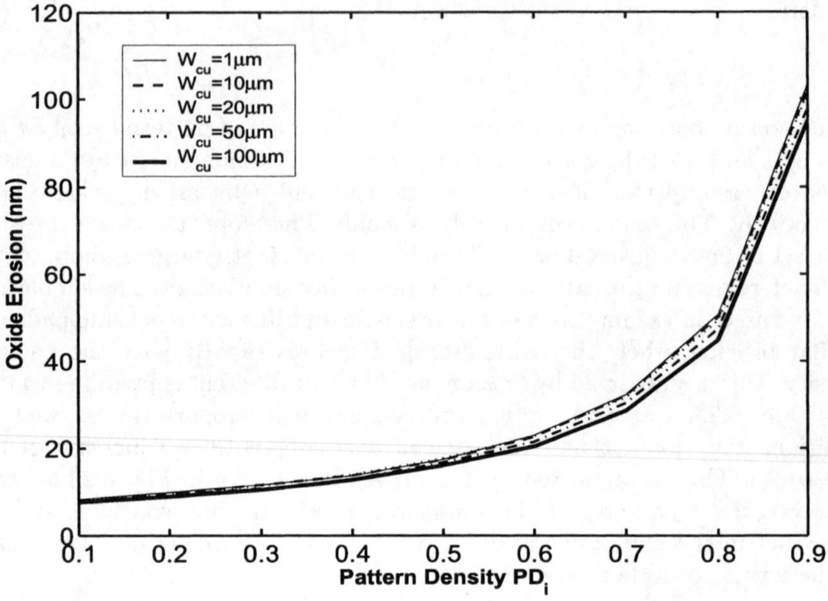

Fig. 6.22. Simulation results of oxide erosion as a function of pattern density based on assumption of linear viscoelastic pad

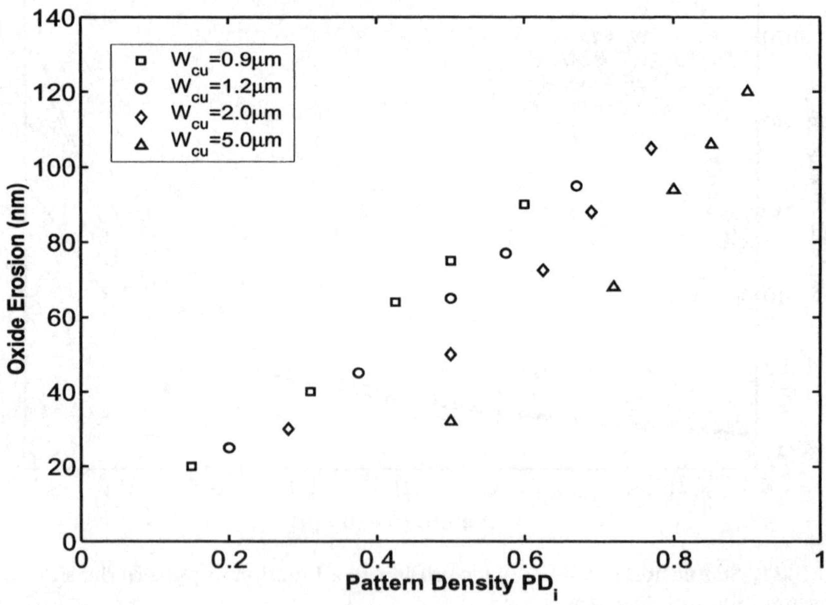

Fig. 6.23. Experimental data of oxide erosion as a function of pattern density [6.3]

There are primarily two reasons for the rapid increase of erosion with pattern densities: one is the longer overpolishing time, and the other is the larger pressure supported by the exposed oxide (or larger oxide removal rate), as shown in Figs. 6.24 and 6.25.

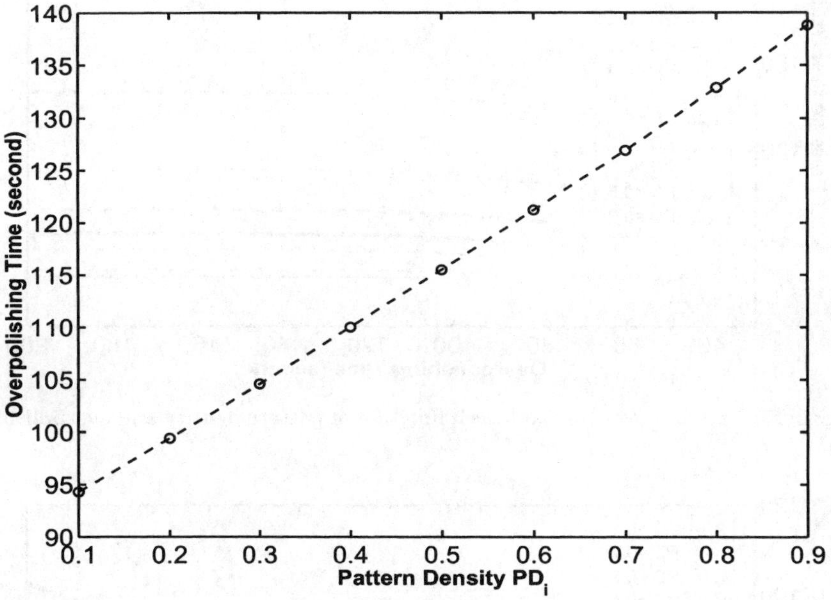

Fig. 6.24. Overpolishing time of the third stage as a function of pattern density

6.8.3 Oxide Erosion as a Function of Time

Fig. 6.26 shows the simulation results of the oxide erosion as a function of planarization time for different pattern densities under Pressure $P_0 = 25$kPa, asperity layer thickness $l = 58\mu$m and line width $W_{cu} = 10\mu$m. It is seen that the oxide erosion increases approximately linearly with time and the increase is quicker for higher pattern densities. This is because the pressure supported by the oxide is almost constant and larger for larger pattern densities, Fig. 6.25. The experimental data [6.4] supporting the simulation results is shown in Fig. 6.27.

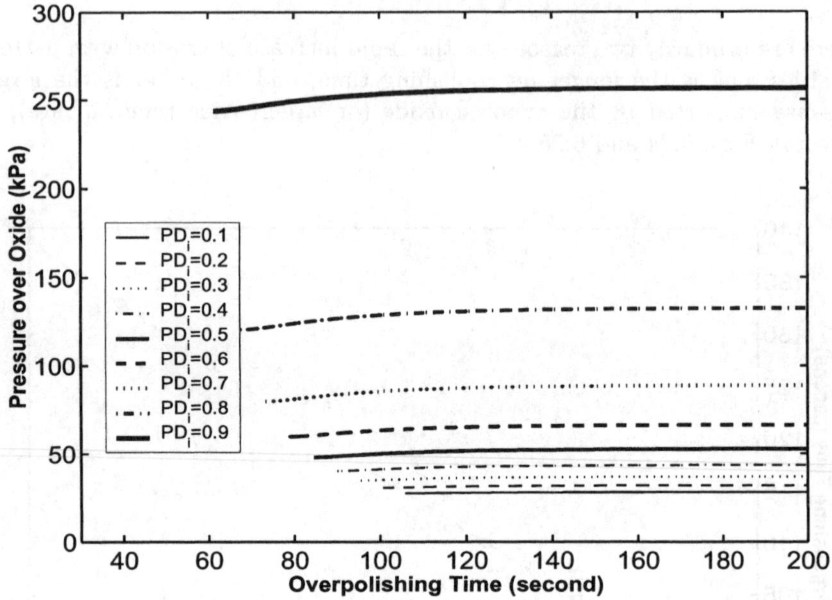

Fig. 6.25. Pressure over the oxide as a function of pattern density and overpolishing time

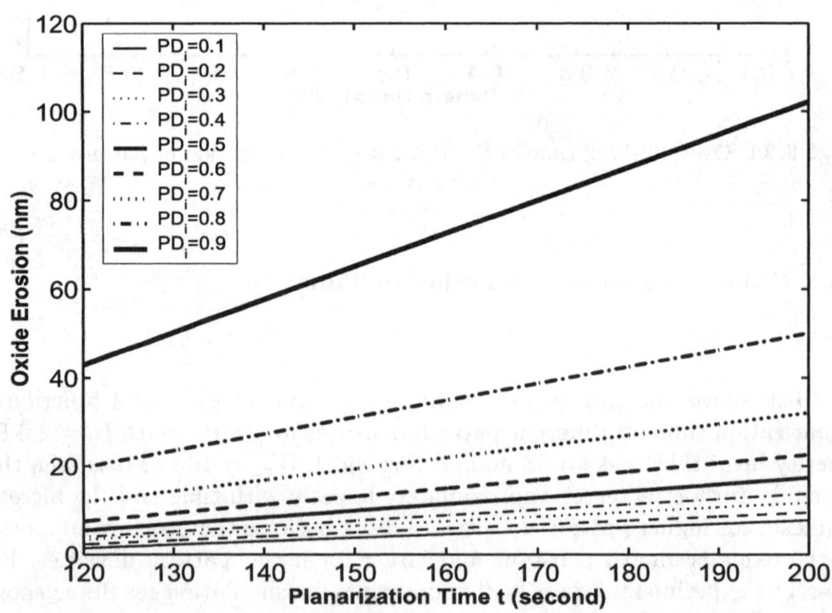

Fig. 6.26. Simulation results of erosion as a function of time based on assumption of linear viscoelastic pad

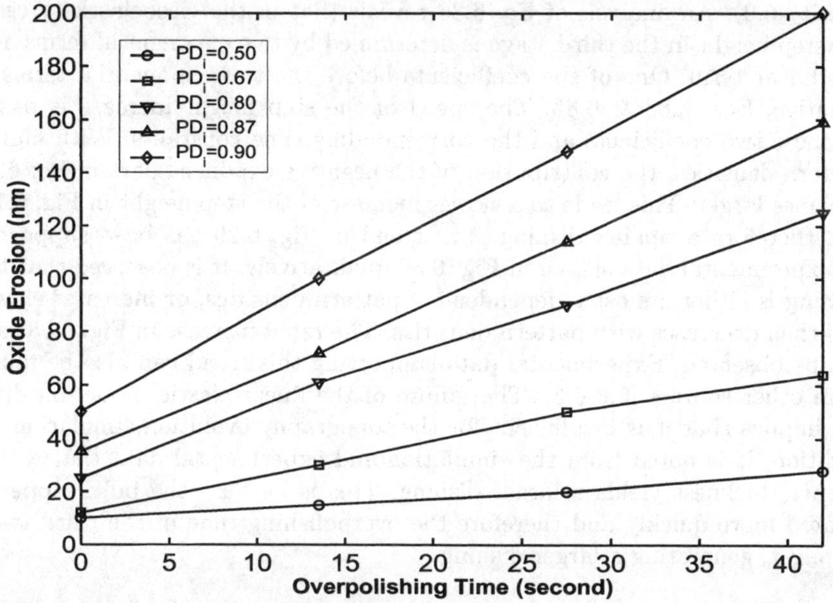

Fig. 6.27. Experimental results of erosion as a function of overpolishing time [6.4]

6.8.4 Dishing as a Function of Pattern Density and Line Width

While the simulation results of erosion based on assumptions of linear elastic and linear viscoleastic pads both support the experimental data qualitatively, this is not the case for dishing. Figs. 6.28 and 6.29 show the different simulation results of dishing as a function of pattern density under pressure P_0= 15kPa, Asperity layer thickness l= 58μm for linear elastic and linear viscoelastic pads, respectively. In the simulation, two copper thickness, one, 1000nm and the other 600nm are used. The unfilled symbols represent results for the thin copper and the filled symbols represent results for the thick copper. Similar to the erosion, the dishing increases rapidly (from 100nm at pattern density 0.1 to 350nm at pattern density 0.9 for W_{cu} = 100μm and copper thickness 600nm) in the linear elasticity case. This trend can also be observed from Fig. 6.14 (a) and Eq. 6.61. This rapid increase can be attributed to the longer overpolishing time and larger material removal rate difference (pressure difference) over the upper and bottom parts of the steps. The simulation results based on the linear viscoelasticity assumption are different. It is observed that in Fig. 6.29 the dishing increases slowly at lower pattern densities (from 158nm at pattern density 0.1 to 190nm at pattern density 0.6 for W_{cu}=100μm and copper thickness 600nm.) and then decreased at higher pattern densities (from 190nm at pattern density 0.6 to 140nm at pattern density 0.9). An analysis of Eq. 6.86 reveals that in the viscoelasticity case, the step height in the third stage is determined by two exponential terms and one linear term. One of the coefficients before the two exponential terms is negative, Eqs. 6.84 & 6.85. The speed of the step height increase is based on these two coefficients and the corresponding time constants. With higher pattern densities, the contribution of the negative exponent part in Eq. 6.86 becomes larger. This leads to a slower increase of the step height in Fig. 6.15, and therefore a smaller dishing. The trend in Fig. 6.29 has been supported by experimental data shown in Fig. 6.30 qualitatively. It is observed that the dishing is either almost independent of pattern densities, or increases slowly and then decreases with pattern densities. The rapid increase in Fig. 6.28 cannot be observed. Experimental data supporting this trend can also be found from other sources [6.2-6.3]. The failure of the linear elastic model on dishing implies that it is insufficient for the topography evolution simulation. In addition, it is noted from the simulation and experimental data that a thin copper thickness yields a larger dishing. This is because the bulk copper is cleared more quickly and therefore the overpolishing time of the third stage is longer, generating a larger dishing.

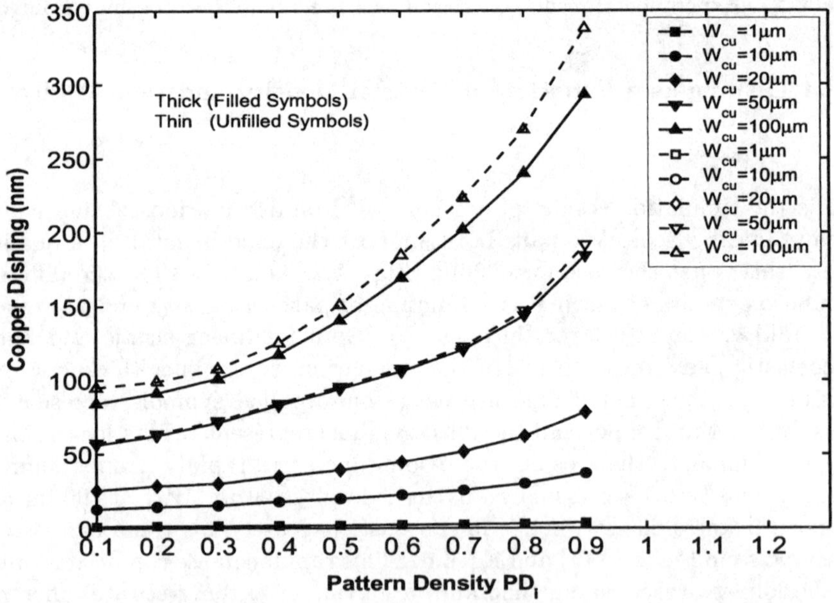

Fig. 6.28. Simulation results of dishing as a function of pattern density and line width based on assumptions of linear elastic pad

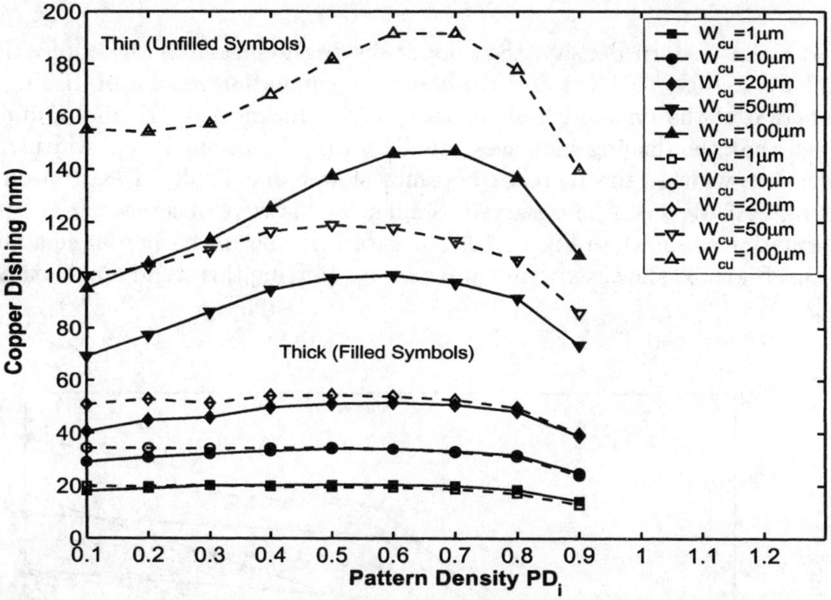

Fig. 6.29. Simulation results of dishing as a function of pattern density and line width based on assumptions of linear viscoelastic pad

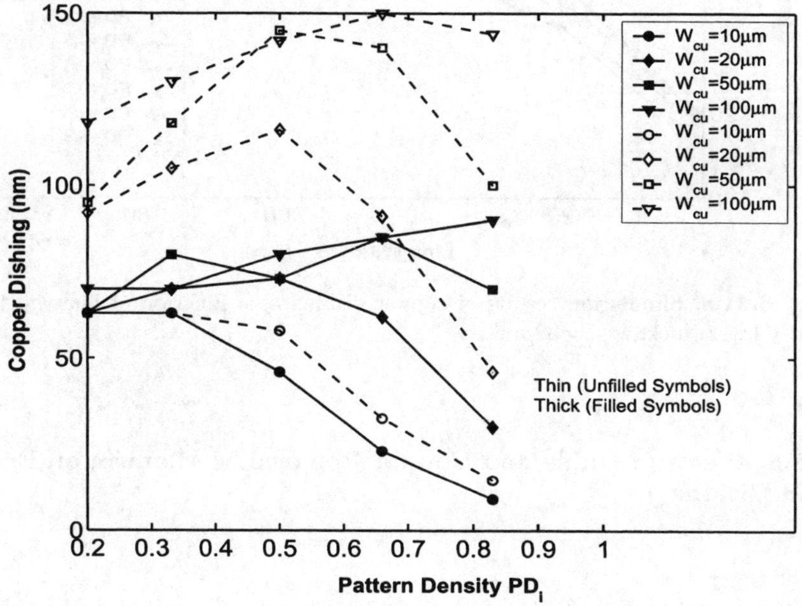

Fig. 6.30. Typical experimental data of dishing as a function of pattern density and line width [6.6]

Besides the pattern density, the copper dishing depends on the copper line width W_{cu}. Figs. 6.31 (a) and (b) show the simulation results of dishing as a function of the line width for asperity layer thickness $l=$ 32 and 58μm. It is seen that the dishing increases rapidly with W_{cu} at small line widths. At larger line widths, the increase becomes slower and finally a saturation of dishing with W_{cu} can be observed. Similar trends were observed for the pad deformation as shown in Figs. 6.10. The saturation occurs earlier for small line widths. Fig. 6.32 shows experimental data supporting this trend schematically [6.6].

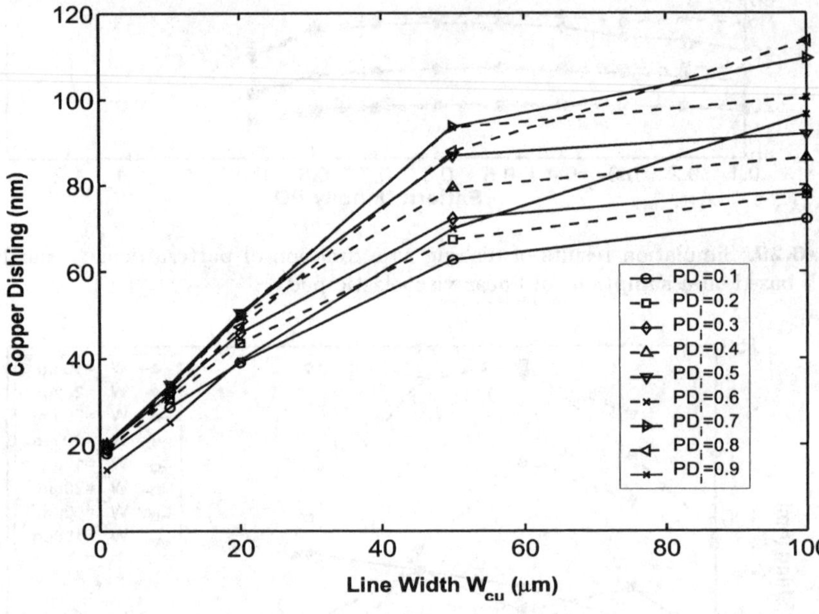

Fig. 6.31.a. Simulation results of copper dishing as a function of line width (Asperity layer thickness $l=$ 32μm)

6.8.5 Effects of Oxide and Copper Removal Mechanism on Erosion and Dishing

In this section, the effects of the material removal properties of oxide and copper on the erosion and dishing will be addressed. All simulations are based on the assumption of a linear viscoelastic pad.

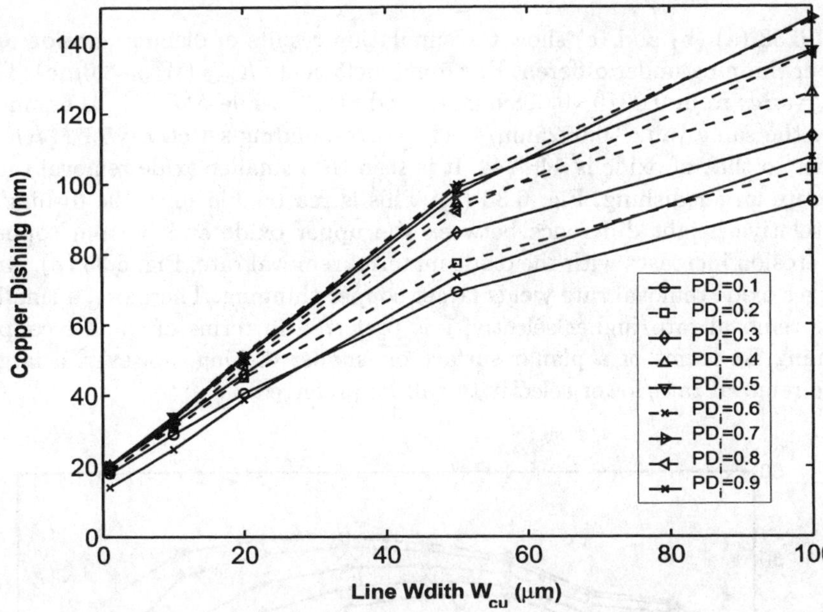

Fig. 6.31.b. Simulation results of copper dishing as a function of line width (Asperity layer thickness $l = 58\mu m$)

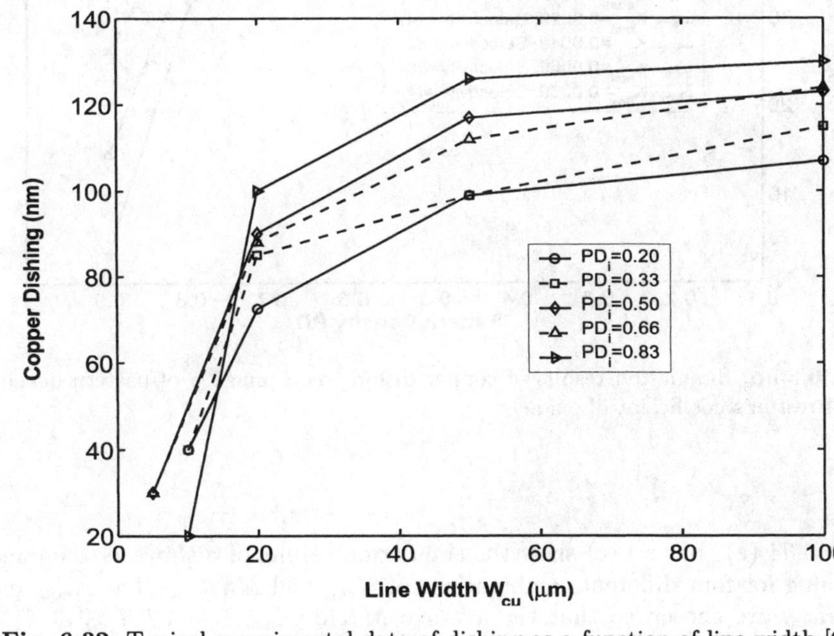

Fig. 6.32. Typical experimental data of dishing as a function of line width [6.6]

Figs. 6.33 (a) (b) and (c) show the simulation results of dishing, erosion and copper thinning under different Preston's coefficients K_{eox} (W_{cu}= 20μm). The K_{eox} varies from 0.0019~0.0089nm/second·kPa^{-1} while MRR_{ox0} is assumed to be the same value -0.0025nm/s. The corresponding selectivity of MRR of copper to that of oxide is 44~244. It is seen that smaller oxide removal rates generate larger dishing, Fig. 6.33 (a). This is reasonable since the dishing is the relative height difference between the upper oxide and bottom copper. The erosion increases with the oxide material removal rate, Fig. 6.33 (b). And a larger oxide removal rate yields larger copper thinning. Therefore, a smaller oxide removal rate/higher selectivity is preferred in terms of smaller copper thinning. In terms of a planar surface or smaller dishing, however, a larger oxide removal rate/lower selectivity will be preferred.

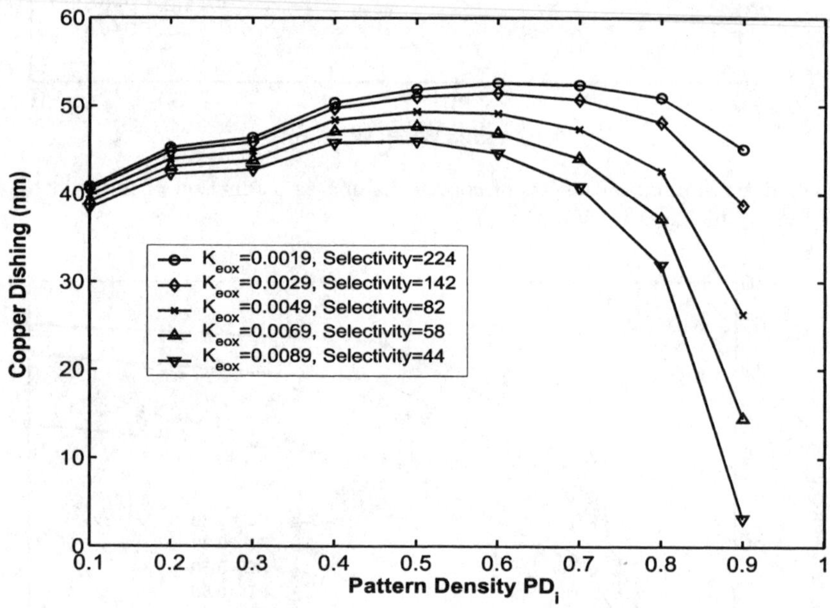

Fig. 6.33.a. Simulation results of copper dishing as a function of pattern density and Preston's coefficient of oxide

Figs. 6.34 (a) (b) and (c) show the simulation results of dishing, erosion and thinning for four different combinations of K_{ecu} and MRR_{cu0}. The K_{ecu} and MRR_{cu0} are chosen so that the average MRR=$K_{ecu}P_0$+ MRR_{cu0} at P_0= 15kPa remains the same. The same average MRR is to ensure that the over-polishing time is comparable to each other, and the final dishing and erosion

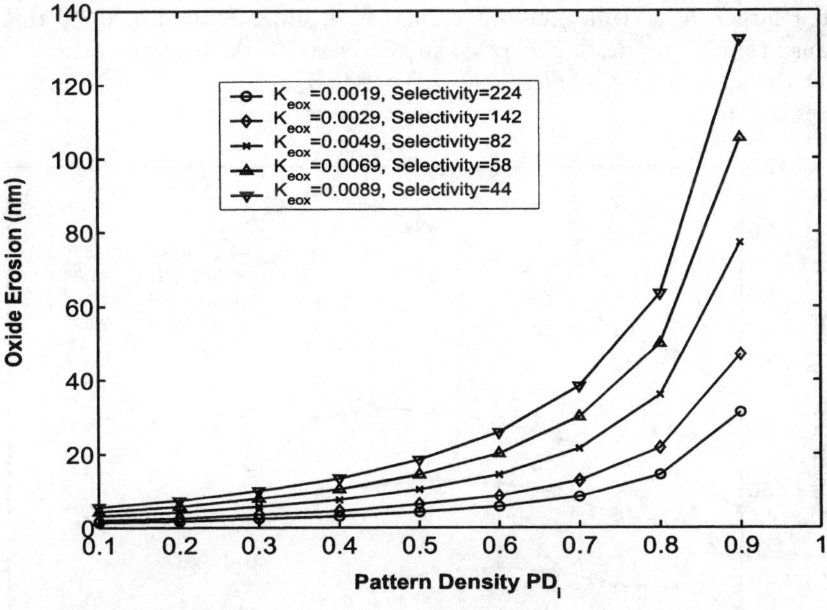

Fig. 6.33.b. Simulation results of oxide erosion as a function of pattern density and Preston's coefficient of oxide

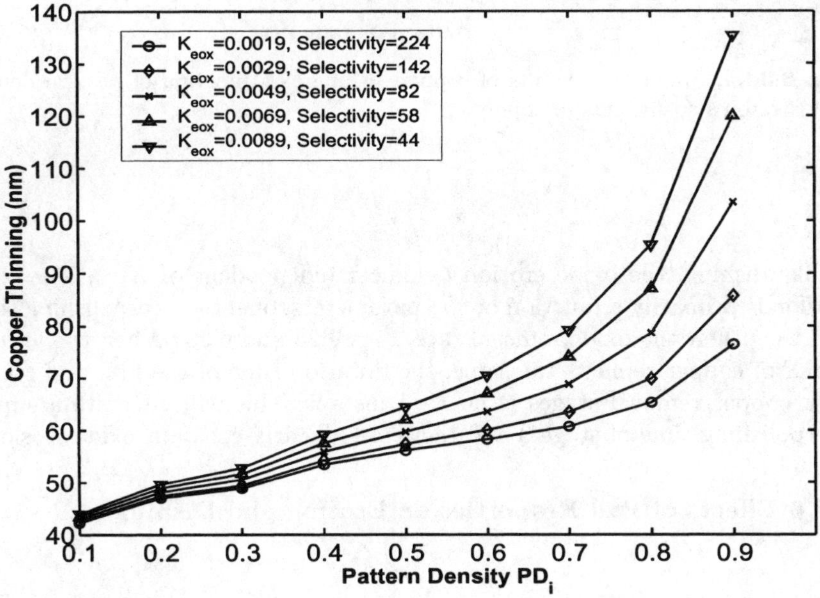

Fig. 6.33.c. Simulation results of copper thinning as a function of pattern density and Preston's coefficient of oxide

is a sole function of copper removal parameters K_{ecu} and MRR_{cu0}. It is seen that a larger K_{ecu} will generate a smaller dishing. From Eq. 6.79, this is because that larger K_{ecu} generates smaller step height increase rates dS/dt under the same average $MRR = K_{ecu}P_0 + MRR_{cu0}$.

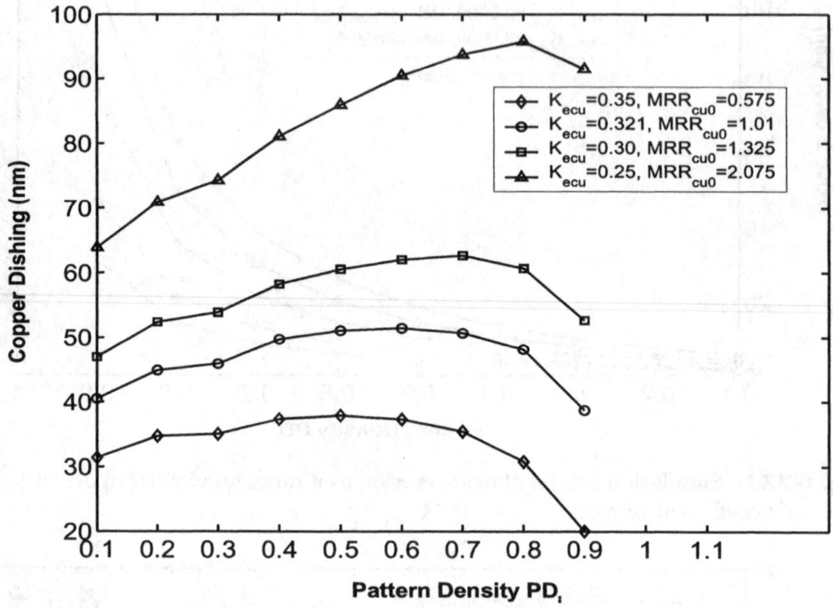

Fig. 6.34.a. Simulation results of copper dishing as a function of pattern density and Preston's coefficients of copper

Unlike dishing, the oxide erosion is almost independent of K_{ecu}. The oxide erosion is primarily a function of two parameters, one, the overpolishing time, and the other, the oxide removal rate, Figs 6.25 and 6.26. When the average MRR of copper remains the same, the duration time of the first and second bulk copper removal stages is nearly constant. This will generate a similar overpolishing time of stage 3 and therefore a nearly constant oxide erosion.

6.8.6 Effects of Pad Properties on Erosion and Dishing

In the earlier sections, the pad is assumed to be IC1000. Fig. 6.35 shows the simulation results of copper dishing, oxide erosion and copper thinning when using other commercial pads. It is seen that IC1400 generates the smallest

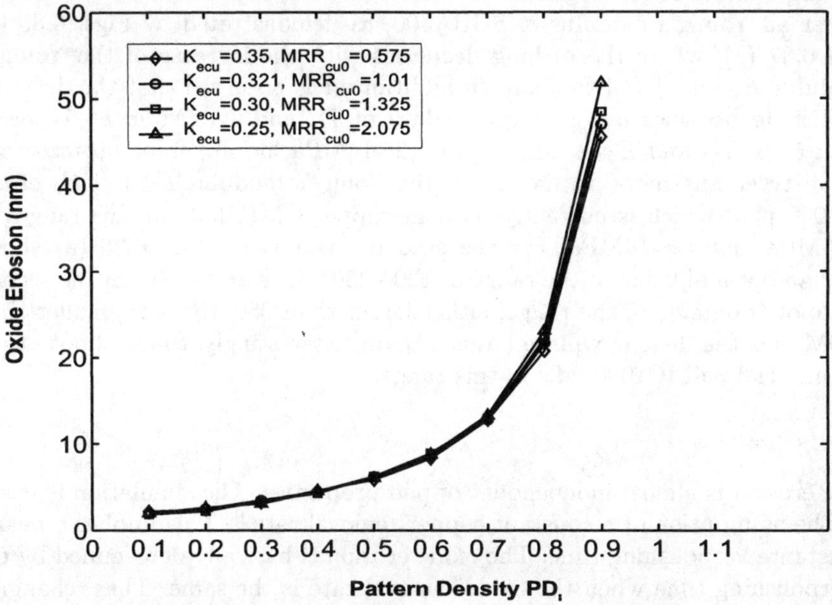

Fig. 6.34.b. Simulation results of oxide erosion as a function of pattern density and Preston's coefficients of copper

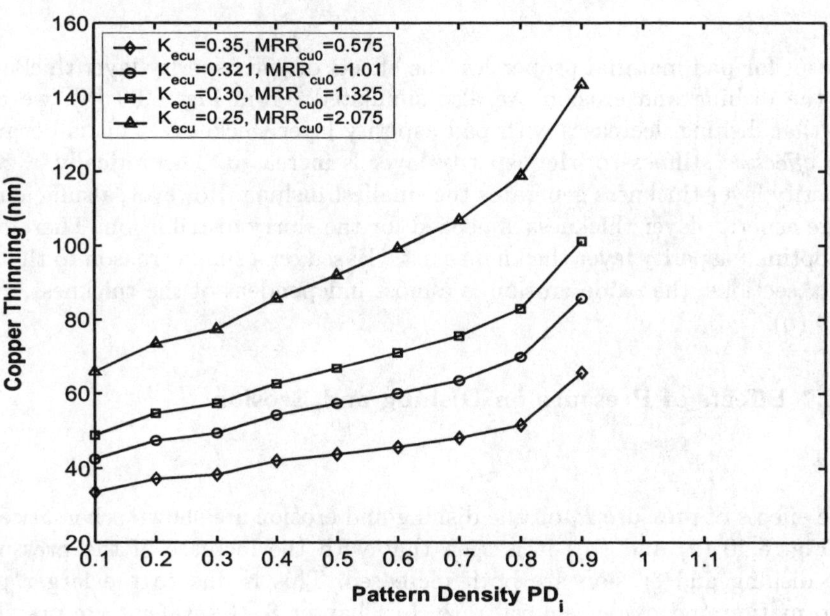

Fig. 6.34.c. Simulation results of copper thinning as a function of pattern density and Preston's coefficients of copper

dishing while SUBA400 generates the largest dishing. This is attributed to the small Young's modulus of SUBA400, as demonstrated by Figs. 6.36 (a) and 6.37 (a), where the dishing decreases with the increase of the Young's modulus E_1 and E_2 of the pad. In addition, it is observed that the decrease of dishing becomes less at larger values of E_1 and E_2. After E_1 is larger than 5∼6MPa and E_2 is larger than 12∼14MPa, no apparent decrease can be observed any more. Interestingly, the Young's modulus E_1 and E_2 of the IC1000 pad, which is currently used in copper CMP, falls in this ranges of 5∼6MPa and 12∼14MPa. For the second dashpot η_2, Fig. 6.38 (a) shows that an optimal value in the range of 120∼250MPa·s exists. When the second dashpot is outside of the range, either larger than 250MPa·s or smaller than 120MPa·s, the dishing will be larger. Again, interestingly, the dashpot of the commercial pad IC1000 falls in this range.

The erosion is almost independent of pad properties. The simulation is based on the assumption of a constant copper removal rate [8]. This implies a nearly constant overpolishing time. The oxide erosion is basically determined by the overpolishing time when the oxide removal rate is the same. Thus, changing pad material properties, which does not change the overpolishing time much, will not cause an apparent change of oxide erosion.

Except for pad material properties, the effects of pad asperity layer thickness on the dishing and erosion are also simulated. From Fig. 6.39 (a), we can see that dishing decreases with pad asperity layer thickness. This is because the *effective* stiffness of the asperity layer is increased. Theoretically, a zero asperity layer thickness generates the smallest dishing. However, a sufficiently large asperity layer thickness is needed for the slurry distribution. Therefore, an optimal asperity layer thickness exists. Based on a similar reason to that in early sections, the oxide erosion is almost independent of the thickness, Fig. 6.39 (b).

6.8.7 Effects of Pressure on Dishing and Erosion

The effects of pressure P_i on the dishing and erosion are shown schematically in Fig. 6.40 (a) and (b). It is seen that with the increase of the pressure, the dishing and erosion are both increased. This is due to the larger pad deformation and oxide removal rate. In Chapter 8, the wafer-scale pressure

---

[8] It will be shown in Chapter 8 that there is a slight increase of material removal rate with pad Young's modulus.

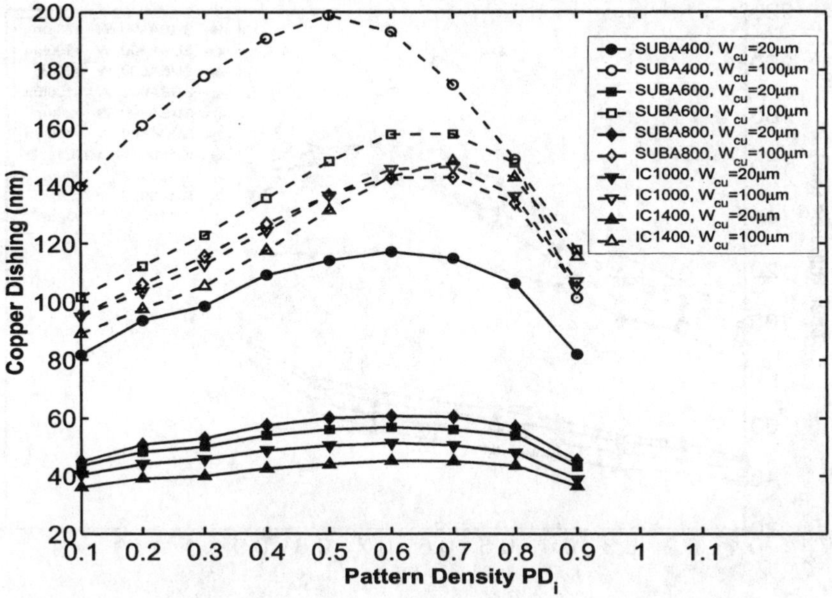

Fig. 6.35.a. Simulation results of copper dishing using commercial pads

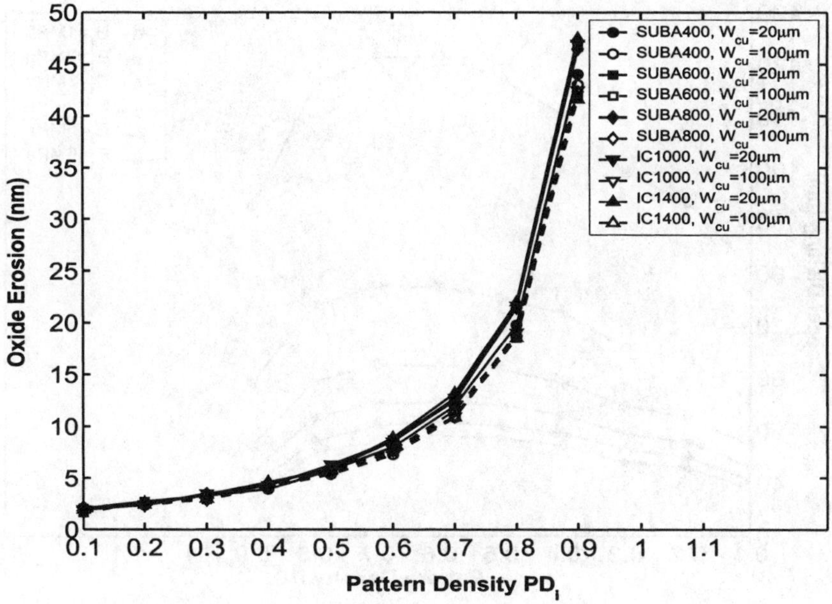

Fig. 6.35.b. Simulation results of oxide erosion using commercial pads

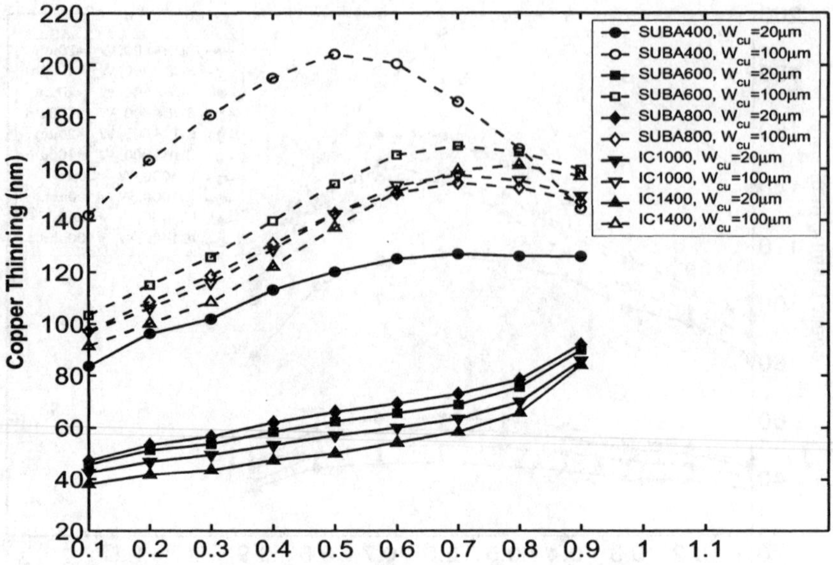

Fig. 6.35.c. Simulation results of copper thinning using commercial pads

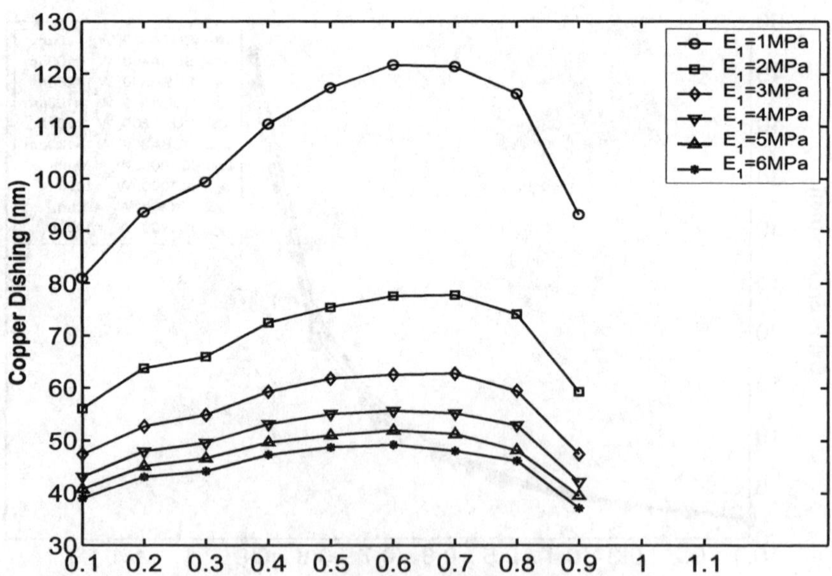

Fig. 6.36.a. Simulation results of copper dishing as a function of E_1

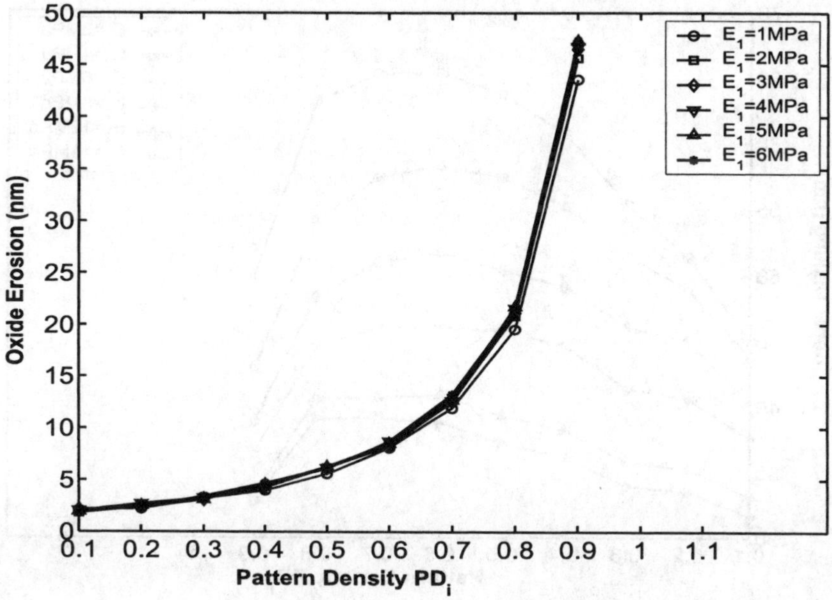

Fig. 6.36.b. Simulation results of oxide erosion as a function of E_1

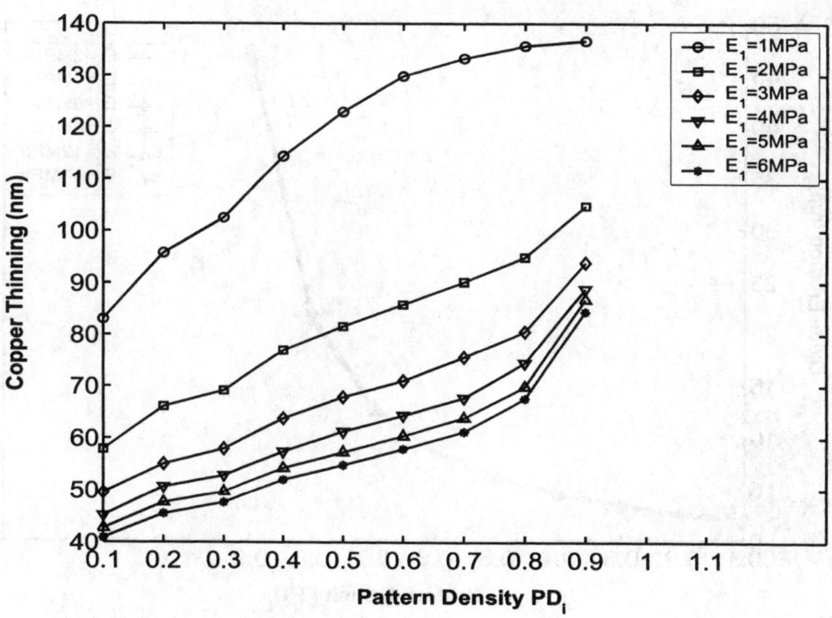

Fig. 6.36.c. Simulation results of copper thinning as a function of E_1

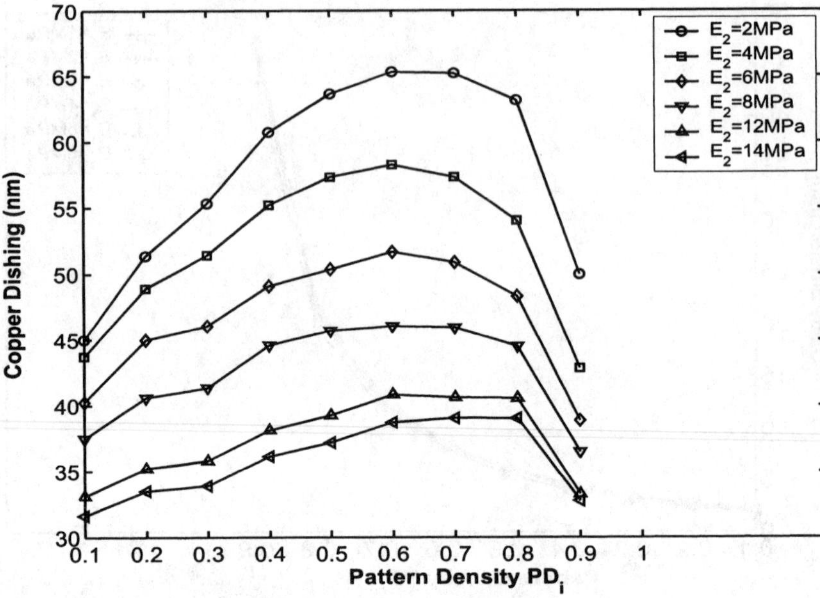

Fig. 6.37.a. Simulation results of copper dishing as a function of E_2

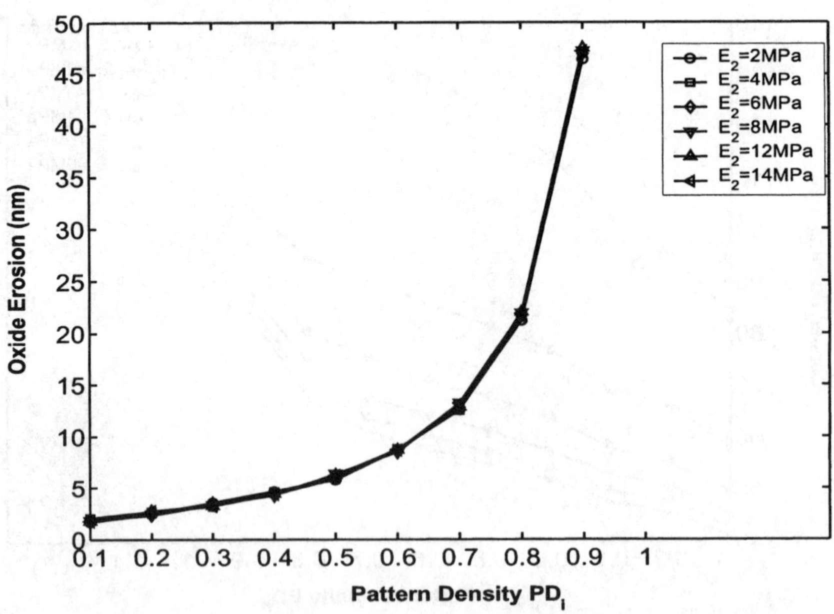

Fig. 6.37.b. Simulation results of oxide erosion as a function of E_2

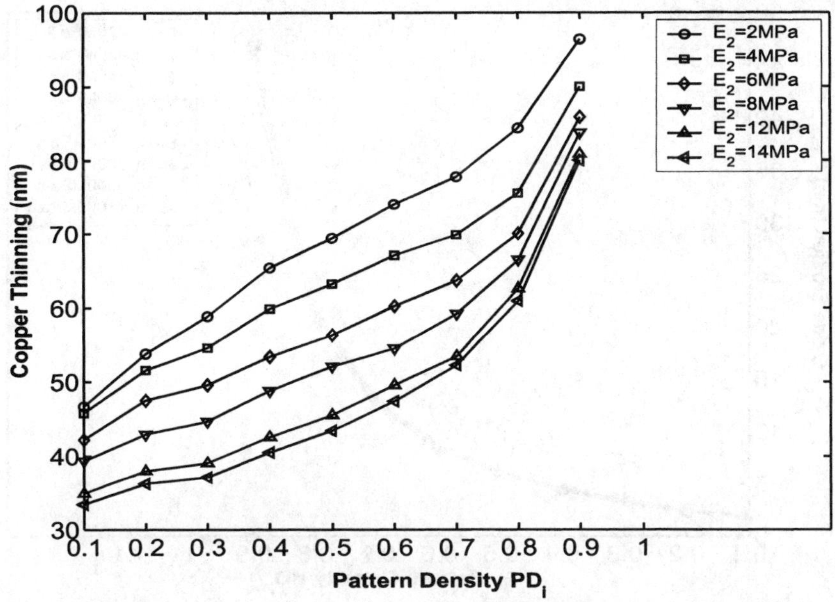

Fig. 6.37.c. Simulation results of copper thinning as a function of E_2

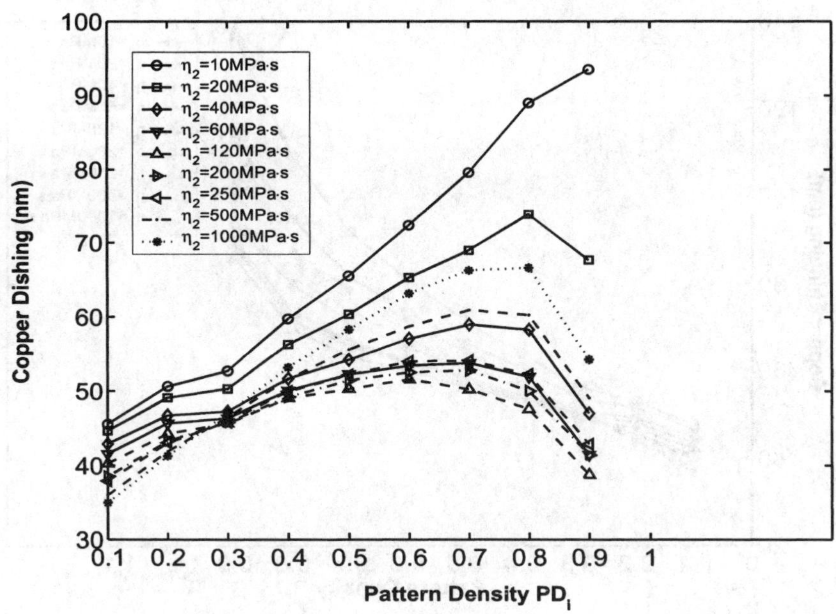

Fig. 6.38.a. Simulation results of copper dishing as a function of η_2

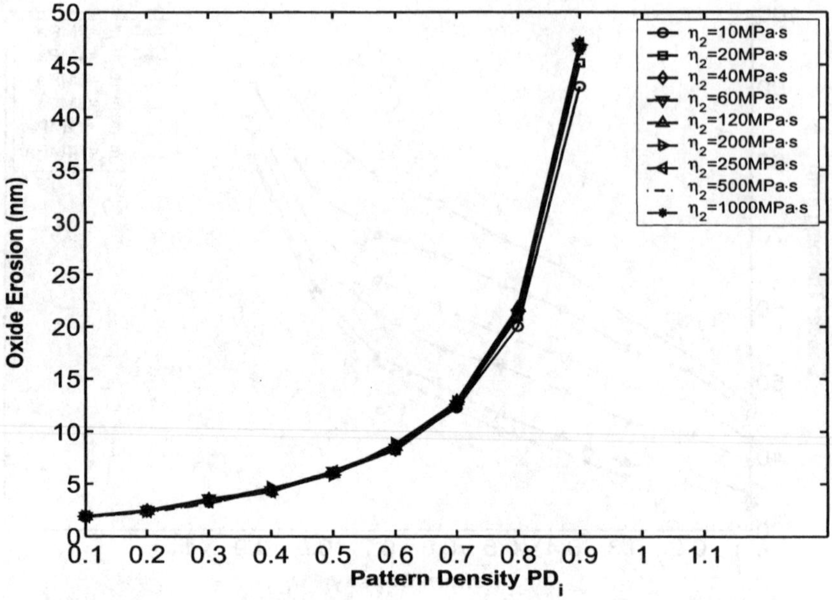

Fig. 6.38.b. Simulation results of oxide erosion as a function of η_2

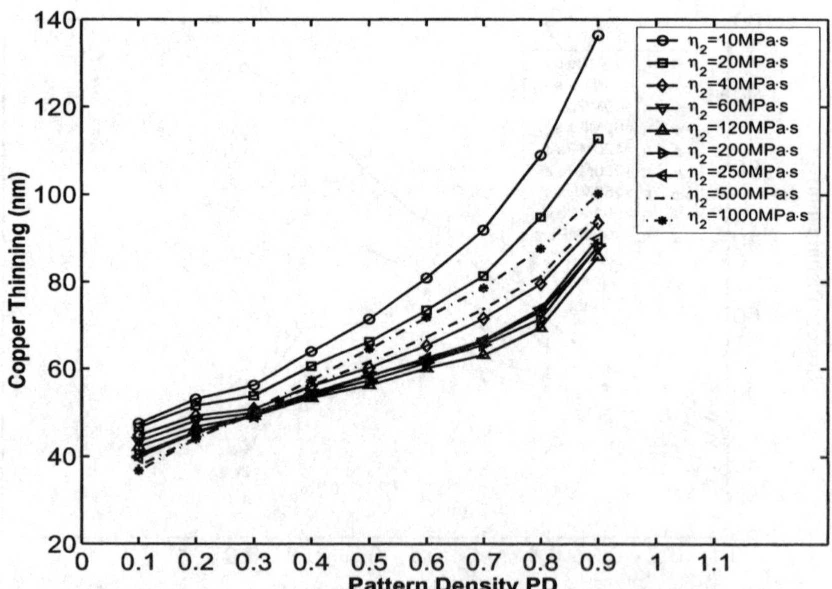

Fig. 6.38.c. Simulation results of copper thinning as a function of η_2

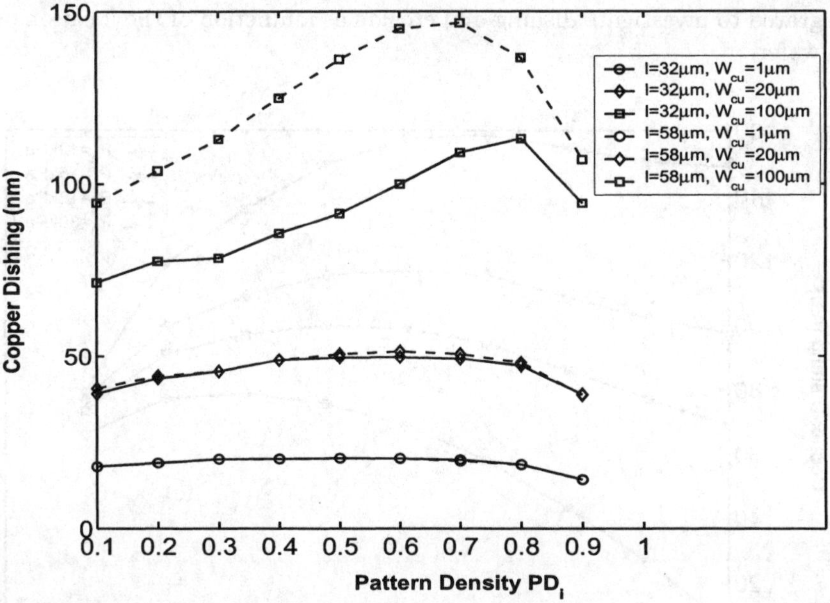

Fig. 6.39.a. Simulation results of copper dishing as a function of asperity layer thickness

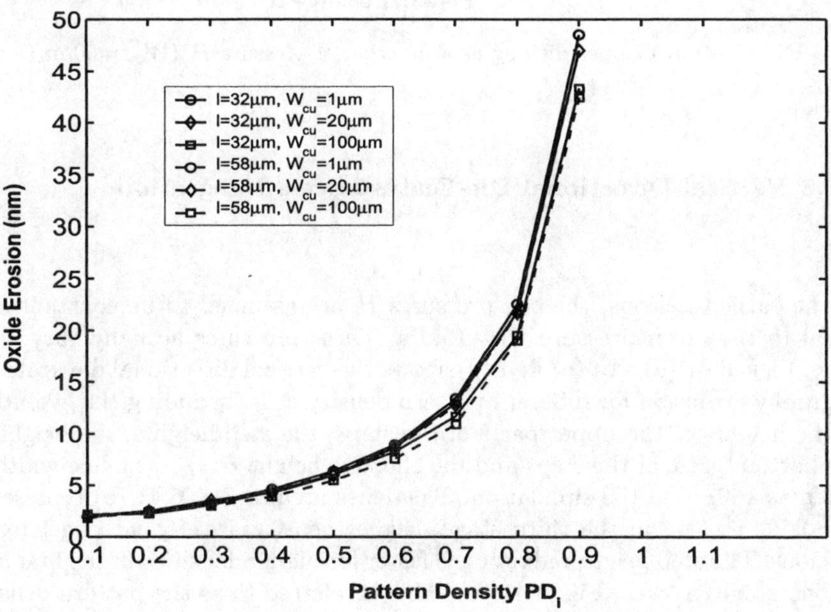

Fig. 6.39.b. Simulation results of oxide erosion as a function of asperity layer thickness

distribution model and the feature-scale topography evolution model will be integrated to investigate dishing and erosion as a function of the location over the wafer.

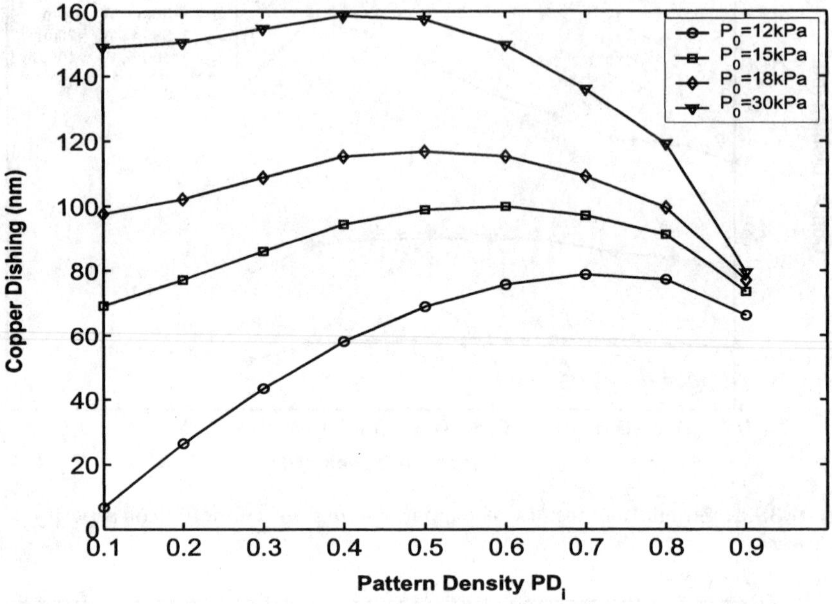

Fig. 6.40.a. Copper dishing as a function of pressure P_i (W_{cu} =50μm)

6.8.8 Vertical Directional Die-Feature Scale Integration

In the earlier sections, the cell pressures P_i are assumed to be constant and equal to the down pressure P_0= 15kPa. These pressures actually vary with time. Figs. 6.41 (a), (b) (c) and (d) shows the vertical directional die-scale topography evolution for different pattern density cells, including the evolution of the heights of the upper parts of the steps, the step heights, the height of the bottom parts of the steps and the effective height H_{eff_i}. The line width is taken as 100μm in the simulation. The dense area in Fig. 6.41 (a) represents the oxide erosion in the third stage. Higher density areas yield larger oxide erosions. The step height reduction is faster for higher densities in the first and second stages as well, Fig. 6.41 (b). This is referred to as the pattern density effects. The step height increases in the third stage. This is represented by the dense area in Fig. 6.41 (b). The step height increases fastest at a certain density between 0.1 and 0.9.

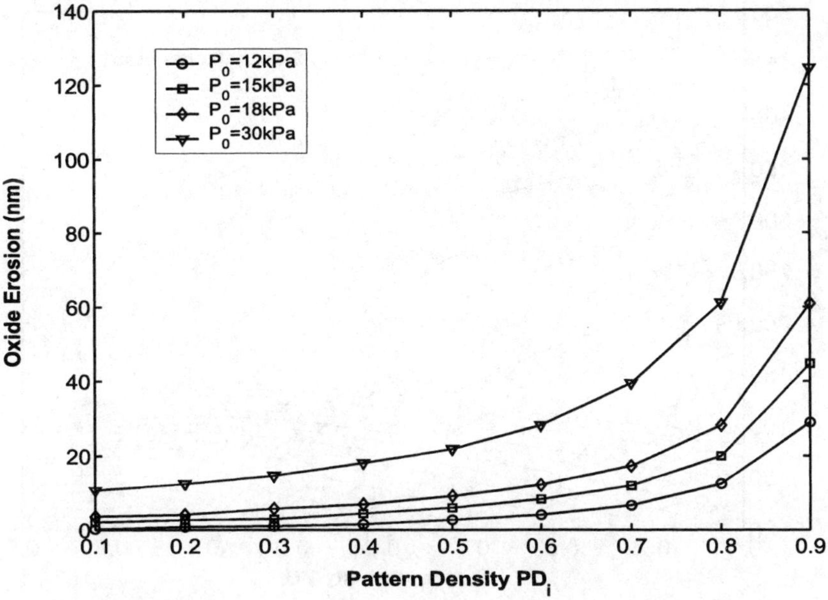

Fig. 6.40.b. Oxide erosion as a function of pressure P_i (W_{cu}=50μm)

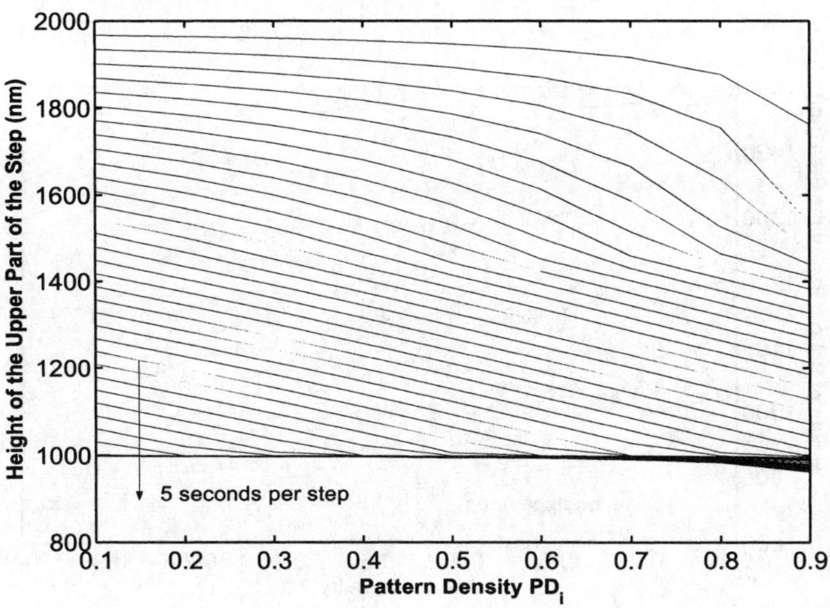

Fig. 6.41.a. Evolution of upper part of the feature step (W_{cu}=100μm)

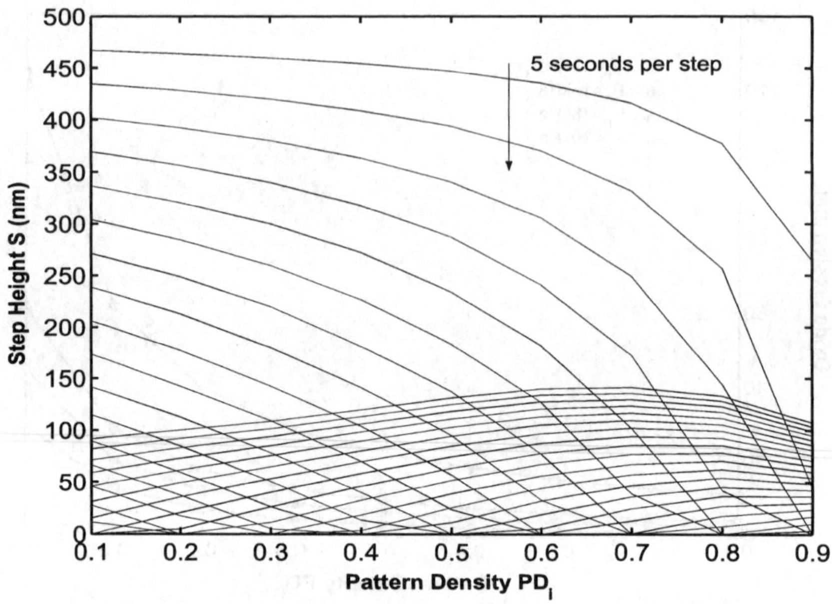

Fig. 6.41.b. Evolution of step height (W_{cu}=100μm)

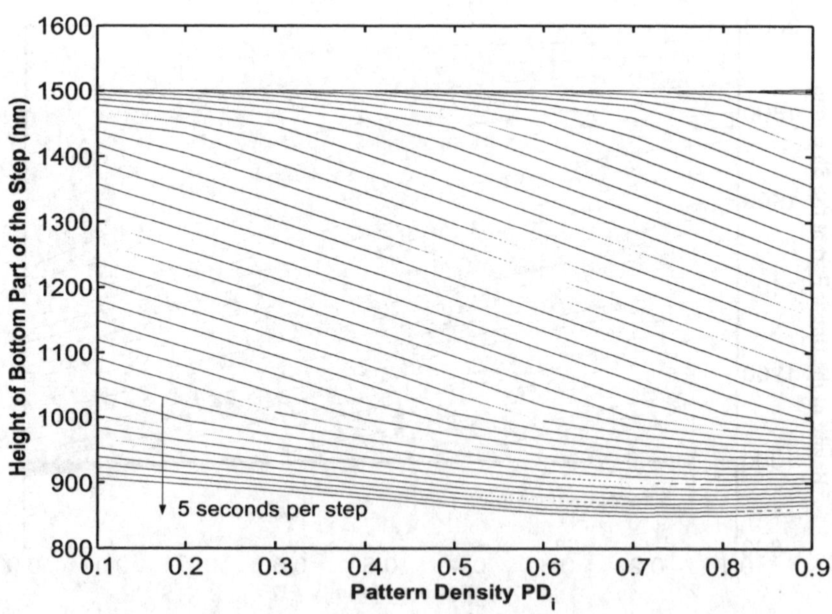

Fig. 6.41.c. Evolution of bottom part of the step (W_{cu}=100μm)

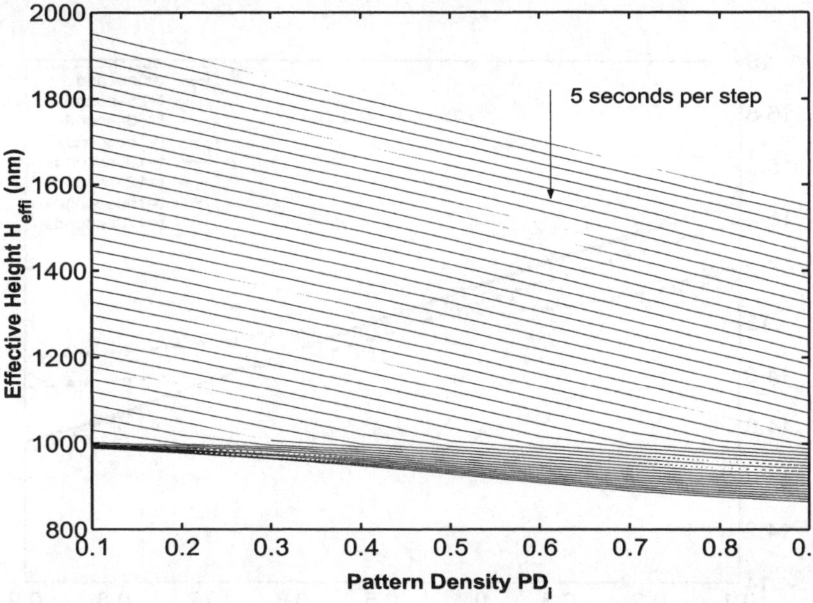

Fig. 6.41.d. Evolution of effective height H_{eff} (W_{cu}=100μm)

From the above simulations, it is seen that the higher pattern density area is removed faster during the planarization. This leads to an uneven cell pressure distribution. From Fig. 6.42, it is seen that a lower pressure P_i is applied on the higher pattern density area. The maximum pressure difference between the higher pattern density cell and lower pattern density cell is around 1kPa at the beginning of the planarization. The pressure difference is smaller at the overpolishing stage since the surface has become smoother, Fig. 6.41 (d). In addition, it is noted that for the same height variance, the pressure difference at die scale is much smaller than that at feature scale; see Figs. 6.16 and 6.17. This is due to the smaller die-scale stiffness K_{1d} (or larger pad substrate thickness than the pad asperity layer thickness). Figs. 6.43 and 6.44 show that a change of the dishing and erosion does occur when the pressure variance among different pattern density cells is considered. However, this change is relatively small, especially, when the pattern density is in the range of 0.4 to 0.6. Therefore, the analytical solution in section 6.7 can be used to obtain simulation results as in earlier sections.

6.8.9 Summary of Simulation Results

From the above simulations, we can get several conclusions:

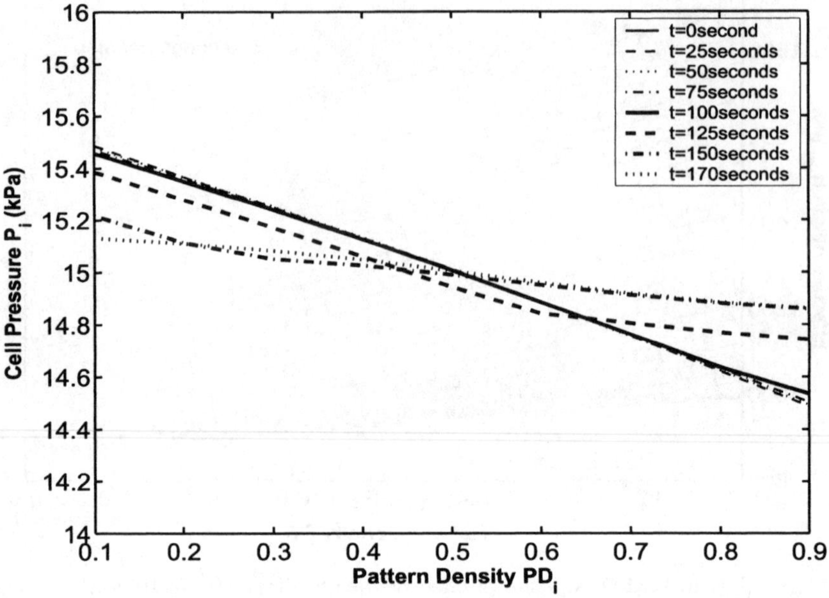

Fig. 6.42. Evolution of pressure P_i (W_{cu}=100μm)

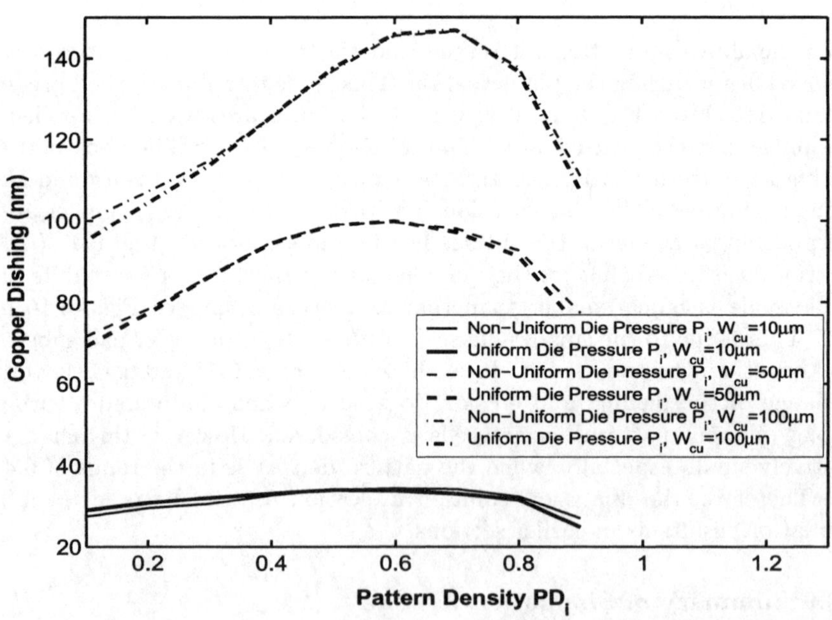

Fig. 6.43.a. Effects of pressure variance on dishing for different copper line widths

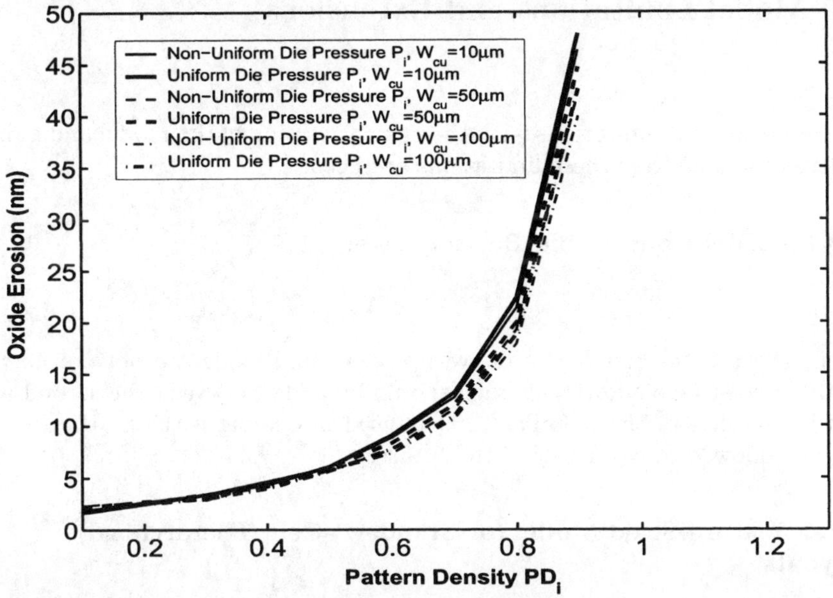

Fig. 6.43.b. Effects of pressure variance on erosion for different copper line widths

i. The oxide erosion is basically determined by the oxide material removal rate and the overpolishing time. The consumable parameters which influence the oxide material removal properties will affect the oxide erosion. The pad parameters, including the pad materials and pad asperity layer thickness l, may not affect the oxide erosion if they do not change the oxide material removal properties.
ii. The copper dishing depends on both the material removal properties and the pad properties. Changing pad parameters, including pad material properties and asperity layer thickness (or roughness), even if that does not change the copper material removal rate, can affect the copper dishing effectively.
iii. At the small line widths, the copper dishing is usually not so big a concern as the oxide erosion. In this case, a small oxide removal rate will be preferred in terms of a smaller copper thinning. From i and ii, adjusting slurry parameters, instead of pad parameters, to obtain a larger selectivity may therefore be a more feasible way to reduce the copper thinning at small line widths.
iv. Both oxide erosion and copper dishing are strongly dependent on the feature pattern density. A dummy filling at the higher density area may be needed to obtain a uniform distribution of dishing and erosion across the die. In section 6.9.4, a dummy filling algorithm will be proposed.

6.9 Model Limitations and Extensions

There are several limitations with the current version of the model and model extensions to address these limitations are needed.

6.9.1 Model Considering Barrier Layer

First, the thin barrier layers between copper and oxide are not considered in the model. One additional stage should be added between the second and third stages. A method similar to that used in dealing with the third stage can be followed to cover this in the future.

6.9.2 Feature-Scale Model for Standard-Cell (Hierarchical) Layouts

Standard-cell, or hierarchical layouts with a certain function are applied in most modern IC designs. While the above feature-scale step height reduction model is concentrated on a single isolated step, it is possible to extend it to standard cell layouts. For certain standard cell-topographies, a look-up table or library of the *effective* stiffness of the asperity layer could be constructed using three-dimensional (3-D) BEM simulation or the three-dimensional contact mechanics model to be discussed in Chapter 7. Then, as in the semi-two dimensional model, the tabulated *effective* stiffness values obtained from the 3-D simulation can be input into the one-dimensional model to evaluate the dishing and erosion as a function of a certain process condition for standard cells. The results can then be incorporated into cell libraries.

6.9.3 Lateral Directional Die-Feature Scale Integration

In the current version of the model, when a feature is studied, the features surrounding it are assumed to have a uniform layout pattern density. In this case, the average die-scale pressure over a feature is approximately equal to the down pressure P_0[9] [10]. It can be input directly into the feature-scale topography evolution model developed earlier to evaluate the dishing and erosion. However, when the layout pattern density is distributed randomly, this is not the case and more detailed lateral die-feature interactions should be included.

To account for the effects of the lateral directional topography, Ouma [6.26] introduced the concept of the effective pattern density for oxide (ILD/IMD) CMP. The key idea is that the *effective* pattern density ρ at location (x, y) is the convolution of the layout density $PD(x, y)$ and a weight function $w(x, y)$ in a pattern density window W. The size of W, which is often referred to as planarization length, is primarily dependent on the process and usually in the range of several minimeters. The center of the pattern density window W is at the location (x, y) of the feature studied. It is like a circular symmetric filter, with larger values at the center and smaller values at the boundary. Mathematically, the *effective* pattern density ρ can be expressed as

$$\rho(x,y) = PD(x,y) \otimes w(x,y) = \iint_W PD\,(\xi,\eta)\, w\,(x-\xi, y-\eta) d\xi d\eta \qquad (6.95)$$

The weight density function is given by Ouma [6.26] using the following function of pad deformation:

$$Df\,(r) = \frac{4\left(1-\upsilon_h^2\right) qR_p}{\pi E_h} \int_0^{\pi/2} \sqrt{1-\frac{r^2}{R_p^2} sin^2\theta d\theta} \qquad (6.96)$$

by considering that a uniform pressure q is applied on the pattern density window W, where R_p is the radius of W, and r the distance to the center of W. This function is obtained using contact mechanics by assuming the polishing platen with the equipment foundation as a half space. E_h is the Young's modulus of the half space and υ_h is the Poisson's ratio.

Considering that in copper CMP the definition of pattern density is equal to 1- the pattern density defined in ILD/IMD CMP, Eq. 6.95 can be written as

[9] More accurately, it should be the wafer-scale pressure at the die location considering the wafer-scale pressure distribution as to be discussed in Chapter 8.

[10] Of course, if considering the evolution of the topography, this pressure will slightly change with time due to the different material removal rates over different pattern density areas, Fig. 6.42. This variance of pressure, (smaller than 1kPa) caused by the vertical topography change is negligible in comparison with that due to the lateral topography variance.

$$1-\rho(x,y) = [1-PD(x,y)]\otimes w(x,y) = \iint_W [1-PD(\xi,\eta)]\, w(x-\xi, y-\eta) d\xi d\eta \tag{6.97}$$

This concept of effective pattern density is primarily derived from intuition [6.26]. Although it works quite well, its physical meaning is not clear. In the following sections, we will first give a rough physical explanation of this concept. (Note: As will be shown in the following section, it is not a strict physical explanation from the viewpoint of mechanics. Our aim is to offer some physical insights into this empirical concept instead of its solid physical proof, which may not exist.) Then a scheme of feature- and die-scale integration for the topography evolution modeling is proposed based on the implication from this explanation.

Fig. 6.44 shows the top view of a die. It is discretized into $n \times n$ small windows with size A_w (usually with a size 20μm$\times$20μm to 40 μm$\times$40μm. The size is not scaled in Fig. 6.44.) The local layout pattern density PD is calculated in these small windows and referred to as PD_k for the kth window. Based on Eq. 6.97, the effective pattern density $\rho_k(x,y)$ of the feature at the kth window can be written as

$$\begin{aligned} 1-\rho_k(x,y) &= \iint_{W_k} [1-PD(\xi,\eta)]\, w(x-\xi, y-\eta)\, d\xi d\eta \\ &= \sum_{j=1}^{M} (1-PD_j)\, w_{kj} \end{aligned} \tag{6.98}$$

where W_k is the effective pattern density window (with the dashed line in Fig. 6.44 as boundary) surrounding the kth feature window, M the number of small windows in W_k, PD_j the layout pattern density in the jth window, and w_{kj} the weight function value at the location of the jth window. From the pad deformation formulation 6.96, w_{kj} can be obtained as:

$$w_{jk} = \frac{\int_0^{\pi/2} \sqrt{1-\frac{r_{jk}^2}{R_p^2} sin^2\theta d\theta}}{\sum_{j=1}^{M} \int_0^{\pi/2} \sqrt{1-\frac{r_{jk}^2}{R_p^2} sin^2\theta d\theta}} \tag{6.99}$$

where r_{jk} is the distance between the kth and jth feature windows. The spread of the weight function is changed with R_p. The MIT mask by Ouma [6.26] is needed to determine/calibrate R_p.

Besides the feature layout density PD, we introduce a concept of pad contact density PCD, which is defined as the ratio of pad-feature contact area to the feature window size A_w, and denotes the PCD at the kth window as PCD_k. It is seen that before the pad touches the bottom of the features, such as that in stage 1 of copper CMP, PCD_k is equal to 1- PD_k. When the pad touches the bottom, PCD will be equal to 1 since the pad touches the features everywhere. From the definition, PD is a feature related only and fixed during the planarization. The PCD not only depends on the feature but also changes with the planarization time.

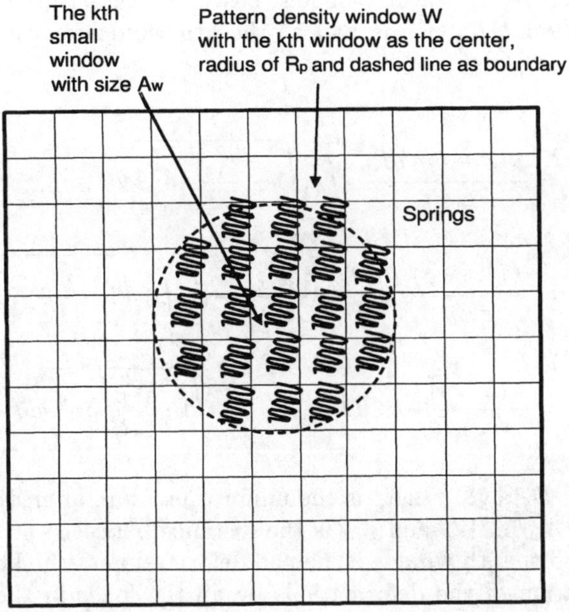

Fig. 6.44. Top view of die-pad contact

One dimensional springs, one over each feature window, can be used to model the part of the pad which is in touch with the wafer feature. Fig. 6.44 shows springs in the pattern density window W_k surrounding the kth window. (Outside of W_k, the contribution of pad-die contact is negligible.) Two different kinds of springs, one considering only the part of the pad which contacts the raised part of the features, and the other one considering the pad in the whole kth window, can be applied. The relationship between their stiffness values is

$$K_{dk} = PCD_k \times K'_{dk} \tag{6.100}$$

where K'_{dk} is the stiffness of the first spring and K_{dk} the stiffness of the second spring.

Now, let us first estimate the stiffness K'_{dk} of the spring at the kth window We assume that a unit pressure is applied on the springs. Thus the load on any feature window j will be equal to $PCD_j \times A_w \times 1 = PCD_j \times A_w$. The deformation of the pad at the kth window is the superposition of the deformation caused by all the loads in the pattern density window W_k. If taking the platen and the equipment foundation as a half-space and considering the load $PCD_j \times A_w$ are distributed uniformly over a pattern density window W_j surrounding the jth feature window, based on contact mechanics, the die-scale deformation Df_{kj} of the pad at the kth window caused by the load $PCD_j \times A_w$ [6.29] is:

$$\begin{aligned} Df_{kj} &= \frac{4\left(1-v_h^2\right) qR_p}{\pi E_h} \int_0^{\pi/2} \sqrt{1-\frac{r_{jk}^2}{R_p^2} sin^2\theta d\theta} \\ &= \frac{4\left(1-v_h^2\right)}{\pi E_h} \frac{PCD_j \times A_w \times R_p}{\pi R_p^2} \int_0^{\pi/2} \sqrt{1-\frac{r_{jk}^2}{R_p^2} sin^2\theta d\theta} \\ &= \frac{4\left(1-v_h^2\right)}{\pi^2 E_h R_p} PCD_j \times A_w \int_0^{\pi/2} \sqrt{1-\frac{r_{jk}^2}{R_p^2} sin^2\theta d\theta} \end{aligned} \tag{6.101}$$

where $q = \left(PCD_j \times A_w\right)/\pi R_p^2$ is the uniform pressure after distributing the load $PCD_j \times A_w$ on W_j and r_{jk} is the distance between the centers of the kth window and the jth window. The pad deformation at the kth window will be the summation of the deformations by all the loads in the window W_k, therefore,

$$Df_k = \frac{4\left(1-v_h^2\right) A_w}{\pi^2 E_h R_p} \sum_{j=1}^{M} \left(PCD_j \int_0^{\pi/2} \sqrt{1-\frac{r_{jk}^2}{R_p^2} sin^2\theta d\theta} \right) \tag{6.102}$$

The deformation Df_k at the kth window is caused by a unit pressure 1. Therefore, we estimate the spring stiffness K'_{dk} as

$$K'_{dk} = 1/Df_k = \frac{\pi^2 E_h R_p}{4A_w\left(1-v_h^2\right)} \frac{1}{\sum_{j=1}^{M} \left(PCD_j \int_0^{\pi/2} \sqrt{1-\frac{r_{jk}^2}{R_p^2} sin^2\theta d\theta} \right)} \tag{6.103}$$

The stiffness K_{dk} will be

$$K_{dk} = PCD_k \times \frac{\pi^2 E_h R_p}{4A_w\left(1-v_h^2\right)} \frac{1}{\sum\limits_{j=1}^{M}\left(PCD_j \int\limits_0^{\pi/2} \sqrt{1-\frac{r_{jk}^2}{R_p^2} sin^2\theta d\theta}\right)} \tag{6.104}$$

We noted that these stiffness values are dependent on the pad contact density of features surrounding the window k. In other words, it is a function of die topography and feature location. In addition, for the pad contacting with a cell with constant PCD as shown in Fig. 6.4, K_{dk} is independent of PCD, i.e., the same die-scale spring K_d stiffness can be used for all cells, as assumed in the simulation in section 6.8.

After obtaining the stiffness, the pressure distribution at the die-scale can now be evaluated as the deformation times the stiffness. In CMP, if taking the wafer features as rigid bodies, the deformation of the pad over them should be equal to each other. This implies that the pressure distribution across the die will be proportional to the stiffness distribution. Under an average down pressure P_0, this deformation can be evaluated approximately as

$$Df = \frac{4\left(1-v_h^2\right)}{\pi^2 E_h R_p} P_0 A_w \sum_{j=1}^{M} \int\limits_0^{\pi/2} \sqrt{1-\frac{r_{jk}^2}{R_p^2} sin^2\theta d\theta} \tag{6.105}$$

It is noted that this deformation is independent of location and feature topography.

Therefore, the pressure P'_k over the raised part in the kth window will be

$$P'_k = Df \times K'_{dk} = P_0 \frac{\sum\limits_{j=1}^{M} \int\limits_0^{\pi/2} \sqrt{1-\frac{r_{jk}^2}{R_p^2} sin^2\theta d\theta}}{\sum\limits_{j=1}^{M}\left(PCD_j \int\limits_0^{\pi/2} \sqrt{1-\frac{r_{jk}^2}{R_p^2} sin^2\theta d\theta}\right)} \tag{6.106}$$

The average die-scale pressure P_k over the kth window is

$$P_k = P'_k \times PCD_k = Df_k \times K_{dk} \tag{6.107}$$

$$= P_0 \times PCD_k \frac{\sum\limits_{j=1}^{M} \int\limits_0^{\pi/2} \sqrt{1-\frac{r_{jk}^2}{R_p^2} sin^2\theta d\theta}}{\sum\limits_{j=1}^{M}\left(PCD_j \int\limits_0^{\pi/2} \sqrt{1-\frac{r_{jk}^2}{R_p^2} sin^2\theta d\theta}\right)} \tag{6.108}$$

We note that for a pattern density cell with a constant layout pattern density PD_i, the average die-scale pressure P_k is independent of PD_i and equal to down pressure P_0. This agrees with that assumed in the simulation in section 6.8.

From Eq. 6.106,

$$\frac{P_0}{P'_k} = \sum_{j=1}^{M} \left(PCD_j \frac{\int_0^{\pi/2} \sqrt{1 - \frac{r_{jk}^2}{R_p^2} sin^2\theta d\theta}}{\sum_{j=1}^{M} \int_0^{\pi/2} \sqrt{1 - \frac{r_{jk}^2}{R_p^2} sin^2\theta d\theta}} \right)$$

$$= \sum_{j=1}^{M} (PCD_j \alpha_{jk}) \tag{6.109}$$

where

$$\alpha_{jk} = \frac{\int_0^{\pi/2} \sqrt{1 - \frac{r_{jk}^2}{R_p^2} sin^2\theta d\theta}}{\sum_{j=1}^{M} \int_0^{\pi/2} \sqrt{1 - \frac{r_{jk}^2}{R_p^2} sin^2\theta d\theta}} \tag{6.110}$$

Note that the summation of α_{jk} in the pattern density window W is equal to 1.

If considering that PCD_j is equal to PD_j in oxide or ILD/IMD CMP (or $1 - PD_j$ in copper bulk removal before the pad touches the bottom of the feature), and taking P_0/P'_k and α_{jk} as the effective pattern density ρ_k and weight function w_{jk}, respectively, the above equation is equivalent to Eq. 6.98. This indicates that the effective pattern density at a feature can truly be written as a convolution of the layout pattern densities of features surrounding it and a weight function.

From this rough physical explanation, several implications can be obtained:

i. The effective pattern density ρ_k is an indicator of the die-scale pressure P'_k over the raised part of the features in CMP, as pointed out by Ouma [6.26]. The pressure over the raised, or contact part of the features, is equal to P_0/ρ_k. The die-scale pressure P_k over the whole kth window, is equal to the the pressure P'_k times the local contact density PCD_k, or $P_0/\rho_k \times PCD_k$.

ii. The effective pattern density is a concept at die-scale, reflecting the effects of lateral direction die-scale topography on the pressure distribution. It is a concept totally different from the feature density PD.
iii. The effective pattern density should be equal to the convolution of the pad contact density PCD and the weight function w instead of the convolution of the feature layout density PD and the weight function w. In oxide or ILD/IMD CMP, PCD is always equal to PD if assuming that the pad never touches the bottom of the features.
iv. In a damascene process such as copper CMP and STI, since the pad contact density PCD changes from PD to 1 from stage 1 to stage 2 and stage 3, the *effective* pattern density calculation should be updated with the planarization.

Based on the above implications, the following scheme of the feature-die scale integration can be proposed:

i. Use the MIT semi-empirical model to evaluate the effective pattern density ρ_k (or 1- ρ_k for damascene process) at the kth window as a convolution of local pad contact density PCD_j (or 1-PCD_j for the damascene process) and weight function w_{kj} in the pattern density window W. This includes the calibration of the size R_p of W_k using the MIT mask [Ouma].
ii. Evaluate the die-scale pressure P_k at the kth window as $P_0/\rho_k \times PCD_k$.
iii. Input the die-scale pressure P_k and local feature density PD_k (Note: not local pad contact density PCD_k) into the feature-scale topography evolution model, evaluating step height reduction.
iv. Update local pad contact density PCD_j from PD_j to 1 once the pad begins to touch the whole jth window. (The time of the touching can be evaluated as a function of P_k and local pattern density PD_k, either analytically or numerically, using the feature-scale model in earlier sections.) Update the effective pattern density ρ_k as the convolution of the updated local density PCD_j and the calibrated weight function w_{kj}.
v. Update the pressure P_k at the kth window as $P_0/\rho_k \times PCD_k$. PCD_k may be updated to 1 already.
vi. Input the updated die-scale pressure P_k and local feature density PD_k (Note: it is a fixed parameter.) into the feature-scale topography evolution model.
vii. Continue the above process until the planarization is finished. The dishing, erosion and other feature-scale topography parameters are obtained across the die.

In the past, the following method has sometimes been applied for the die-feature scale integration:

i. Use the MIT semi-empirical model to evaluate the effective pattern density ρ_k at the kth window as a convolution of local pattern density PD_k and weight density function w_{kj}.
ii. Take the pressure over the feature windows as down pressure P_0 and replace the local density PD_k in the feature-scale model with the effective pattern density ρ_k to evaluate the topography evolution across the die.

The second method is equivalent to method 1 for oxide (ILD/IMD) CMP if we assume that the pad is hard and does not touch the bottom of features before full planarization. First, this is because there is simply a linear relationship between step height reduction and pressure P_k. Second, this is because that PCD is equal to PD. If neglecting the MRR_0 term in the revised Preston's equation, the feature step height reduction should be proportional to the pressure P_k over the raised part of the feature and can be written as

$$S = K_e P_k / PD_k \times t \tag{6.111}$$

Using the first method, P_k is equal to $\frac{P_0}{\rho_k PCD_k} = \frac{P_0}{\rho_k PD_k}$, therefore,

$$S = K_e P_0 / \rho_k \times t \tag{6.112}$$

From the above two equations, it is seen that replacing the feature density PD_k with the effective density ρ_k and taking the pressure P_k as P_0 using method 2 yield the same model as that using method 1.

However, there is no physical basis for directly applying the second method to the damascene process since there is a no-linear relationship between step height and pressure P_k and PCD_k is not equal to PD_k in stages 2 and 3 of the damascene process.

A final point is that the above die-scale topography evolution model is still a semi-empirical model, replying on the calibration of the size of the *effective* pattern density window and a continuous update of the die-scale pressure P_k. The computation time will be much faster than a pure physical die-scale topography evolution model. This kind of model can help generate/verify pattern density related design rules.

6.9.4 Dummy Filling

To improve the performance predictability and yield, the die-level manufacturing variation due to CMP should be reduced. The major reason for die-level variation in CMP is the pattern density variation, as shown in earlier simulations. Therefore, one of the methods to reduce the non-uniformity is to insert dummy features in the low density areas wherever there exist large enough slack areas to make the pattern density uniform across the die.

As discussed in Chapter 2, dummy filling algorithms based on linear optimization have been proposed by Tian et al. [6.34] for ILD/IMD CMP. Tian et al. [6.35] also proposed an algorithm for the damascene process. The validity of the model-based dummy filling algorithms relies on correctly understanding the physics of the process. Tian et al. [6.35] use a die-scale topography model by replacing the feature layout density with the *effective* pattern density. As discussed in the last section, the validity of the model is in doubt. In this section, we propose a filling algorithm based on the physical understanding of the die-scale topography evolution discussed in last section.

Various kinds of objective functions may be used in the dummy filling. It can be uniform topography such as minimum variation of copper dishing, oxide erosion and copper thinning across the die. It can also be minimum filling. Here, we use mini-variation of copper dishing as an example; thus, we have

Minimize:

$$d_H - d_L \tag{6.113}$$

subject to

$$0 \leq PD'_k \leq PD_k + PD^a_k \quad 1 \leq k \leq N \tag{6.114}$$

where d_H is the maximum dishing and d_L the minimum dishing across the die after the planarization, PD'_k the layout pattern density at the kth feature window after dummy filling, PD_k the original layout pattern density at the kth window, PD^a_k the maximum incremental change of the pattern density that can be introduced at the kth window based on design rules, and N the number of feature windows in the die.

The key of the algorithm is the model to evaluate the dishing d_k as a function of the layout pattern density PD_k. The die-feature scale integrated model proposed in the last section can evaluate the dishing for arbitrary feature layout distribution. However, it relies on a continuous update of the effective pattern density ρ_k and the pressure P_k supported by the kth feature window. Although it is accurate and sufficiently fast in comparison with a pure physical topography evolution model, the computation time is a still concern if integrating it into the dummy filling algorithm. Therefore, an analytical model, which can capture the major feature of the above model, even with less accuracy, will be preferred.

A closer look at the model in the last section reveals that across the whole die, there are three stages of planarization:

i. At the beginning of the planarization, only the upper part of the features are in contact with the pad. The pressure P_k is determined by the *effective* pattern density as P_0PD_k/ρ_k. We refer to this as stage 1. Stage 1 ends when one of the feature windows across the die begins to fully touch the pad.
ii. At the end of the planarization, all features are in full touch with the polishing pad. The *effective* pattern density is 1 for all feature windows and the pressure P_k is constant and equal to the down pressure P_0. We call this stage 3.
iii. Between the above stage 1 and stage 3, only part of the feature windows are in full contact with the pad and the pressure P_k at the kth window decreases smoothly from P_0PD_k/ρ_k to P_0.

Since the pressure P_k remains unchanged for all feature windows at stages 1 and 3, an analytical topography evolution equation exists for them. By replacing the P_i with P_0PD_k/ρ_k in stage 1 and P_0 in stage 3, the equations derived in section 6.4 can be applied to obtain the analytical solutions. The pressure P_k at the second stage varies with time. Therefore, the analytical solution is not available. However, we may avoid this problem by assuming a constant pressure P_k which is equal to the average of the start pressure P_0PD_k/ρ_k and the end pressure P_0, that is, $P_k = (P_0PD_k/rho_k + P_0)/2$. The analytical solution of the topography evolution for stage 2 exists under this assumption.

Based on the above discussion, we propose the following algorithm of dummy filling:

Minimize

$$d_H - d_L \tag{6.115}$$

subject to

$$t_{1k} = t_{stage1}(\rho_k, P_0) = t_{stage1}(PD_k, P_0 PD_k/\rho_k) \tag{6.116}$$
$$t_{2k} = t_{stage2}\left[t_{1k}, (P_0 PD_k/\rho_k + P_0)/2\right] \tag{6.117}$$
$$d_k = S_{stage3}(t_{2k}, P_0, PD_k) \tag{6.118}$$
$$0 \leq PD_k' \leq PD_k + PD_k^a \tag{6.119}$$
$$1 \leq k \leq N \tag{6.120}$$

where t_{1k} is the end time of stage 1, t_{stage1} is the an analytical equation to evaluate the t_{1k} as a function of the *effective* pattern density ρ_k and down pressure P_0, t_{2k} the end time of stage 2, t_{stage2} an analytical function to evaluate the t_{2k} as a function of t_{1k} and the average down pressure $P_k = (P_0 PD_k/\rho_k + P_0)/2$, d_k the final dishing, and S_{stage3} the step height evolution function to evaluate the dishing as a function of t_{2k}, the feature layout density PD_k and down pressure P_0.

The final dishing at any location is primarily determined by two parameters: one, the layout feature density and the other, the start time of stage 3. The above scheme captures both of them, although with approximations: first, the start time of the third stage is basically determined by the *effective* pattern density at stage 1, and second, the layout feature density, instead of the *effective* pattern density, determines the dishing at stage 3.

In addition, considering that S_{stage3} is a nonlinear function of pattern density PD_k, the above dummy filling problem is a nonlinear optimization problem, which is much more complicated than linear optimization in terms of computer implementation.

6.10 Conclusion

In this chapter, a CMP model for the topography evolution in the copper damascene process has been developed. Pad models based on both linear elasticity and linear viscoelasticity assumptions are used in the model development. The die and feature scale topography interactions are implemented as well. In comparison with the previous linear elastic models, the simulation results in this work shows that a viscoelastic model of pads can evaluate the pattern density dependency of dishing more accurately, implying the previous models based on the linear elasticity assumption are insufficient. This model can help provide a better understanding of the formation mechanism of erosion and dishing in the copper damascene process.

It is noteworthy that the model developed in this chapter is applicable to other damascene processes such as the shallow trench isolation process. The model of the first stage of planarization can also be used to model the ILD/IMD planarization process.

6.11 References

[6.1] S. P. Murarka, in: *Metallisation*, pp. 100, Butterworth-Heinemann, Boston, U. S. A., 1993.

[6.2] J. M. Steigerwald, R. Zirpoli, S. P. Murarka, D. Price and R. J. Gutmann, "Pattern geometry effects in the chemical-mechanical polishing of inlaid copper structures," *Journal of the Electrochemical Society,* Vol. **141**, pp. 2842-2848, 1994.

[6.3] Z. Stavreva, D. Zeidler, M. PlÖtner, G. Grasshoff and K. Drescher, "Chemical-mechanical polishing of copper for interconnect formation," *Microelectronic Engineering*, Vol. **33**, pp. 249-257, 1997.

[6.4] Z. Stavreva, D. Zeidler, M. Plötner and K. Drescher, "Influence of process parameters on chemical-mechanical polishing of copper," *Microelectronic Engineering*, Vol. **37/38**, pp. 143-149, 1997.

[6.5] M. Fayolle and F. Romagna, "Copper CMP evaluation: planarization issues," *Microelectronic Engineering*, Vol. **37/38**, pp. 135-141, 1997.

[6.6] V. Nguyen, H. VanKranenburg and P. Woerlee, "Dependency of dishing on polish time and slurry chemistry in Cu CMP," *Microelectronic Engineering*, Vol. **50**, pp. 403-410, 2000.

[6.7] J. T. Pan, P. Li, K. Wijekoon, S. Tsai and F. Redeker, "Copper CMP integration and time dependent pattern effect," *1999 International Interconnect Technology Conference*, pp. 164-166, San Francisco, CA, U. S. A., May 1999.

[6.8] J-Y. Lai, N. Saka and J-H. Chun, "Evolution of copper-oxide damascene structures in chemical mechanical polishing, I. Contact mechanics modeling," *Journal of the Electrochemical Society*, Vol. **149**, pp. G31-G40, 2002.

[6.9] J-Y. Lai, N. Saka and J-H. Chun, "Evolution of copper-oxide damascene structures in chemical mechanical polishing, II. Copper dishing and oxide erosion," *Journal of the Electrochemical Society*, Vol. **149**, pp. G41-G50, 2002.

[6.10] S. Hymes, T. Brown, P. LeFevre, B. Mikkola, I. Emesh, R. Bajaj, K. Smekalin, F. Redeker, T. Park, T. Tugbawa, D. Boning and J. Nguyen, "Evolution of topography during 1st step CMP of Cu-plated damascene structures," *Electrochemical Society Meeting*, Vol. **99-37**, pp. 149-157, Honolulu, HA, U. S. A., Oct. 1999

[6.11] K. Wijekoon, S. Mishra, S. Tsai, K. Puntambekar, M. Chandrachood, F. Redeker, R. Tolles, B. Sun, L. Chen, T. Pan, P. Li, S. Nanjangud, G. Amico, J. Hawkins, T. Myers, R. Kistler, V. Brusic, S. Wang, I. Cherian, L. Knowles, C. Schmidt and C. Baker, "Development of a production worthy copper CMP process," *1998 SEMI/IEEE Advanced Semiconductor Manufacturing Conference and Workshop*, pp. 354-363, Sept. 1998.

[6.12] Q. Luo and S. V. Babu, "Dishing effects during chemical mechanical polishing of copper in acidic media," *Journal of the Electrochemical Society*, Vol. **147**, pp. 4639-4644, 2000.

[6.13] J. Luo and D. A. Dornfeld, "Material removal mechanism in chemical mechanical polishing: theory and modeling," *IEEE Transaction: Semiconductor Manufacturing*, Vol. **14**, pp. 112-133, 2001.

[6.14] J. Luo and D. A. Dornfeld, "Effects of abrasive size distribution in chemical mechanical planarization: modeling and verification," *IEEE Transaction: Semiconductor Manufacturing*, Vol. **16**, pp. 469-476, 2003.

[6.15] J. Luo and D. A. Dornfeld, "Material removal regions in chemical mechanical planarization for sub-micron integrated circuit fabrication: coupling effects of wafer-pad contact area, abrasive size distribution and slurry chemicals," *IEEE Transaction: Semiconductor Manufacturing*, Vol. **16**, pp. 45-56, 2003.

[6.16] J. Luo and D. A. Dornfeld, "Material removal saturation in chemical mechanical planarization," *ESRC Report, University of California at Berkeley*, Berkeley, CA, U. S. A., 2001.

[6.17] J. Luo and D. A. Dornfeld, "Optimization of CMP from the viewpoint of consumable effects," *Journal of the Electrochemical Society*, Vol. **150**, pp. G807-G815, 2003.

[6.18] B. Stine, D. Ouma, R. Divecha, D. Boning, J. Chung, D. L. Hetherington, I. Ali, G. Shinn, J. Clark, O. S. Nakagawa and S.-Y. Oh, "A closed-form analytical model for ILD thickness variation in CMP processes," *1997 CMP-MIC Conference*, pp. 266-273, Santa Clara, CA, U. S. A., Feb. 1997.

[6.19] E. Tseng, C. Yi and H. C. Chen, "A mechanical model for DRAM dielectric chemical-mechanical polishing process," *1997 CMP-MIC Conference*, pp. 258-265, Santa Clara, CA, U. S. A., Feb. 1997.

[6.20] T. H. Smith, S. J. Fang, D. S. Boning, G. B. Shinn and J. A. Stegani, "A CMP model combining density and time dependencies," *1999 CMP-MIC Conference*, pp. 97-104, Santa Clara, CA, U. S. A., Feb. 1999.

[6.21] J. Grillaert, M. Meuris, E. Vrancken, N. Heylen, K. Devriendt, W. Fyen and M. Heyns, "The use of a semi empirical CMP model for the optimization of the STI module," *1999 CMP-MIC Conference*, pp. 105-112, Santa Clara, CA, U. S. A., Feb. 1999.

[6.22] E. Tseng, M. Meng and S. C. Peng, "Modeling and discussions for STI CMP process," *1999 CMP-MIC Conference*, pp. 113-119, Santa Clara, CA, U. S. A., Feb. 1999.

[6.23] T. Tugbawa, T. Park, D. Boning, T. Pan, P. Li, S. Hymes, T. Brown and L. Camilletti, "A mathematical model of pattern dependencies in Cu CMP processes," *Electrochemical Society Meeting*, Honolulu, HA, U. S. A., Oct. 1999.

[6.24] T. Tugbawa, T. Park, B. Lee, D. Boning, P. Lefevre and J. Nguyen, "Modeling of pattern dependencies of abrasive-free copper chemical mechanical polishing processes," *2001 VMIC Conference*, pp. 113-122, Santa Clara, CA, U. S. A., Nov. 2001.

[6.25] O. G. Chekina, L. M. Keer and H. Liang, "Wear-contact problems and modeling of chemical mechanical polishing," *Journal of the Electrochemical Society*, Vol. **145**, pp. 2100-2106, 1998.

[6.26] D. O. Ouma, "Modeling of chemical mechanical polishing for dielectric planarization," *Ph.D. Dissertation, Department of Electrical Engineering and Computer Science, MIT*, MA, U. S. A., 1999.

[6.27] S. R. Runnels, F. Miceli and I. Kim, "Validation of a large area three-dimensional erosion simulator for chemical mechanical polishing," *Journal of the Electrochemical Society*, Vol. **146**, pp. 4619-4625, 1999.

[6.28] Y. M. Haddad, *Viscoelasticity of Engineering Materials*, Chapman and Hall, London, U. K., 1995.

[6.29] H-Y. Kim, H-J. Kim and H.-D. Jeong, "Development of an abrasive embedded pad for dishing reduction and uniformity enhancement," *Journal of the Korean Physical Society*, Vol. **37**, No. 6, pp. 945-951, 2000.

[6.30] T. H. Park, "Characterization and modeling of pattern dependencies in copper interconnects for integrated circuits," *Ph.D. Dissertation, Department of Electrical Engineering and Computer Science, MIT*, Cambridge, MA, U. S. A., 2002.

[6.31] J. F. Luo, Y. J. Liu and E. J. Berger, "Analysis of two dimensional thin structures (from micro- to nano-scales) using the boundary element method," *Computational Mechanics*, Vol. **22**, pp. 404-412, 2003.

[6.32] Y. Moon, "Mechanical aspects of the material removal mechanism in chemical mechanical polishing (CMP)," *Ph.D. Dissertation, Department of Mechanical Engineering, University of California at Berkeley*, Berkeley, CA, U. S. A., 1999.

[6.33] K. L. Johnson, *Contact Mechanics*, Cambridge University Press, Cambridge, U. K., 1985.

[6.34] R. Tian, D. F. Wong and R. Boone, "Model-based dummy feature placement for oxide chemical-mechanical polishing manufacturability," *IEEE Transactions: Computer-Aided Design of Integrated Circuits and Systems*, Vol. **20**, pp. 902-910, 2001.

[6.35] R. Tian, X. Tang and M. D. F. Dong, "Dummy-feature placement for chemical-mechanical polishing uniformity in shallow-trench isolation process," *IEEE Transactions: Computer-Aided Design of Integrated Circuits and Systems*, Vol. **21**, pp. 63-71, 2002.

7

THREE-DIMENSIONAL FEATURE-SCALE MODELING OF CMP

7.1 Introduction

In Chapter 6, a one and semi-two dimensional model based on linear elasticity and viscoelasticity is developed to investigate feature- and die-scale topography evolution. The influences of lateral directional parameters such as line width on the topography evolution in the vertical direction can be evaluated using the look-up table based semi-two dimensional model. The advantage of the model is that the computation time is minimal. It can be integrated into a design for manufacturing software to investigate the effects of process parameters on circuit performance. However, it is not able to predict exactly the two and three dimensional shape evolution of the features during the CMP process, which may be a concern in some applications. For instance, an exact three-dimensional shape of the copper dishing may help to extract the accurate resistor of a copper line under a certain process; the edge rounding of the alignment mark generated by CMP may create problems for the lithography process following CMP. A three-dimensional model with capability in topography evolution will help obtain the optimal design of alignment marks.

In this chapter, a three dimensional (3-D) feature-scale topography evolution model based on the framework of Fig. 6.2 is discussed. The pressure distribution model is the key part in the framework. While a full finite element or boundary element model based on linear elasticity or linear viscoelasticity can be developed, in this chapter, a linear elasticity based contact mechanics model is used. The contact mechanics based governing equation is simpler than that in linear elasticity and its numerical implementation is simpler. Considering that the thickness of the asperity layer is in the range of 10$\sim$100μm, this contact mechanics based model may be only accurate for feature lines smaller than 10μm. The numerical implementation of the model is the focus of this chapter.

7.2 Contact Mechanics Based Governing Equation

The contact mechanics based governing equation which can be applied in modeling three-dimensional topography evolution is an integral equation. It has been presented in Chapter 2 as Eq. 2.20 already [7.1]:

$$\overline{u}_z(x,y) = \frac{1-\upsilon_p^2}{\pi E_p} \iint\limits_D \frac{P(\xi,\eta)}{r(x,y,\xi,\eta)} d\xi d\eta \tag{7.1}$$

This equation relates the deformation $\overline{u}_z$ of the pad at an arbitrary location (x, y) to the pressure distribution $P(\xi,\ \eta)$ applied in an arbitrary domain D over the pad, where E_p and υ_p are the Young's modulus and Poisson's ratio of the polishing pad, respectively, and r (x, y, ξ, η) the distance between locations (x, y) and (ξ, η).

The above contact mechanics equation is based on linear elasticity. Following the numerical implementation in this chapter and the boundary conditions of the one-dimensional model in Chapter 6, a 3-D model based on linear viscoelasticity may be developed to cover the case of a linear viscoelastic pad in the future.

7.3 Numerical Solution of the Governing Equations

In CMP, the top view of the domain of interest might look like that shown in Fig. 7.1. Features with different geometry and size are in contact with the pad. Knowing part of the pressure distribution and pad deformation over the domain as boundary conditions, the press distribution and pad deformation over the whole domain can be evaluated using Eq. 7.1. For example, if we assume a uniform pressure distribution over a rectangle feature A as shown in Fig. 7.1, the pad deformation at any location (x,y) due to this pressure can be obtained immediately by substituting the locations of (x,y), the pressure and geometry parameters of rectangle A into the integral in Eq. 7.1. If we know the pressure distributions over all the features, the deformation will be the summation of the deformation caused by the pressures at each feature. Similarly, once we know the deformation of the pad at any location (x, y) of the domain, the pressure distribution over feature A can be determined by Eq. 7.1.

In general, the analytical solution to integral equation 7.1 for arbitrary pressure distribution and topography of the domain D is difficult to obtain and the numerical method which reduces the integral equation to linear algebraic equations has to be used. In order to solve the integral equation numerically, the domain D can be discretized into a series of elements over which displacements and pressures are written in terms of their values at a series of nodal points, Fig. 7.2. Writing the discretized form of Eq. 7.1 for every nodal point and applying boundary conditions, a system of linear algebraic equations is obtained and can be solved for the unknown values.

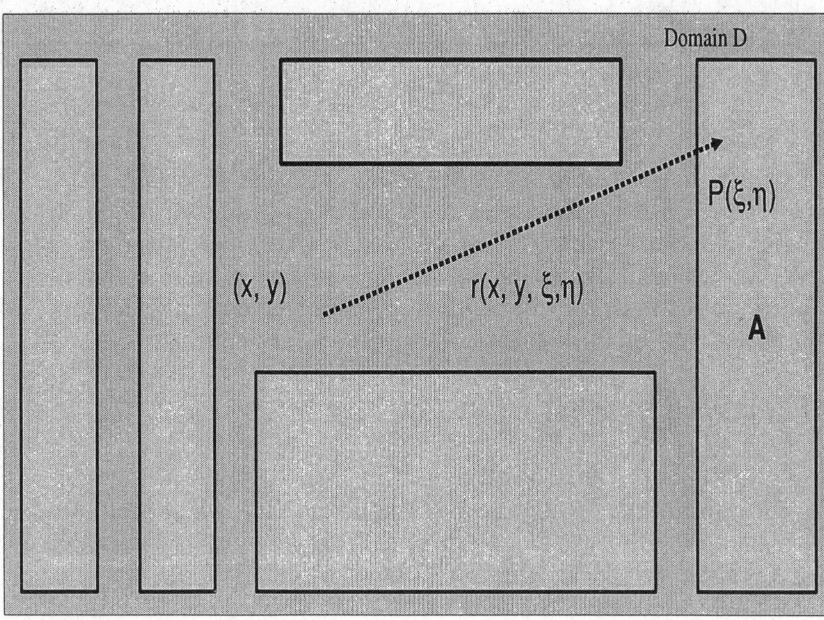

Fig. 7.1. Top view of a pad domain D with wafer features

Two different kinds of elements can be used to mesh the domain, one, constant elements, and the other, quadratic elements. The numerical implementation using constant elements is easier. It was first discussed by Johson [7.1]. Later,

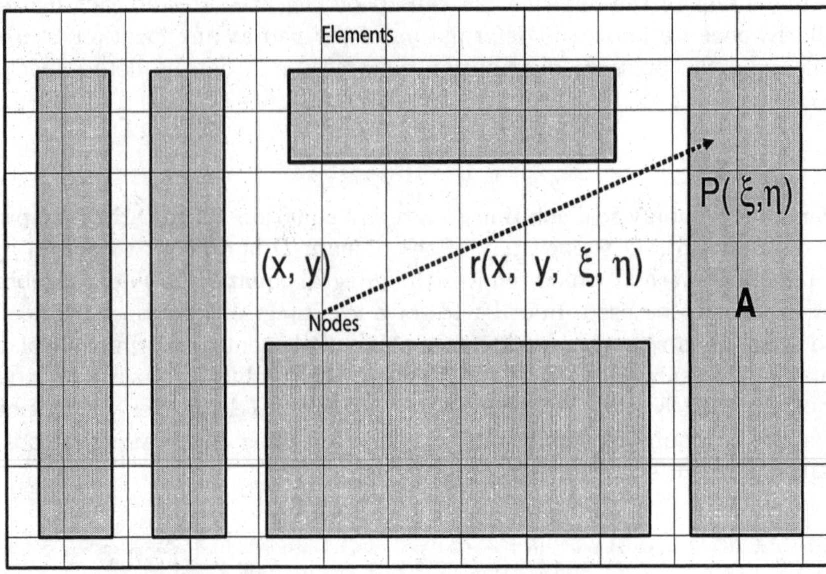

Fig. 7.2. Mesh of domain D with elements

Yoshida [7.2] applied it in 3D topography evolution modeling of the CMP process. They do not discuss the numerical implementation using quadratic elements. Quadratic elements may be needed when more accurate solutions are needed, although its numerical implementation is more complicated. In the following sections, we will address the major numerical issues involved in solving Eq. 7.1 using both constant and quadratic elements.

7.3.1 Constant Element

For a constant element, there is only one node point, which can be taken as the center of the element. The pressure and displacement/deformation distribution over the element is assumed to be constant. If M elements/nodes are used to mesh the domain, based on Eq. 7.1, the displacement/deformation $\overline{u}_{zj}$ at the jth node or the jth element with center location (x, y) can be written approximately as:

$$\overline{u}_{zj}(x,y) = \frac{1-v_p^2}{\pi E_p} \sum_{k=1}^{M} P_k \iint\limits_{E_k} \frac{1}{r(x,y,\xi,\eta)} d\xi d\eta = \sum_{k=1}^{M} (e_{jk} P_k) \qquad (7.2)$$

where P_k is the pressure over the kth element E_k, and $e_{jk} = \frac{1-v_p^2}{\pi E_p} \iint\limits_{E_k} \frac{1}{r(x,y,\xi,\eta)} d\xi d\eta$ is the deformation of element j due to a unit pressure on element k. After writing Eq. 7.2 for every nodal point and applying the boundary conditions at each node[1], a system of $M \times M$ linear algebraic equations is obtained.

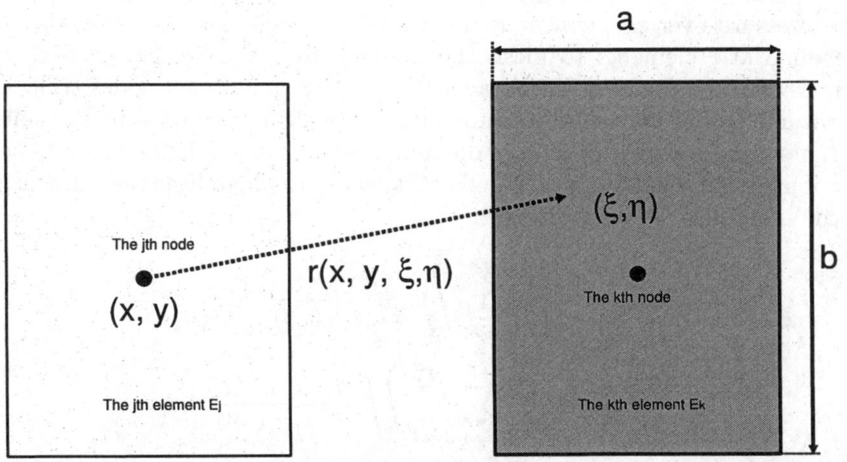

Fig. 7.3. Constant element

To form the system matrix of the linear algebraic equations, the value of $e_{jk}(x, y)$ must be obtained. For a rectangle constant element with width a and length b, as shown in Fig. 7.3, $e_{jk}(x, y)$ can be evaluated analytically. It has been given by Johnson [7.1]:

$$e_{jk}(x,y) = \frac{1-\upsilon_p^2}{\pi E_p} \left\{ \begin{array}{l} (x+a)\ln\left\{ \frac{(y+b)+\left[(y+b)^2+(x+a)^2\right]^{1/2}}{(y-b)+\left[(y-b)^2+(x+a)^2\right]^{1/2}} \right\} \\ +(y+b)\ln\left\{ \frac{(x+a)+\left[(y+b)^2+(x+a)^2\right]^{1/2}}{(x-a)+\left[(y+b)^2+(x-a)^2\right]^{1/2}} \right\} \\ +(x-a)\ln\left\{ \frac{(y-b)+\left[(y-b)^2+(x-a)^2\right]^{1/2}}{(y+b)+\left[(y+b)^2+(x-a)^2\right]^{1/2}} \right\} \\ +(y-b)\ln\left\{ \frac{(x-a)+\left[(y-b)^2+(x-a)^2\right]^{1/2}}{(x+a)+\left[(y-b)^2+(x+a)^2\right]^{1/2}} \right\} \end{array} \right\} \quad (7.3)$$

Note that in the above equation, the coordinate of the center of the element $E_k$is (0, 0).

For an arbitrary shape of the element, the integral of e_{jk} can be evaluated using numerical integration.

[1]Either pressure or displacement at a node must be known as the boundary condition.

7.3.2 Quadratic Element

The pressure over an element is not necessarily constant. We may want to use quadratic elements to mesh the domain for accuracy, Fig. 7.4. A two dimensional quadratic element can either have 8 nodes or 9 nodes [7.3]. Let us assume that 9-node elements are used. If M elements with a total number of H nodes are used to mesh the domain, based on Eq. 7.1, the displacement $\overline{u}_{zj}$ of node j at location (x, y) can be written approximately as the summation of the integrals over each element:

$$\begin{aligned}
\overline{u}_{zj}(x,y) &= \frac{1-\upsilon_p^2}{\pi E_p}\sum_{k=1}^{M}\iint\limits_{E_k}\frac{P(\xi,\eta)}{r(x,y,\xi,\eta)}d\xi d\eta \\
&= \frac{1-\upsilon_p^2}{\pi E_p}\sum_{k=1}^{M}\sum_{m=1}^{9}\iint\limits_{E_k}\frac{\phi_m(\xi,\eta)\,P_m}{r(x,y,\xi,\eta)}d\xi d\eta \\
&= \frac{1-\upsilon_p^2}{\pi E_p}\sum_{k=1}^{M}\sum_{m=1}^{9}P_m\iint\limits_{E_k}\frac{\phi_m(\xi,\eta)}{r(x,y,\xi,\eta)}d\xi d\eta \qquad (7.4)
\end{aligned}$$

where m is the local index of nodes over an element, j the global index of a node over the whole domain, k the index of the element, P_m the pressure at the mth node and ϕ_m the shape function of the element E_k corresponding to the mth node [7.3]. As in the case of the constant element, an $H \times H$ system of linear equations can be formed by writing Eq. 7.4 and applying boundary conditions to each nodal point.

From Eq. 7.4, it is seen that the integral over the kth element E_k is separated into 9 integrals $\iint\limits_{E_k}\frac{\phi_m(\xi,\eta)}{r(x,y,\xi,\eta)}d\xi d\eta$, with each corresponding to one element node. They can be evaluated accurately using numerical Gaussian integration [7.4] when node j is not on element k. When node j is one of the nodes of element k, however, the denominator in the integral will be equal to zero when (ξ, η) coincides with (x, y). This leads to a singular integral and it cannot be evaluated accurately by the normal Gaussian integration. Fortunately, among the 9 integrals, only one is singular. To demonstrate this idea more clearly, let $I(m)$ denote the global index of the mth nodes in the domain. When $I(m) \neq j$, the shape function $\phi_m = 0$ at the jth node (x, y). This cancels out the singular term $r(x, y, \xi, \eta)$ in the denominator. And therefore the mth integral will be evaluated accurately using the Gaussian integration. However, when $I(m) = j$, $\phi_m = 1$ at (x, y). The singularity in the denominator cannot be cancelled and the regular Gaussian integration will fail to evaluate the

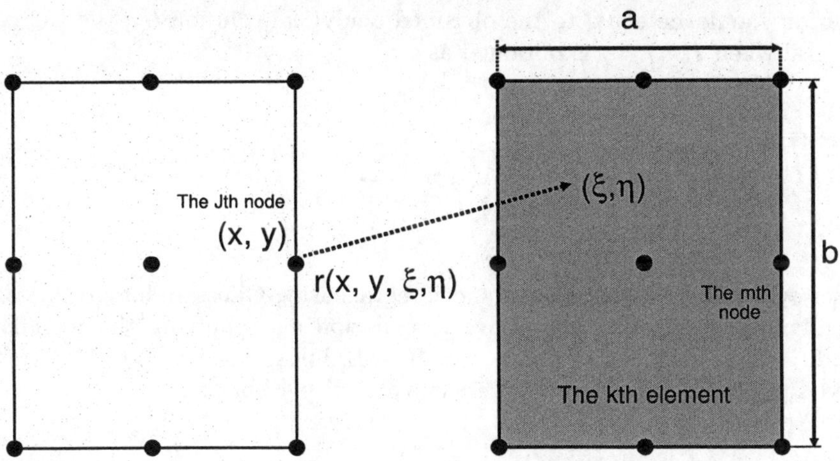

Fig. 7.4. Quadratic element

integral. In this case, the following semi-analytical method can be used, which is similar to the 'rigid-body translation' method used to handle the singular integrals in boundary integral equations in BEM.

Let us assume that a constant unit pressure 1 is applied on the quadratic element. Then the displacement of node j due to the unit pressure can be evaluated analytically using Eq. 7.1, even if node j is one of nodes on the element. The analytical solution depends on the shape of the element. For a rectangular element with width a and height b, the solution is given by Eq. 7.3. We assume that it is U_j.

Besides the analytical solution, the numerical solution of the displacement U_j under the unit pressure can also be attempted using the following formulation:

$$U_j(x,y) = \frac{1-\upsilon_p^2}{\pi E_p} \sum_{m=1}^{9} \iint_{E_k} \frac{1 \times \phi_m(\xi,\eta)}{r(x,y,\xi,\eta)} d\xi d\eta$$

$$= \frac{1-\upsilon_p^2}{\pi E_p} \left[\begin{array}{l} \sum\limits_{m=1;I(m)\neq j}^{9} \iint\limits_{E_k} \frac{\phi_m(\xi,\eta)}{r(x,y,\xi,\eta)} d\xi d\eta \\ + \iint\limits_{E_k} \frac{\phi_{m|I(m)=j}(\xi,\eta)}{r(x,y,\xi,\eta)} d\xi d\eta \end{array} \right] \quad (7.5)$$

As discussed earlier, the integrals when $I(m) \neq j$ in above equation can be evaluated using Gaussian integration. Considering that the above numerical solution should be equal to the obtained analytical solution U_j, the singular integral when $I(m) = j$ is obtained as

$$\iint\limits_{E_k} \frac{\phi_{m|I(m)=j}(\xi,\eta)}{r(x,y,\xi,\eta)} d\xi d\eta = U_j - \sum_{m=1;I(m)\neq j}^{9} \iint\limits_{E_k} \frac{\phi_m(\xi,\eta)}{r(x,y,\xi,\eta)} d\xi d\eta \tag{7.6}$$

After evaluating all the non-singular integrals using Gaussian integration and singular integrals using the above formulation and applying the boundary conditions at each nodal point, the $H \times H$ linear algebra Eq. 7.4 can be solved to obtain the unknown pressures and displacements.

7.4 Boundary Conditions in CMP

7.4.1 Displacement Condition

The boundary conditions of pressure and displacement at each node need to be known to solve the unknown pressure and displacement using Eqs. 7.2 and 7.4. In CMP, the wafer can be taken as a rigid body in comparison with the polishing pad asperity layer. Therefore, the relative pad deformation over a feature will follow the feature topography. For instance, at the beginning of the planarization, the surface of a feature is flat. Then the relative pad asperity layer deformation over that feature will be zero. This indicates that the boundary conditions of the pad over the features are displacement conditions. The pressure distributions are unknown values and need to be solved.

However, it is noted that the known node displacements over features are relative displacements depending on the feature surface topography. In Eqs. 7.2 and 7.4, the displacement is an absolute value. Therefore, a reference displacement across the whole domain has to be added into the relative displacement. The reference displacement is an unknown value. Therefore, there will be $M+1$ (or $H+1$ for quadratic element) instead of M (or H) unknown variables in Eqs. 7.2 and 7.4. One more equation based on force equilibrium has to be added into the system equation so that it is solvable. For the case of constant elements, the equation is

$$\sum_{k=1}^{M} P_k A_k = P_{app} \tag{7.7}$$

where P_{app} is the force applied on the domain and A_k the area of the kth element. For the case of quadratic elements, the equation is

$$\sum_{k=1}^{M}\sum_{m=0}^{9} P_m \iint_{E_k} \phi_m d\xi d\eta = P_{app} \tag{7.8}$$

All the integrals in Eq. 7.8 can be evaluated numerically.

7.4.2 Pressure Condition

For the part of the pad asperity layer which is not in touch with the features, say, the asperity above the copper lines at stage 1 of copper polishing, the pressure over them is zero. The unknown boundary values to be solved are displacements or deformations.

7.5 Topography Evolution

The whole planarization can be separated into N time steps. Using the framework of Fig. 6.2, the boundary conditions for the contact mechanics model are continuously updated using the material removal rate equation. For example, at time step t_i, the displacement boundary condition over the contact parts/features is determined by the feature topography. With this boundary condition known, the pressure distribution over the contact part/features can be obtained. Inputting this pressure distribution into a material removal rate equation such as Preston's equation updates the surface shape of the features, which is used as boundary conditions to evaluate the pressure distribution at next time step.

Similar to that in the one dimensional model, part of the pad which does not contact the wafer at the beginning of the planarization t_i may contact the wafer some time during the planarization, the boundary conditions over that part of the pad should be updated from a zero pressure condition to a displacement condition.

Note that the selection of the time step is critical. If it is too small, the computation time is huge. If it is too large, the convergence is a concern. A large step time may update the feature surface into one which cannot be fully contacted by pads, leading to negative contact pressures. The selection of the time step should consider the pad material, material removal properties (Preston's coefficient) and mesh size. Generally, When the pad is stiffer or the Preston's coefficient is larger, the time step should be smaller to ensure the convergence.

7.6 Simulation Example

A simulator is written using C++ to implement the above numerical model. At the current stage, the simulator can only handle the planarization of a single material. We use an isolate copper step feature as shown in Fig. 7.5 to test this simulator. The step height is assumed to be 500nm. The height of the upper part of the step is 2μm and the height of the bottom part is 1.5μm. The size of the upper part of the step is 2μm$\times$2μm. The ratio of the bottom part area to the area of the whole domain, or, the pattern densities PD are 2/3 for both cases. The same copper material removal properties as in Chapter 6 is used in the simulation, i.e., K_{ecu}=0.321nm/s·kPa^{-1} and MRR_{cu0}=1.01nm/s. The pad is assumed to be linear elastic material with first Young's modlulus E_1= 5.2MPa and Poisson's ratio 0.5, as commercial pad IC1000. An down pressure 10kPa is assumed. All together 48 quadratic elements are used for the simulation.

Figs. 7.6 (a) and (b) show the simulation results of pressure distribution and pad deformation (or location) over the feature at the beginning of the planarization. The pad has a full contact with the upper part of the step. It can be seen that the pressure distribution over the top surface is very no-uniform. At the edge, the pressure is much larger than that in the center. This will lead to an edge rounding, as will be shown later. The bottom part is not in touch with the pad. Therefore the pressure distribution there is zero. Fig. 7.6 (b) shows the pad locations after deformation. The pad has a full contact with the flat top surface of the feature while there is a relative -2nm deformation above the bottom of the step.

Figs. 7.7 (a) and (b) show the snapshot of the pressure distribution and deformation after 0.5 second of the planarization. We can see that an edge rounding

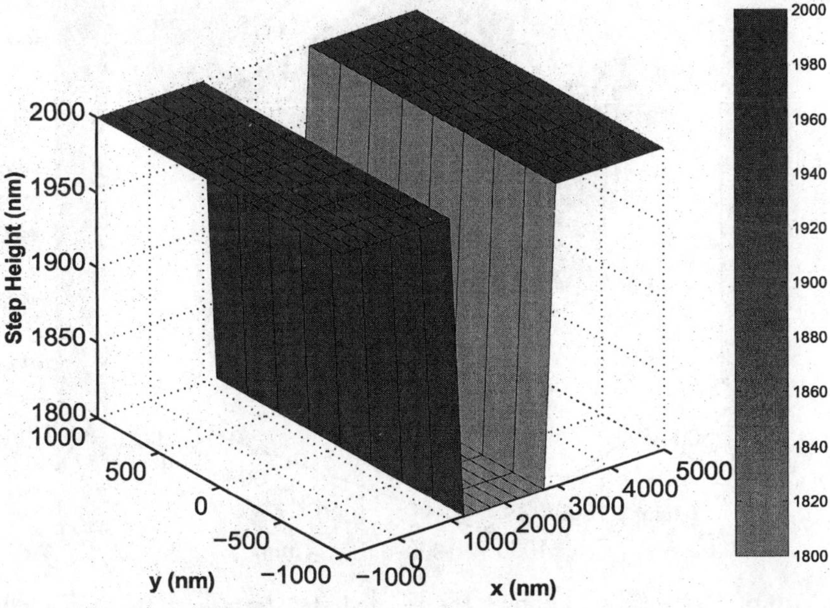

Fig. 7.5. An isolated step feature with step height 500nm

Fig. 7.6.a. Pressure distribution over the isolated step feature at the beginning of planarization

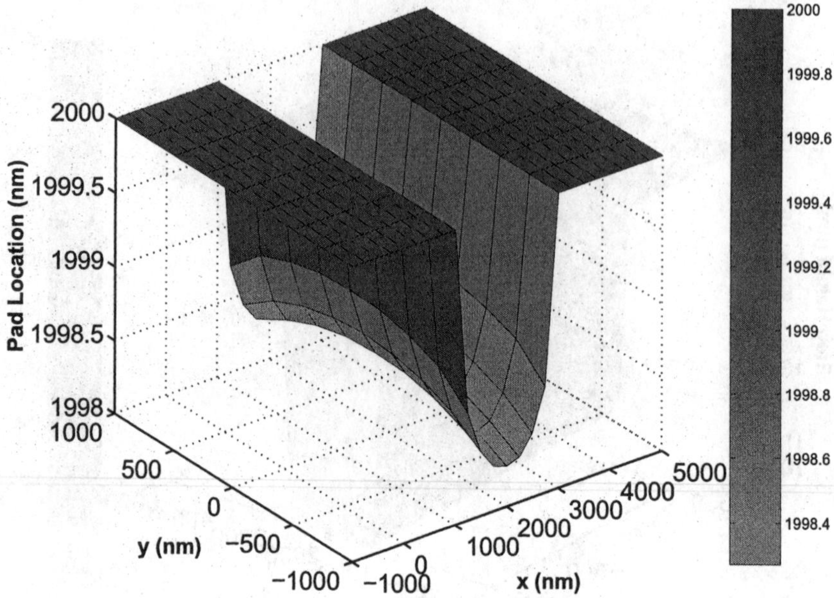

Fig. 7.6.b. Pad deformation over the isolated step feature at the beginning of planarization

at the upper part of the step occurs. The pressure distribution becomes more uniform due to the edge rounding.

Figs. 7.8 (a) and (b) show the pressure distribution and pad deformation (or location) at the third second of the planarization. There is more edge rounding in comparison with that at second 3. The pressure distribution has become uniform with the edge rounding. It is 10kPa/PD=10/(2/3)kPa=15kPa. Before the pad touchs the bottom of the step, the shape of the upper part's surface will remain the same. We can see that for a feature size as small as 1μm, the edge rounding is small (around 2nm) and can reach its stable status quickly.

After around 85 second of planarization, the pad touches the bottom part of the step, Fig. 7.9 (a) and (b). It is seen that the pad touches the bottom of the step and deforms there, Fig. 7.9 (b). The pressure distribution over the top part of the step becomes non-uniform again. Part of the down force now is supported by the bottom of the step. The pressure at the top of the step continuously decreases to 10kPa while the pressure at the bottom increases to 10kPa, Figs. 7.10 and 7.11. Because the studied feature is an isolated line, the

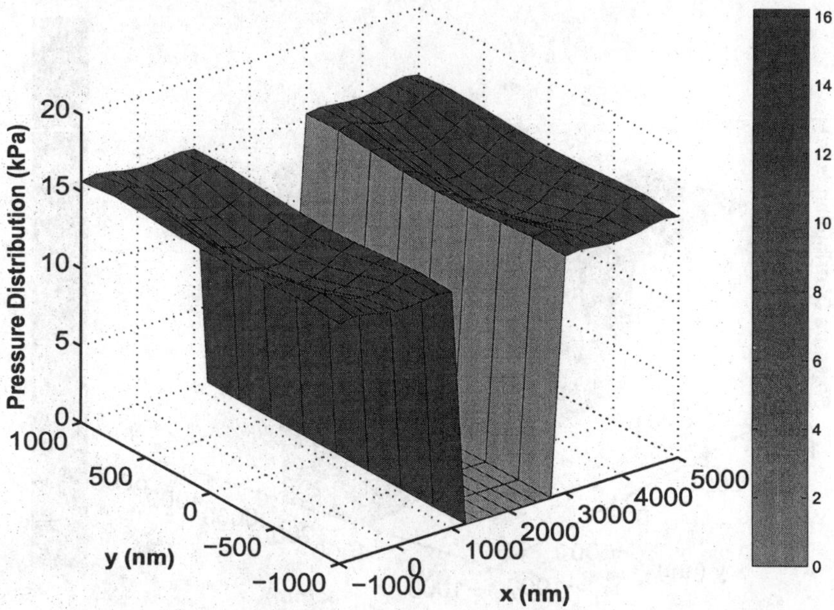

Fig. 7.7.a. Pressure distribution over the smaller isolated step feature at 0.5 second of planarization

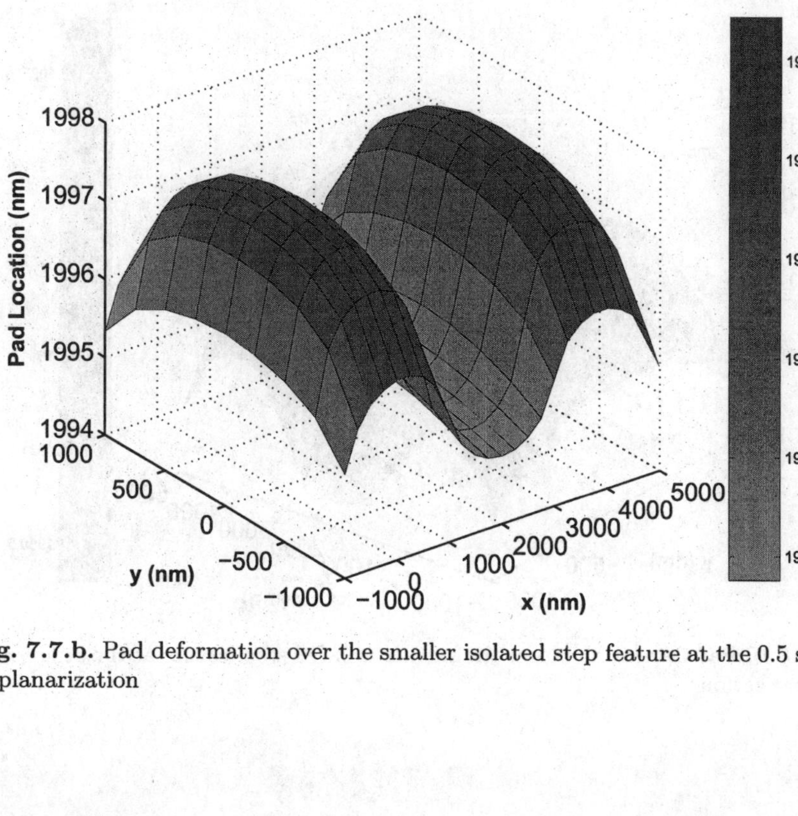

Fig. 7.7.b. Pad deformation over the smaller isolated step feature at the 0.5 second of planarization

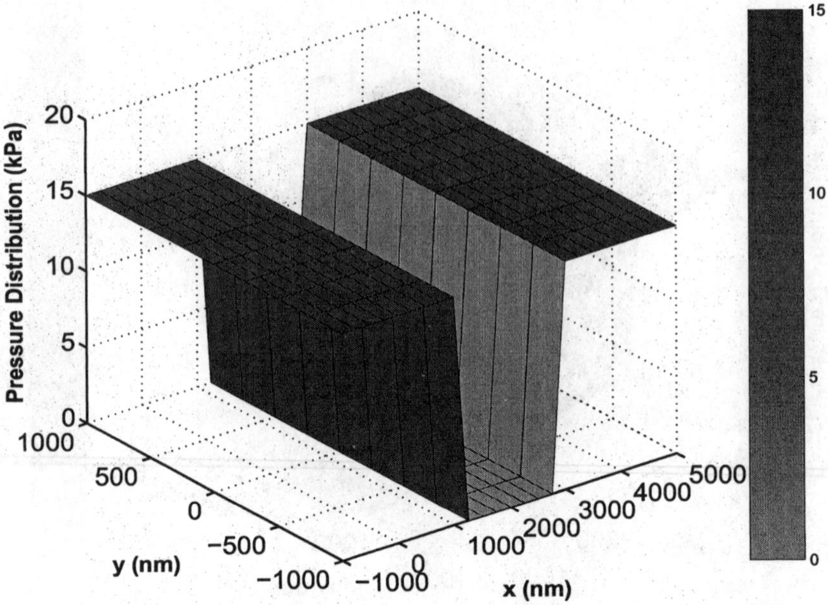

Fig. 7.8.a. Pressure distribution over the smaller isolated step feature at the 3rd second of planarization

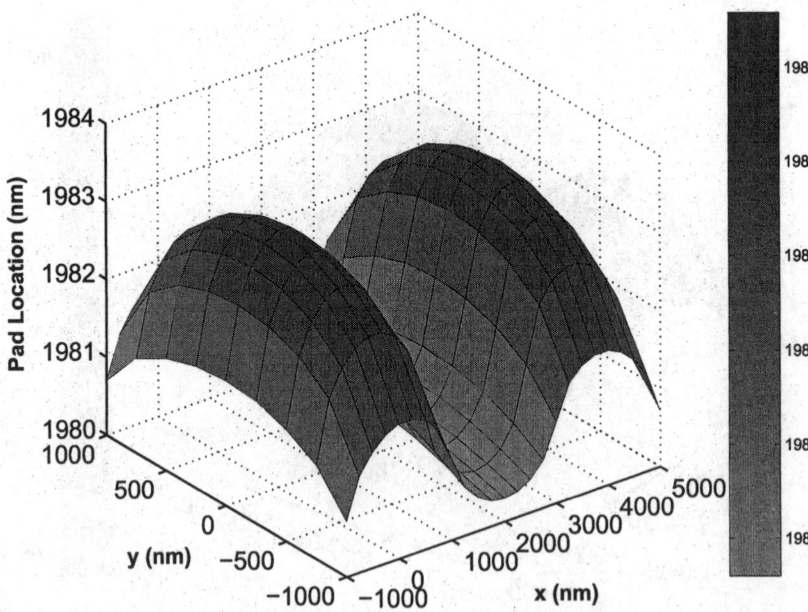

Fig. 7.8.b. Pad deformation over the smaller isolated step feature at the 3rd second of planarization

pressure over the boundary of the domain is larger than that in the center if the whole domain is flat. The pressure difference decreases continuously with the planarization. A stable bow with a center 20nm higher than the boundary (which balances the pressure singularity at the boundary), is finally formed with a uniform pressure 10kPa across the whole domain.

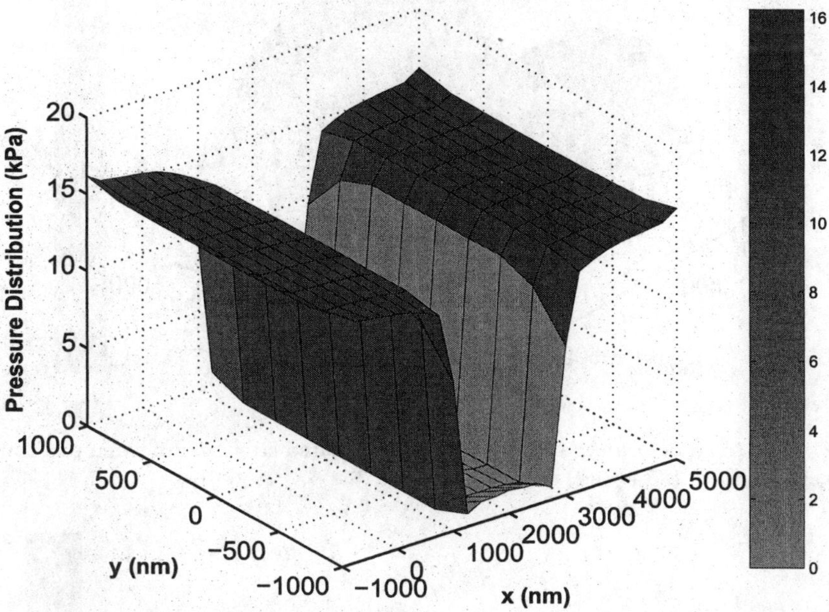

Fig. 7.9.a. Pressure distribution over the smaller isolated step feature when pad touchs the step bottom (85second)

When the line size is larger, say, 10μm, the pad deformation over the bottom part of the step will be larger. The edge rounding will be larger as well. And it takes a longer time to form a stable edge rounding.

7.7 Conclusion and Future Work

In this chapter, the numerical details of a 3-D topography evolution model for single material planarization are discussed. In the future, a model following this idea may be developed for the damascene process. The linear viscoelasticity may be integrated into the topography evolution model as well.

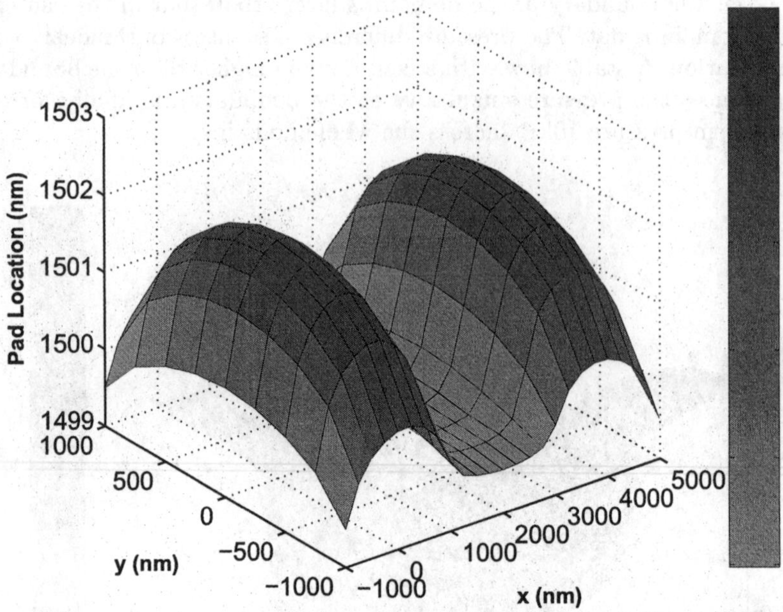

Fig. 7.9.b. Pad deformation over the smaller isolated step feature when pad touchs the step bottom (85second)

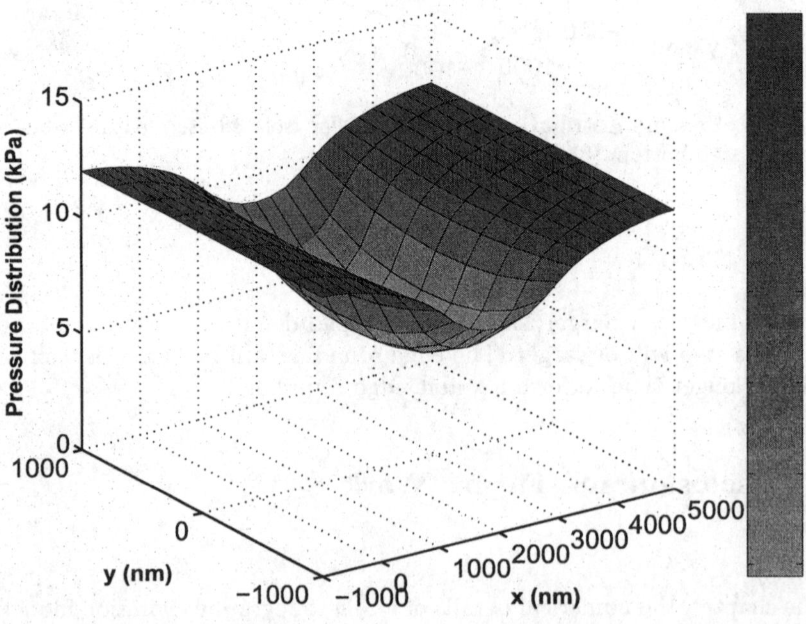

Fig. 7.10.a. Pressure distribution over the smaller isolated step feature after pad touchs the step bottom (90second)

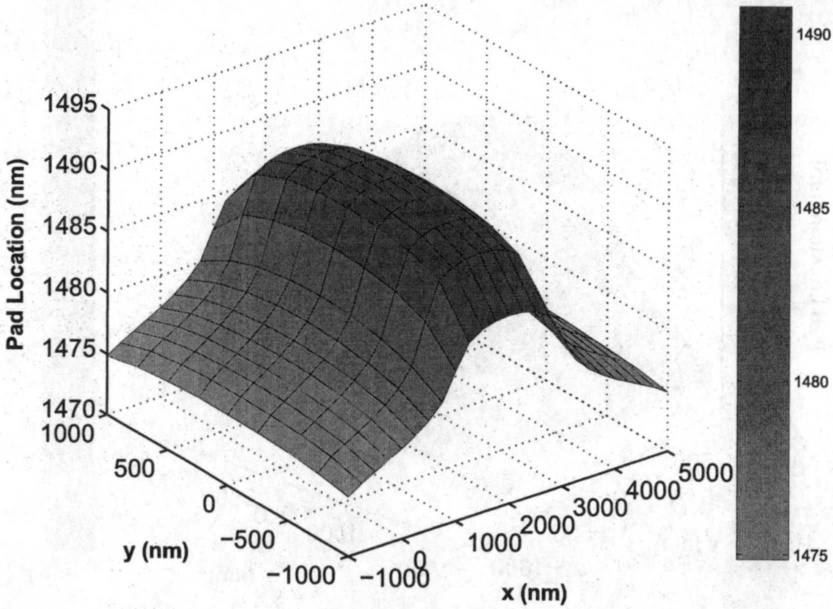

Fig. 7.10.b. Pad deformation over the smaller isolated step feature after pad touchs the step bottom (90second)

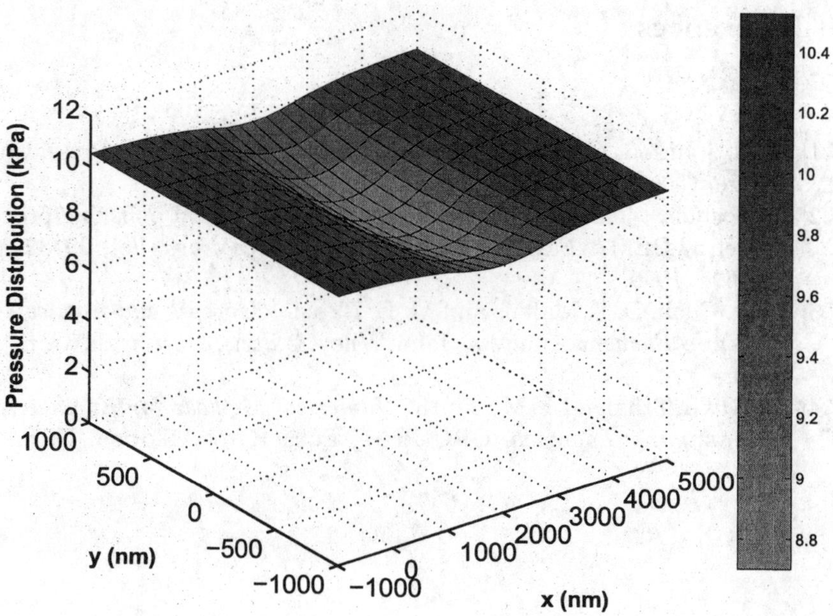

Fig. 7.11.a. Pressure distribution over the smaller isolated step feature after pad touchs the step bottom (95second)

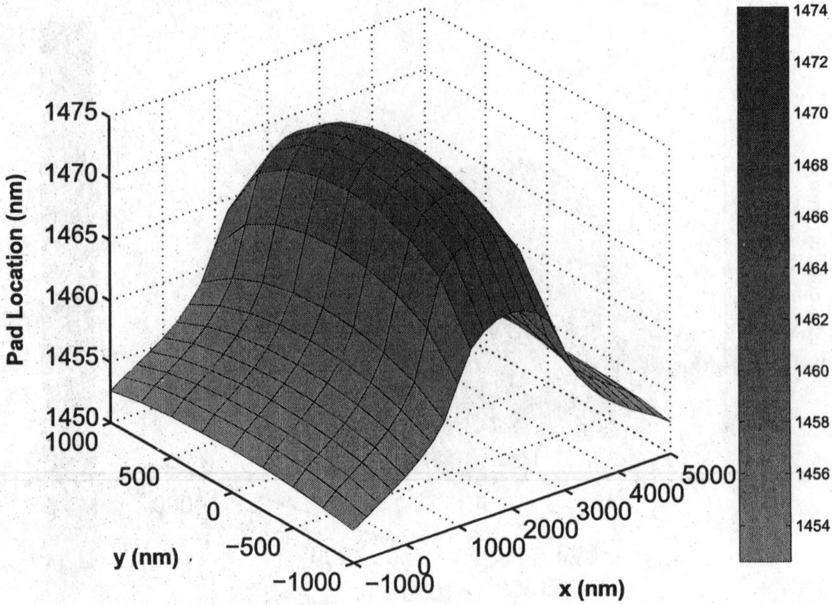

Fig. 7.11.b. Pad deformation over the smaller isolated step feature after pad touchs the step bottom (95second)

7.8 References

[7.1] K. L. Johnson, *Contact Mechanics*, Cambridge University Press, Cambridge, U. K., 1985.

[7.2] T. Yoshida, "Three-dimensional chemical mechanical polishing process model by BEM," *Electrochemical Society Proceedings*, Vol. **99-37**, pp. 593-604, 1999.

[7.3] R. D. Cook, D. S. Malkus and M. E. Plesha, *Concepts and Applications of Finite Element Analysis*, John Wiley & Sons, New York, U. S. A., 1989.

[7.4] D. V. Griffiths and I. M. Smith, *Numerical Methods for Engineers: A Programming Approach*, CRC Press, Boca Raton, Florida, U. S. A., 1991.

8

WAFER-SCALE MODELING OF CMP

8.1 Introduction

The with-in wafer non-uniformity (WIWNU) of the material removal rate (MRR) has long been a concern in CMP. For instance, the rapid variation of the material removal rate at the wafer edge, also known as edge effect, requires an exclusion of the wafer edge after CMP. This reduces the yield of the process. In the shallow trench isolation and copper damascene process, the uneven material removal rates across the wafer cause an over-polishing in the regions where the material is removed faster. This causes a degradation of the circuit performance there. The WIWNU will also bring a systematic variation of the circuit performance across the wafer. A better understanding of the mechanisms driving the WIWNU will help to increase the yield and improve the circuit performance.

While the maximum, minimum, and average material removal rate MRR_{max}, MRR_{min} and MRR_{avg} over the wafer surface are known, the WIWNU can be written as:

$$\text{WIWNU} = \frac{MRR_{max} - MRR_{min}}{2MRR_{avg}} \times 100\%. \tag{8.1}$$

WIWNU can also be written as

$$\text{WIWNU} = \frac{6\sigma_{MRR}}{2MRR_{avg}} = \frac{3\sigma_{MRR}}{MRR_{avg}} \times 100\% \tag{8.2}$$

where σ_{MRR} is the standard deviation of the MRR variation across the wafer. The with-in wafer non-planarity WIWNP is defined as the variation of surface height after CMP. It can be related to the WIWNU by assuming a flat surface before planarization:

$$\begin{aligned}\text{WIWNP} &= H_{max} - H_{min} = (MRR_{max} - MRR_{min}) \times T \\ &= (MRR_{max} - MRR_{min}) \times \frac{TH}{MRR_{avg}} \\ &= 2\text{WIWNU} \times TH \end{aligned} \tag{8.3}$$

where H_{max} is the maximum wafer-scale surface height after planarization, H_{min} the minimum surface height, T the total planarization time, and TH the thickness to be removed. From this equation, we can see that the surface non-planarity is proportional to the WIWNU and the thickness [1] to be removed. For the same thickness, a process with smaller WIWNU generates a more planar surface.

The WIWNU can be attributed to the non-uniform distribution of a number of parameters such as the temperature and slurry flow. However, it is believed that the major contributors are the distributions of the pressure P and velocity V. Substitution of the revised Preston's Equation of material removal rate $MRR= K_e PV + MRR_0$ into Eq. 8.1 yields

$$\begin{aligned}\text{WIWNU} &= \frac{K_e (PV)_{max} + MRR_0 - K_e (PV)_{min} - MRR_0}{K_e (PV)_{avg} + MRR_0} \\ &= \frac{K_e [(PV)_{max} - (PV)_{min}]}{MRR_{avg}}. \end{aligned} \tag{8.4}$$

From Eq. 8.4 we see there are two methods to optimize WIWNU. One is to reduce the non-uniformity of the pressure and velocity distributions, or reduce the value of $(PV)_{max}$-$(PV)_{min}$ in Eq. 8.4. This can be realized through the optimization of machine design, specially, the polishing head and platen design. The second method is to reduce the sensitivity of MRR to the pressure and velocity distributions. This can be realized by adjusting the fitting parameters K_e and MRR_0, which are determined by the consumables. In this chapter, models are developed from both sides. First, models are proposed for the prediction of velocity distributions, for both rotational type and linear

[1] In Eq. 8.3, the material removal rate is assumed to be independent of planarization time. Based on this equation, the non-planarization will become larger and larger with the planarization time, or thickness to be removed. However, note that, as with the polishing of the top surface of a step in Chapter 7, the MRR distribution is actually continuously updated with the surface topography. The final distribution of MRR across the whole wafer will become uniform everywhere. After that, the non-planarity will not increase any more.

type of CMP machine, and pressure distributions, in both solid-solid contact and solid-fluid-solid contact modes of polishing. These models can help to optimize the WIWNU from the the machine configuration point of view. Then, a systematic method using particle scale *MRR* models developed in Chapters 3, 4 and 5 is proposed to optimize the non-uniformity from the viewpoint of consumable effects.

8.2 CMP Configurations and Velocity Distribution

The velocity distribution is a function of the configuration of the CMP machine. There are typically two CMP configurations: one, a rotational type and the other, a linear type. In the rotational type machines, which have been shown schematically in Chapter 1, the wafer is affixed to a wafer carrier (via back-pressure), and pressed faced-down on a rotating platen holding a polymeric polishing pad, as illustrated in Fig. 1.1. The CMP equipment by Applied Materials, Ebarra Tech. and Novellus falls into this category. The velocity distribution over the wafer-pad interface in this configuration is shown schematically in Fig. 8.1. In the linear type, a polishing head is rotated above a continuous belt made of polymer materials, which is driven by motors to move linearly. The configuration of a linear type CMP machine is shown in Fig. 8.2 schematically. The movement of the wafer relative to the pad/belt can be decomposed into two components: one, a rotation of the wafer around its center and, the other, a linear translation of the wafer. This is apparent for the linear type of configuration, where the speed of the rotation is equal to the polishing head rotation speed and the speed of the translation is equal to the linear velocity of the belt; see Fig. 8.2. It is not so apparent for the rotational type machine. The following analysis shows how the relative velocity can be decomposed into these two parts for a rotational type of machine.

Let us consider a point p on the wafer surface in a rotational type of machine; see Fig 8.1. The relative velocity of the wafer to the platen over there can be expressed in vector form as

$$\begin{aligned}\mathbf{V}(p) &= \boldsymbol{\Omega}_w \times \mathbf{r}_w - \boldsymbol{\Omega}_p \times \mathbf{r}_p \\ &= \Omega_w(\vec{k} \times \mathbf{r}_w) - \Omega_p(\vec{k} \times \mathbf{r}_p) \\ &= -\Omega_p(\vec{k} \times \mathbf{r}_p) + \Omega_p(\vec{k} \times \mathbf{r}_w) + (\Omega_w - \Omega_p)(\mathbf{k} \times \mathbf{r}_w) \\ &= \Omega_p[\mathbf{k} \times (\mathbf{r}_w - \mathbf{r}_p)] + (\Omega_w - \Omega_p)(\mathbf{k} \times \mathbf{r}_w) \\ &= -\Omega_p(\mathbf{k} \times \mathbf{e}) + (\Omega_w - \Omega_p)(\mathbf{k} \times \mathbf{r}_w) \end{aligned} \tag{8.5}$$

where Ω_p, Ω_w are the rotation speeds of the platen/pad and wafer, respectively, r_p , r_w are the distances between the point p and the pad center O′ and the wafer center O, respectively and e the offset distance between the pad and wafer centers. All symbols with an arrow denote their vector forms. The **k** represents a unit vector in the z direction perpendicular to the rotation plane.

The above equation shows apparently that the relative movement of the wafer over the pad is composed of two parts: (1) a *translation* with constant velocity $\Omega_p \times \boldsymbol{e}$ independent of the location of p, where Ω_p is the rotation speed of pad/platen and e the offset between the centers of the platen and wafer; and (2) a *rotation* with constant rotation speed $\Omega_w - \Omega_p$,where Ω_w is the rotation speed of the wafer/polishing head. Therefore, in terms of velocity distribution, the rotational type of machine with a wafer rotation speed Ω_w and a platen rotation speed Ω_p can be represented by a linear type of machine with an *effective* wafer rotation speed $\omega = \Omega_w - \Omega_p$ and an *effective* belt speed $V_T = \Omega_p \times \boldsymbol{e}$, Fig. 8.1 (b) & Fig. 8.2. Usually, the relative velocity V between wafer and pad/belt refers to the translation velocity V_T.

The rotation component of the relative velocity is the contributor of the velocity non-uniformity for both rotational and linear types of machines. A smaller rotation speed and a larger translation speed are preferred in terms of smaller velocity non-uniformity. For a rotational type of CMP configuration, this indicates a larger platen velocity Ω_p, a larger wafer-platen center offset e and closer platen and head rotation speeds. When the rotational speed is zero [2], the relative movement is a pure translation and therefore there is no velocity non-uniformity. However, there is a tradeoff for the linear configuration. While a smaller rotational speed benefits the non-uniformity, with no wafer rotations, the material removal at each single point p cannot be averaged out. Therefore, a sufficiently large rotation speed will be needed. With this tradeoff in mind, an optimal velocity combination can be obtained with the following procedures:

i. For a linear type of machine: Let the linear velocity be as large as possible and the rotational speed of the head be as small as possible. However, the rotational speed should not be zero to ensure that the material removal is averaged out for each point.

[2] In the case of the rotational type of machine, this implies that the rotation speeds of the head and the platen are equal to each other.

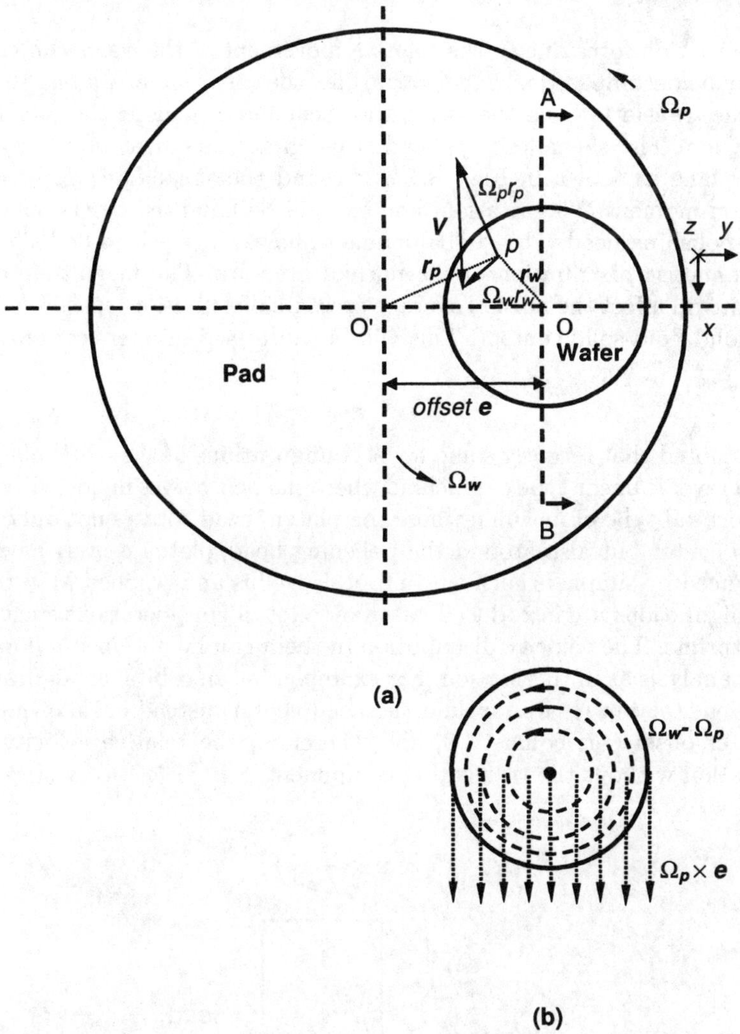

Fig. 8.1. Top view of a rotational type CMP machine and velocity distribution over the wafer-pad interface

ii. For a rotational type of machine: Let the platen rotation speed be as large as possible and the rotation speed difference of the platen and head be as small as possible [3].

---

[3]However, it is noteworthy that a uniform velocity distribution is not always desirable. For a particular process and machine with a smaller polishing rate at the wafer edge, a different head and platen velocity, which leads to higher velocity at the wafer edge, may be applied to bring up the edge polishing rate.

The friction force due to the relative movement of the wafer can correspondingly be decomposed into two parts, one, the shear force f_1 along the direction of the translation and the other, the shear force f_2 along the direction of the rotation. The shear force f_1 leads to a gimbal moment of the wafer in the x-z plane as shown in Figs. 8.1 & 8.2 and the shear force f_2 leads to a rotation moment. The rotation moment will be balanced by the motor driving the polishing head. The gimbal moment, however, needs to be balanced by an asymmetric distribution of the normal pressure. The magnitude of the moment will affect the normal average pressure distribution, specially, in the case of solid-fluid-solid contact. This will be addressed in later sections.

It is noted that recently some novel configurations of the CMP machine, such as a reverse linear type of machine where the belt moves in both directions and an orbital type of machine where the platen/head rotates not only around its own center but also around the polishing head/platen center, have been announced by suppliers such as NuTool, Novellus and Applied Materials. These configurations are introduced based on a lot of considerations such as smaller footprints. The velocity distribution in them can be obtained following a similar analysis as in this section. For example, for an orbital configuration where the pad/platen rotates around the head center instead of its own center, the center offset e is equal to 0, and, therefore, the relative velocity is a pure rotation without the translation component.

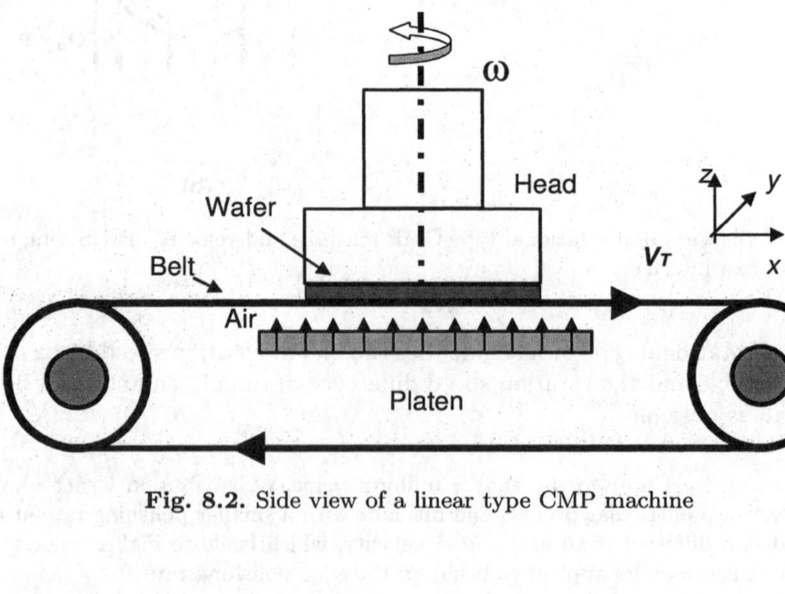

Fig. 8.2. Side view of a linear type CMP machine

8.3 Pressure Distribution

In the following sections, the pressure distribution over the wafer-pad/belt interface is modeled. Two process conditions may exit. One is solid-solid contact between wafer and pad, when the down pressure is larger and the relative velocity is smaller. The other is solid-fluid-solid contact, when the down pressure is smaller and the relative velocity is larger. We first consider the case of the solid-solid contact. This is preferred due to its better planarization ability. Then the solid-fluid-solid contact case is considered. This contact mode may be favorite in low-k CMP. In CMP of low-k dielectrics, usually soft polymer materials, an aggressive contact between low-k material and pad/belt will cause delamination of the metals and dielectrics. The existence of a thin fluid film over the wafer-pad interface may reduce the delamination due to a less aggressive contact between the pad, abrasives and wafer.

8.3.1 Solid-Solid Contact

In the solid-solid contact mode, the governing equation of the wafer and pad deformation is

$$\sigma_{ij} = \frac{2Ev}{(1-2v)(1+v)}\delta_{ij}\varepsilon_{kk} + 2\frac{E}{1+v}\varepsilon_{ij} \tag{8.6}$$

where σ_{ij}, ε_{ij}, E and v are the stress tensor, strain tensor, Young's modulus and Poisson's ratio, and δ_{ij} is the Kronecher delta symbol. The wafer and pad are assumed to be homogeneous isotropic materials. The normal pressure/stress distribution $P(x)$ on the wafer surface arises mainly from two sources; 1) down pressure exerted on the wafer top surface and 2) friction force due to the relative motion of wafer and pad. As mentioned in the last section, for both linear and rotational type machines, the friction force can be separated into two parts: one, a force in the direction of the translation, and the other, a force in the direction of the rotation. The effect of the friction force due to the relative rotation is equal to a pure shear moment, which is balanced by the motor driving the polishing head. It does not have influence on the normal pressure. The friction force in the translation direction has an influence on the pressure distribution. It will be shown by analytical analysis that its effects on the normal pressure distribution can be neglected for the solid-solid contact mode.

Rigid Flat Punch Indentation Model

First, a contact mechanics based rigid flat punch indentation model by Johnson [8.1] is used to model the normal pressure distribution over the wafer-pad interface. This model is a two-dimensional plain strain model with the assumption that the punch does not tilt and therefore the interface remains parallel to the undeformed surface of the solid. The wafer can be modeled as the rigid flat punch and the platen with the whole machine foundation can be modeled as the elastic half-space, as shown in Fig. 8.3.

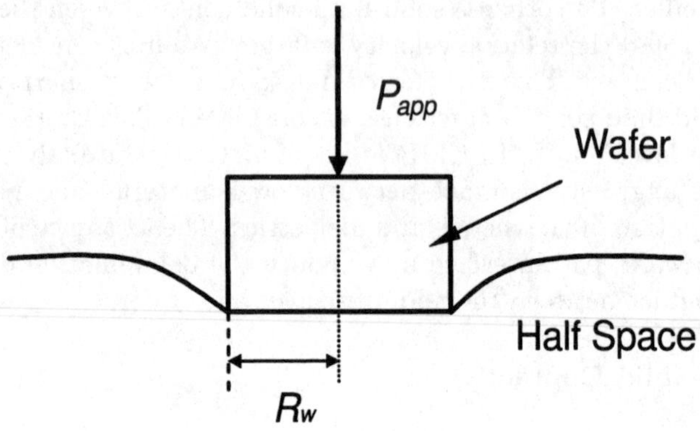

Fig. 8.3. Rigid flat punch model of wafer-platen contact

The pressure distribution functions for frictionless punch and sliding punch are proposed as:

$$P(x) = \frac{P_{app}}{\pi(R_w^2 - x^2)^{1/2}} \tag{8.7}$$

and

$$P(x) = \frac{P_{app}\cos(\pi\gamma)}{\pi(R_w^2 - x^2)^{1/2}}\left(\frac{R_w + x}{R_w - x}\right)^{\gamma} \tag{8.8}$$

respectively, where P_{app} is the force applied on the wafer top surface, R_w the radius of the wafer, $\cot \pi\gamma = -\frac{2(1-\upsilon_h)}{f(1-2\upsilon_h)}$, f the friction coefficient of the wafer-pad interface and υ_h the Poisson's ratio of the foundation. Fig. 8.4 (a) shows the simulation results of the pressure distribution across a 300mm wafer under different friction coefficients and Poisson's ratios. P_0 in the figure is the average pressure across the wafer ($P_0 \times 2R_w = P_{app}$), which is equal to the down pressure. This pressure distribution is asymmetric when the friction coefficient is larger than zero. It is also observed that the pressure increases rapidly at the edge of the wafer. This is due to the fact that the edge is a singular point in Eqs. 8.7 and 8.8. The material removal usually increases rapidly at the wafer edge. This can be attributed to the pressure singularity. Considering the average effects of the rotation component of the velocity distribution, the average pressure distribution should be given by the following equation:

$$\text{Avg}\left[P\left(x\right)\right] = \left[P\left(x\right) + P\left(-x\right)\right]/2 \tag{8.9}$$

Fig. 8.4 (b) shows schematically the simulation results of the average pressure distribution. There is no apparent change when the friction coefficient is changed from 0 to 1. When the friction coefficient is changed from 1 to 2, the average pressure change is more apparent but is still as small as $0.1P_0$. In addition, it is noted that when the Poisson's ratio v_h of the half space goes to 0.5, the Poisson's ratio of pad materials, the effect of friction coefficients becomes minimal. Based on the experimental data of Moon [8.2], the friction coefficients in CMP are usually in the range of 0~1. This indicates that the effects of the shear stress may be ignored in the normal pressure evaluation.

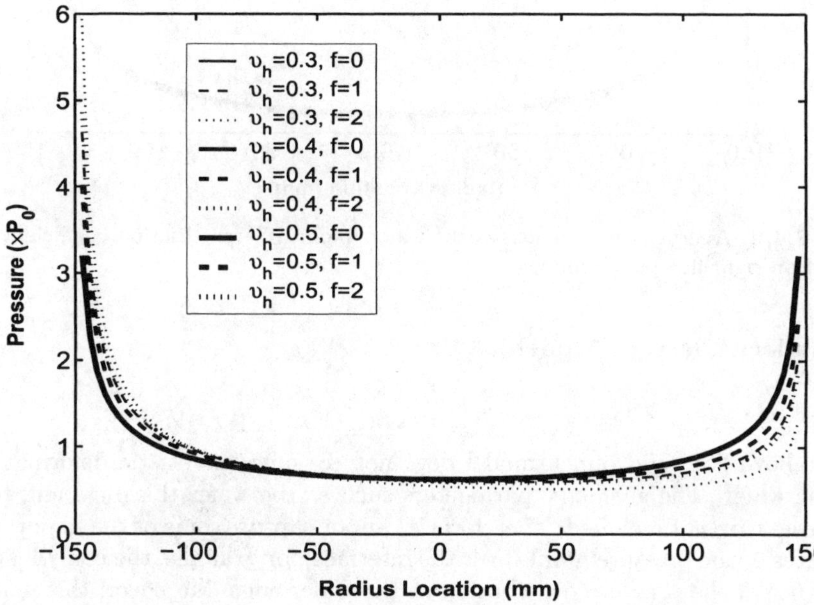

Fig. 8.4.a. Pressure across the wafer under different friction coefficients f based on rigid flat punch model

Following a similar idea of the flat punch indentation model, Fu and Chandra [8.2] developed formulations to predict the pressure distribution for a wafer surface with a curvature. For details, the reader is referred to [8.3]. A three dimensional model as that in Chapter 7 may be developed for wafer-scale pressure distribution as well.

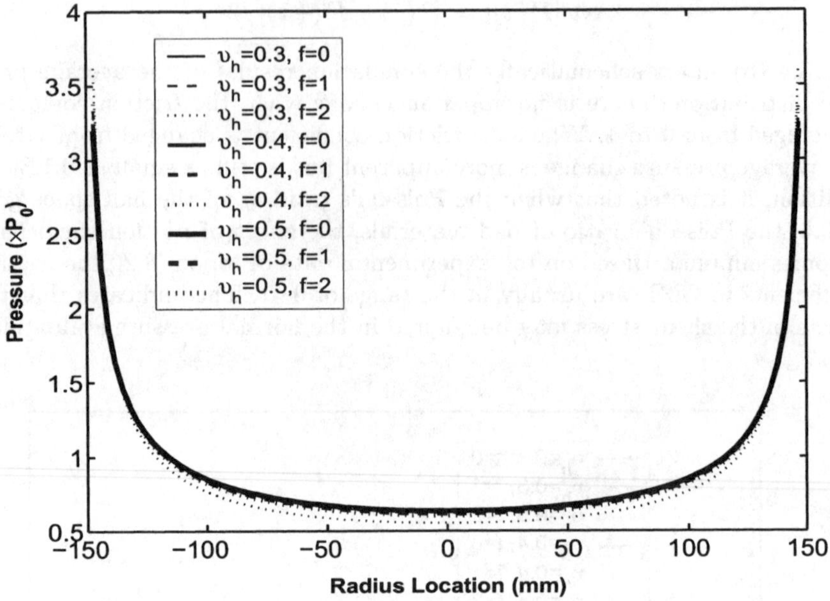

Fig. 8.4.b. Average pressure across the wafer under different friction coefficients f based on rigid flat punch model

Boundary Element Model

The above flat rigid punch model does not consider the elastic deformation of the wafer. The geometry parameters such as the wafer thickness and pad thickness are not included. Therefore, it cannot capture some of the important features of the pressure profile over the interface, for example, the complicated variance of the pressure distribution at the wafer edge. Moreover, this model cannot accommodate flexibility on a polishing head design. A numerical model is therefore preferred. Instead of the conventional finite element model, a two dimensional boundary element model is chosen in this work because of its lower requirements on computer memory and faster computation speed in thin structure analysis [8.4]. Another advantage of a boundary element model is its minimum remeshing requirement for geometry optimization.

The boundary element model is shown schematically in Fig. 8.5 and is the cross section along A-B shown in Fig. 8.1. The wafer is pressed over the pad by a pressure P_0 distributed uniformly over the wafer top surface. Here, the wafer diameter is 300mm. The wafer thickness 1mm. The size of the platen/pad is set as 900mm. This parameter is selected for the rotation type machine. For a linear type of machine, it may be smaller. As discussed in Chapter 6, at the die and wafer scale, the whole pad substrate thickness should be considered in modeling wafer-pad contact. It is set as 2mm. Realizing from the punch model that the pressure increases rapidly at the wafer edge, a retaining ring can be put around the wafer to adjust the pressure distribution. The ring size is fixed as 10mm. A gap exists between the wafer and the retaining ring. In our simulation, it ranges from 0 to 2mm. The pressure over the retaining ring is adjusted from 0 to $4P_0$. The ring is assumed to be thin in the current model and therefore the pressure over the ring can be applied directly over the pad surface. The wafer and pad Young's modulus E_p are set as 100GPa and 1~ 5.2MPa (see Table 6.1 of commercial pad Young's modulus), respectively, which are close to those seen in practice. The Poisson's ratio υ_p is set as 0.5 for polymer materials.

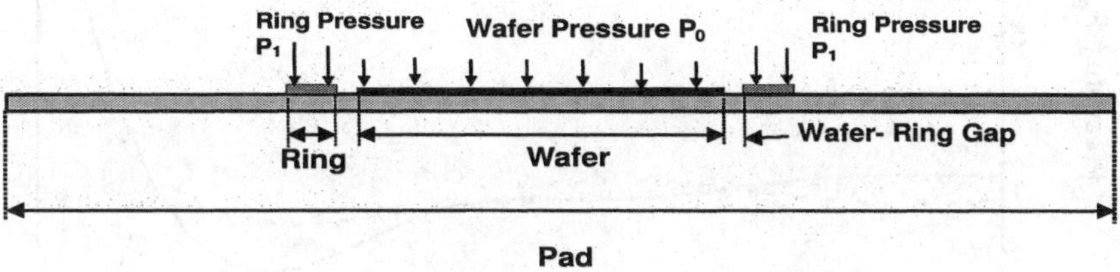

Fig. 8.5. Boundary element model for pressure distribution analysis

The friction force over the wafer-pad interface is assumed to be zero. This zero friction assumption is based on two considerations. First, from the rigid flat punch model, the friction coefficient does not have an apparent influence on the average normal pressure distribution. Therefore, a zero friction assumption should not cause much change of the results. Second, the contact model with non-zero friction coefficient is very complicated and cannot be implemented by the current boundary element code yet.

Fig. 8.6 shows the simulation results of pressure distribution at the wafer edge under a zero ring-wafer gap and different ring pressures. The simulation result without a ring (ring pressure is zero) correlates qualitatively well with the experimental material removal non-uniformity shown in Fig. 8.7 (a) [8.5]. In the experiment, no retaining ring is used. The platen/pad rotation speed is close to the head/wafer rotation speed. Therefore, the velocity non-uniformity is minimal and the material removal non-uniformity can be attributed to the pressure non-uniformity. The correlation of the simulation and experimental results implies that solid-solid contact is the dominant contact mode under the experimental conditions. From Figs. 8.6 and 8.7, it is seen that while the pressure distribution on the center part of the wafer is quite uniform, large variation is observed over the edge: decreased first and then increased rapidly at the very edge. This feature of pressure variation over the edge cannot be captured by the earlier rigid punch model.

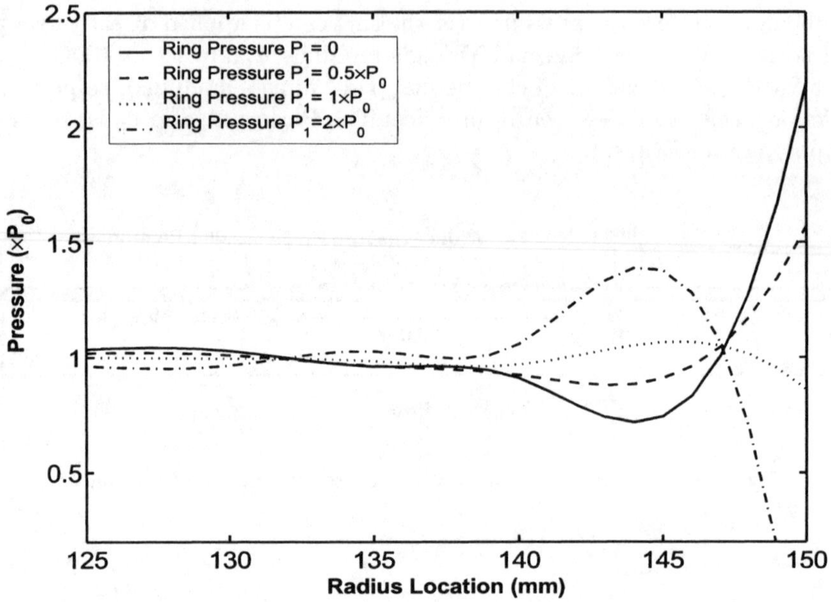

Fig. 8.6. Pressure distribution under different ring pressures when the gap is zero

To remove the edge effects, it is natural to add a retaining ring around the wafer. This ring can be considered as an extension of the wafer and therefore the edge effects may be moved from the wafer to the ring. Based on the simulation results, for the case of the zero ring-wafer gap, when the ring pressure is the same as the pressure over the wafer top surface, the non-uniformity is the smallest, Fig. 8.6. A retaining ring pressure smaller than the wafer pressure P_0 is not sufficient to drop the edge pressure. However, a ring pressure larger than the wafer pressure, say, $2\times P_0$ will drop the edge pressure too much, leading to an edge slow process. The simulation results here imply that an optimal ring pressure must be applied to obtain a uniform edge profile. Fig. 8.7. (b) shows experimental results of *MRR* non-uniformity at the edge of the wafer under different ring pressures [8.6]. The change of

MRR non-uniformity with ring pressures correlates with the simulation results qualitatively well.

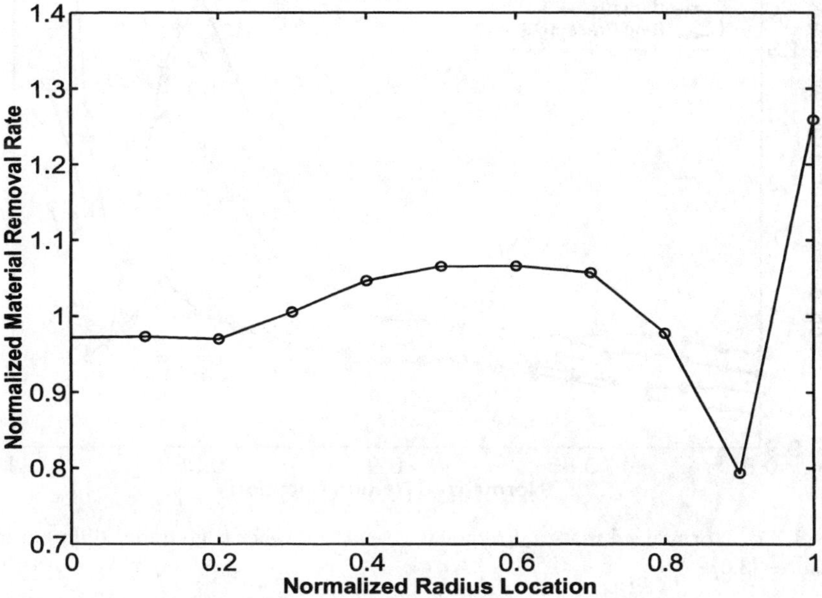

Fig. 8.7.a. Normalized material removal rate across the wafer [8.5]

The zero gap between the ring and wafer is an ideal case. Practically, the gap can never be zero considering the limitation of the manufacturing capability. Therefore, it is needed to simulate the effects of this gap on the pressure profile. Fig. 8.8 shows the simulation results for two gaps. When the gap is 1mm, the optimal ring pressure is $2P_0$. When the gap is 2mm, the optimal ring pressure is $4P_0$. A larger gap needs a larger ring pressure to compensate the edge effects. However, the process condition and machine design may require that this pressure is not too large in comparison with the wafer pressure. Therefore, a gap between 0 and 1mm may be optimal. Based on the same reason, the edge effects under a smaller down pressure will be easier to be compensated using the retaining ring.

In summary, the simulation results here indicate that adding a retaining ring is an effective method to reduce the edge pressure variance. The gap size and ring pressure can be adjusted simultaneously to obtain an optimal pressure

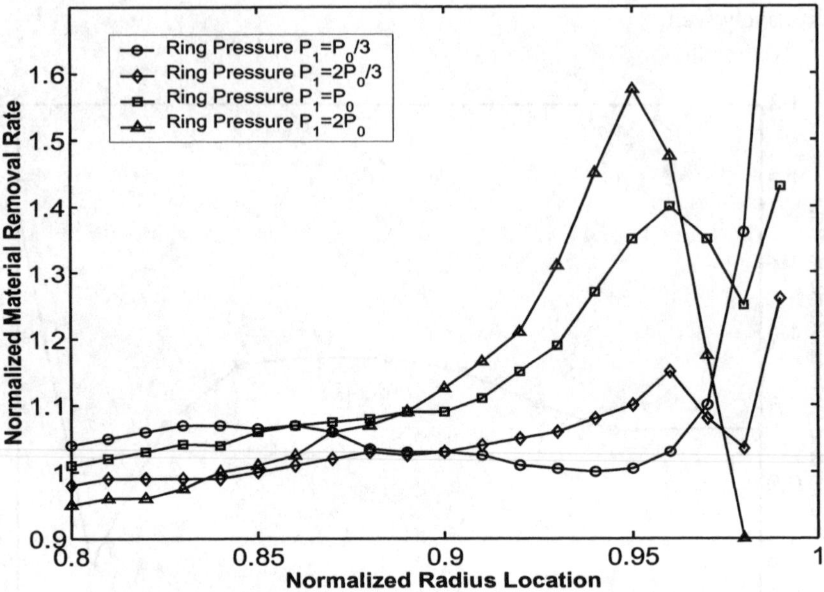

Fig. 8.7.b. Normalized material removal rate at the wafer edge under different ring pressures [8.6]

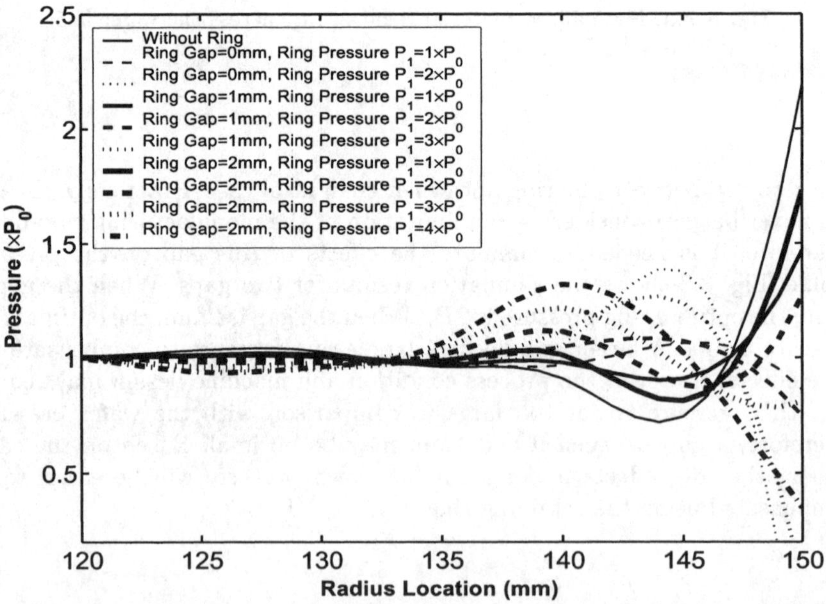

Fig. 8.8. The effects of the retaining ring location and retaining ring pressure on the edge pressure profile

profile. A similar model can be implemented using BEM to see the effects of ring width, ring height and ring materials on the edge pressure improvements. They are not included here.

At the end of this section, let us see the effects of pad material properties on pressure distribution. Fig. 8.9 shows that in the range of commercial pad Young's modulus 1~5MPa, the pressure distribution is almost independent of pad materials. It is often observed that a hard pad yields a smaller non-uniformity. In the later sections, we will show that instead of a smaller pressure distribution variance, this is due to the smaller sensitivity of material removal on the pressure distribution for a harder pad.

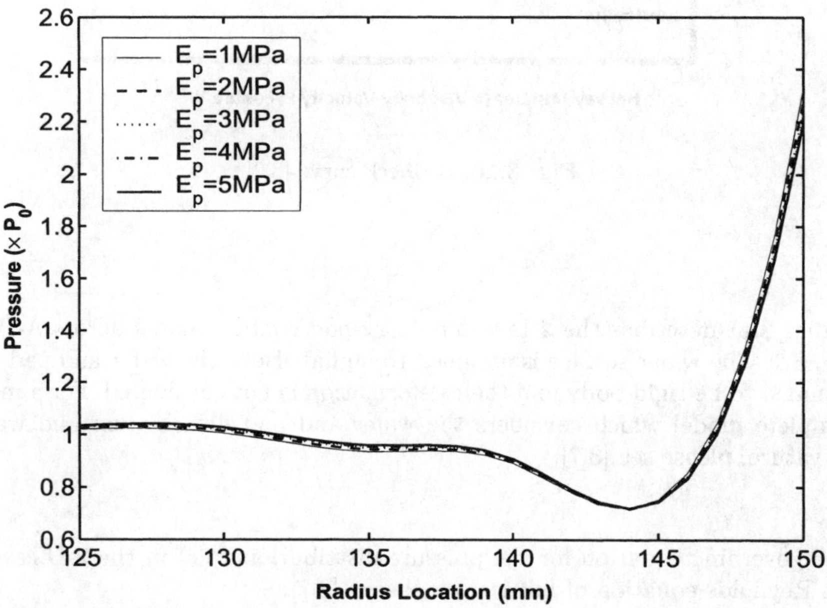

Fig. 8.9. Pressure distribution as a function of pad Young's modulus E_p

8.3.2 Solid-Fluid-Solid Contact

In some process conditions, say, when the velocity V of wafer is large, down pressure P_0 is small or the slurry viscosity μ is large, a fluid film may be formed over the wafer-pad interface; see the hydrodynamic region in Fig. 8.10 of the Stribeck Curve [8.2]. The existence of a fluid film is undesirable in terms of planarization ability. However, when low-k materials are polished, the fluid film may be preferred to avoid delamination due to aggressive contact between wafer and pad in the solid-solid contact mode. Therefore, in this section, a two dimensional (2-D) model is developed to evaluate the pressure distribution in the case of the solid-fluid-solid contact.

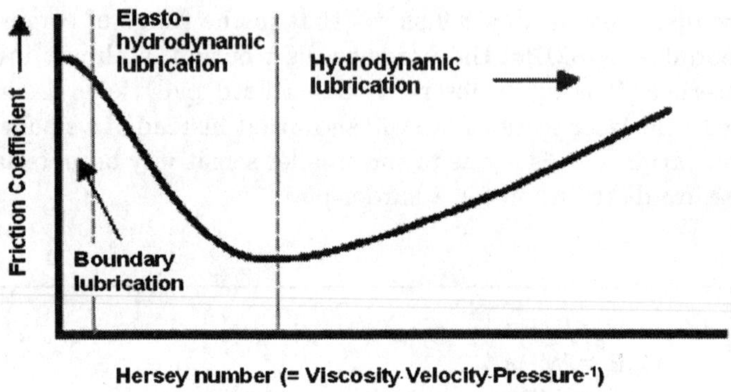

Fig. 8.10. Stribeck curve [8.2]

Figure 8.11 describes the 2-D wafer-slurry-pad contact model across A-B in Fig. 8.1. The wafer surface is assumed to be flat. Both the wafer and pad are assumed to be rigid body and their deformation is not considered. For a more complete model which considers the wafer and pad deformation and wafer curvature, please see [8.7].

The governing equation for the pressure distribution $P(x)$ in the 2-D case is the Reynolds equation of lubrication theory [8.7]:

$$\frac{dh^3(x)\frac{dP(x)}{dx}}{dx} = 6\mu V \frac{dh(x)}{dx} + Q(x) \tag{8.10}$$

where V is the translation component V_T of the wafer-pad relative velocity, μ the slurry viscosity, $P(x)$ the pressure and $h(x)$ is the thickness of the slurry film at the location x and $Q(x)$ is the rate of the slurry flow from the bottom of the polishing pad. For a conventional machine, Q is zero. In a novel orbital CMP configuration by Novellus System, the slurry is distributed from the bottom of the pad considering that non-uniform slurry distribution may be one of the contributors of the WIWNU. In this book, $Q(x)$ is assumed to be uniform across the wafer-pad interface. It is shown later that the magnitude of the uniform Q does not affect the pressure distribution.

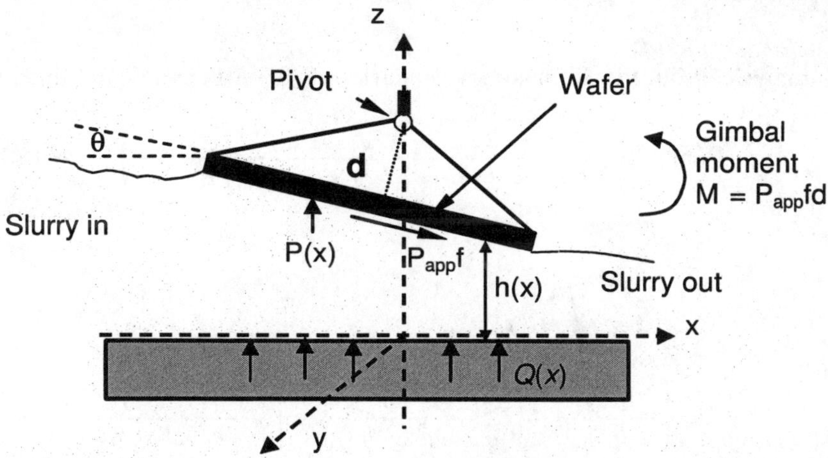

Fig. 8.11. Two-dimensional model of wafer-slurry-pad contact in hydrodynamic lubrication mode

Without considering the wafer and pad deformation, the slurry film thickness can be written as

$$h(x) = h_0 + \theta x \tag{8.11}$$

where h_0 is the slurry film thickness at the center of the wafer, and θ is the slope of the wafer attack angle. The boundary conditions to be satisfied are

$$P(-R_w) = P(R_w) = 0 \tag{8.12}$$

$$\frac{1}{2R_w} \int_{-R_w}^{R_w} P(x)dx = P_{app} \tag{8.13}$$

and

$$\int_{-R_w}^{R_w} P(x)xdx = M = P_{app}fd \tag{8.14}$$

where Eq. 8.12 indicates atmospheric pressure at the inlet and outlet, Eq. 8.13 is from the force equilibrium condition with the down force $P_{app} = P_0 \times 2R_w$ the force supported by the film, and Eq. 8.14 is from the moment equilibrium condition with f the friction coefficient over the wafer-slurry interface, d the distance from wafer surface to the pivot point and $M = P_{app}fd$ the moment around the pivot generated by the friction force in the $x-z$ plane. The friction coefficient f is a function of slurry viscosity μ times velocity V divided by pressure P as shown in Fig. 8.10 of the Stribeck Curve (referred to the Hersey number).

The analytical solution for the above equations 8.10~8.14 can be obtained as:

$$P(x) = C_w \left\{ \frac{2}{1+g(\theta,h_0)\,x} + \frac{g^2(\theta,h_0)\,R_w^2 - 1}{[1+g(\theta,h_0)\,x]^2} - 1 \right\} \tag{8.15}$$

and

$$\begin{aligned} M = P_{app} f d &= \int_{-R_w}^{R_w} P(x) x dx \\ &= C_w \left\{ \frac{6R_w}{g(\theta,h_0)} + \left[R_w^2 - \frac{3}{g^2(\theta,h_0)} \right] \ln \left[\frac{1+g(\theta,h_0)\,R_w}{1-g(\theta,h_0)\,R_w} \right] \right\} \end{aligned} \tag{8.16}$$

where

$$C_w = \frac{P_{app} g(\theta,h_0)}{2\ln\left[\frac{1+g(\theta,h_0)R_w}{1-g(\theta,h_0)R_w}\right] - 4R_w g(\theta,h_0)} \tag{8.17}$$

and $g(\theta,h_0) = \theta/h_0$ a function of the unknown values of the attack angle θ and film thickness h_0. It is difficult to obtain $P(x)$ as a function of M analytically. One way is to solve equation Eq. 8.16 to obtain $g(\theta,\ h_0)$ as a function of M. Then substitute $g(\theta,\ h_0)$ into Eq. 8.15 to evaluate the $P(x)$ as a function of M. However, the nonlinear Eq. 8.16 cannot be solved analytically. The numerical solution can be attempted but the convergence is a concern. An alternative way is to vary $g(\theta,\ h_0)$ in a reasonable range. Substituting $g(\theta,\ h_0)$ into Eqs. 8.15 and 8.16 respectively yields a moment M and a corresponding pressure distribution $P(x)$. This obtained $P(x)$ is then the pressure distribution under the obtained moment M. If the machine design parameter d, which is usually in the range of 20mm-30mm, is known, the relationship between friction coefficient f and the pressure distribution is also obtained. This method is much easier and is used in this work. The disadvantage is that film thickness parameters θ and h_0 cannot be obtained explicitly.

From the above equations, it is also seen that when the Q is uniform, the pressure distribution is independent of film flow rate Q. For an arbitrary function $Q(x)$, it may affect the pressure distribution, and numerical methods can be attempted to obtain the M-$P(x)$ relationship. From Eqs. 8.15 and 8.17, the film viscosity μ and the relative velocity V of the wafer do not contribute directly to the pressure distribution. However, from the lubrication theory, the friction coefficient f is a function of the film viscosity μ and the relative velocity V, as shown by the Stribeck curve in the hydrodynamic lubrication region of Fig. 8.10. Therefore, their effects on the pressure distribution are reflected through the friction coefficient f or the moment M.

Figure 8.12 shows simulation results of the pressure distribution under different moments M or friction coefficients f for a 300mm wafer. The number before the term $(30/d)$ is the friction coefficient for a machine with parameter d= 30mm. For a machine with an arbitrary design parameter d, the corresponding coefficients should be times $30/d$. It is seen that when the friction coefficient is small, the pressure profile is symmetric. With the increase of the friction coefficient, the maximum pressure moves from the center to the edge of the wafer and the pressure profile becomes more and more asymmetric. The asymmetric distribution of the pressure is to balance the anti-clockwise moment M in the $x - z$ plane.

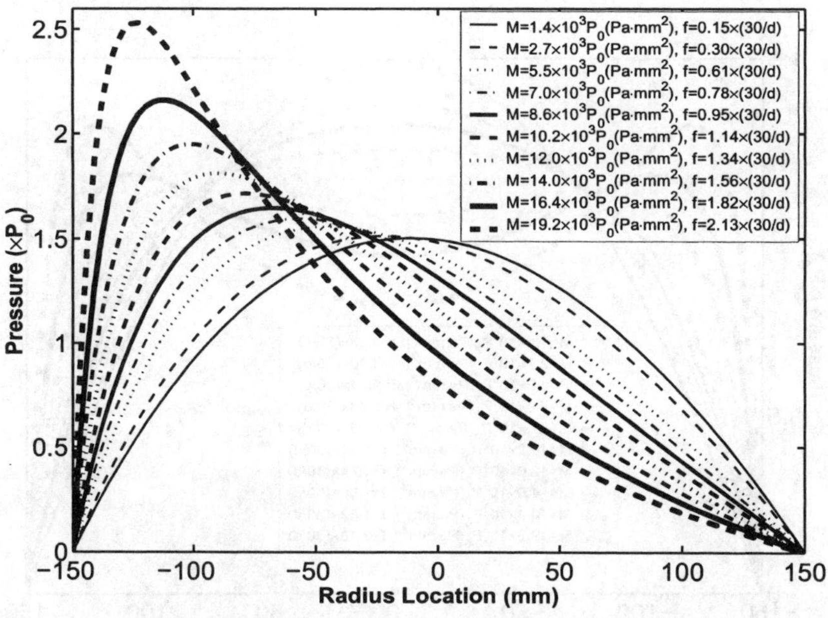

Fig. 8.12. Asymmetric pressure distribution under different moments M or friction coefficients f for a 300mm wafer

The rotation of the wafer around its center in the $x - y$ plane will average out the asymmetric distribution of the pressure. Fig. 8.13 shows the simulation results of the average pressure for a 300mm wafer. It is seen that when the moment M or the friction coefficient f is small, the pressure distribution is like a bow, center higher and edge lower. With the increase of the moment, the pressure distribution profile becomes smoother in the center, with the maximum pressure moving to the edge of the wafer. When the moment is larger than a certain value, a reverse bow will be formed at the center. In terms of smaller non-uniformities, M in the range of 10.2~16.4×10^3 P_0(Pa· mm^2) is preferred. This corresponds to friction coefficients f in the range of 1.14~1.82 for a d of 30mm and 1.14~1.82x(30/d) for an arbitrary d. Similar simulations are done for 200mm (8 inch) and 150mm (6 inch) wafers. The pressure profiles are shown in Figs. 8.14 and 8.15 respectively. Suggested ranges of friction coefficients are given as 0.82~1.21x(30/d) and 0.57~0.91x(30/d). For a smaller wafer size, the optimal friction coefficients are smaller. The above results yield a preferred range of friction coefficients for a machine design parameter d in the range of 10~85mm as shown in Fig. 8.16. Combining Fig. 8.16 with Fig. 8.10 of the Stribeck curve of friction coefficients can yield the optimal process window of down pressure P_0 and relative velocity V.

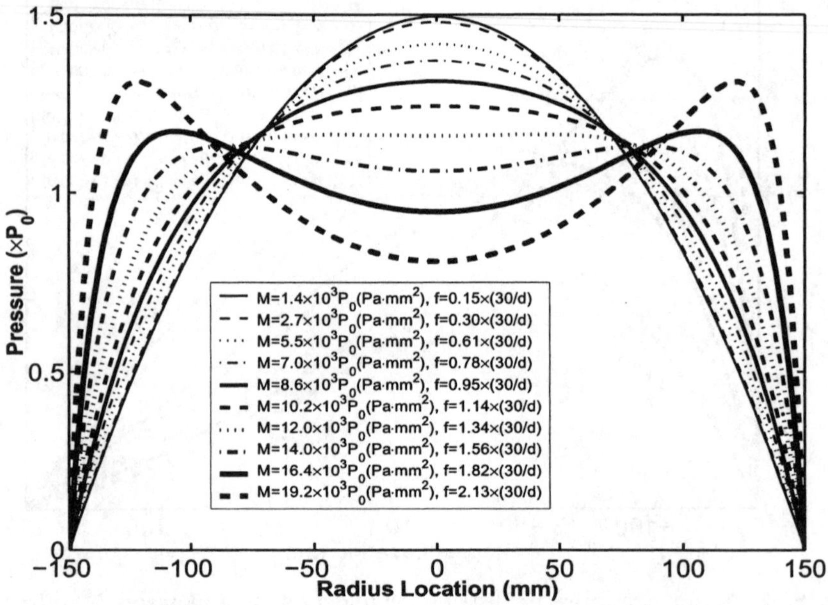

Fig. 8.13. Average pressure distribution under different moments M or friction coefficients f for a 300mm (12 inch) wafer

There are two usages of Fig. 8.16. One is as mentioned above. When the machine design parameter d is known, the process parameters, including the pressure P_0, velocity V and consumable parameters, such as slurry viscosity μ, can be adjusted so that the friction coefficient f falls in the desired region. When the process is fixed, (For example, the P_0, V and μ have been optimized

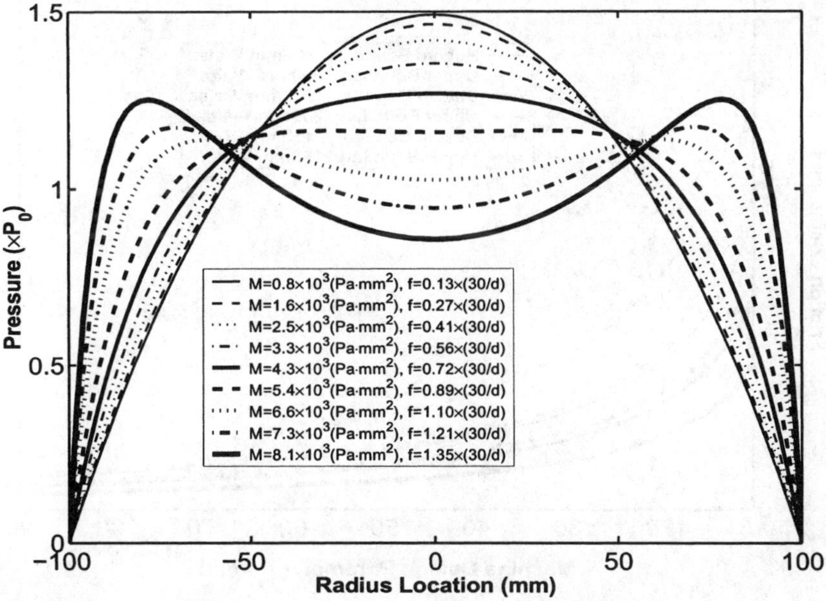

Fig. 8.14. Average pressure distribution under different moments M or friction coefficients f for a 200mm (8 inch) wafer

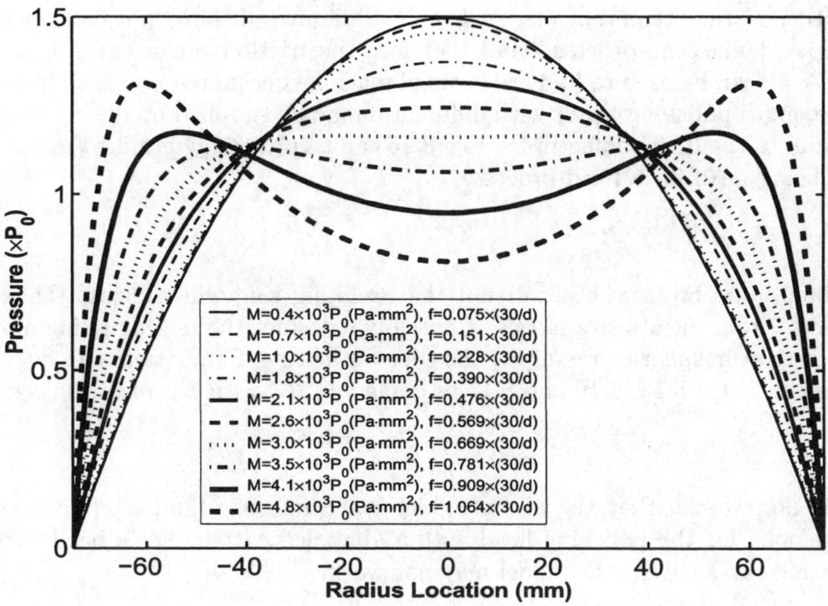

Fig. 8.15. Average pressure distribution under different moments M or friction coefficients f for a 150mm (6 inch) wafer

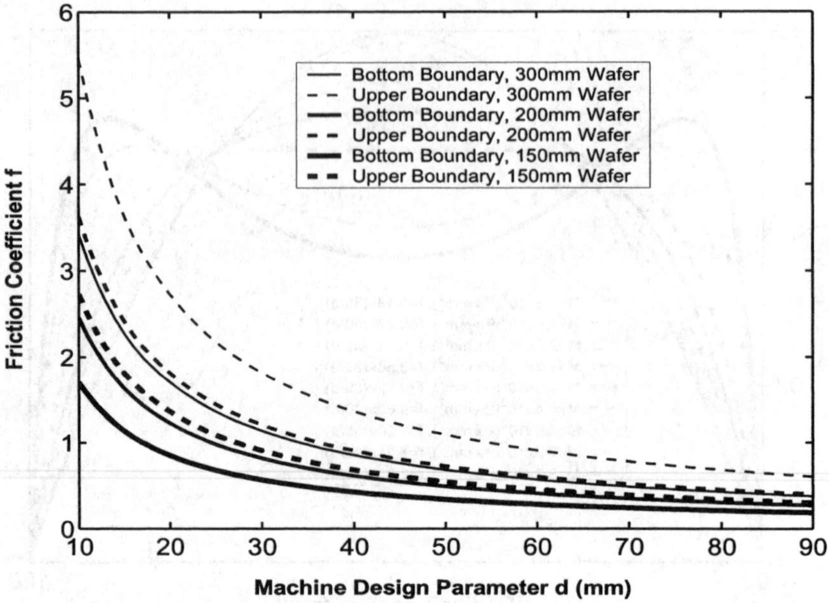

Fig. 8.16. Suggested range of friction coefficients f for different machine design parameter d

based on other requirements, such as to minimize dishing and erosion in a copper damascene process.) and therefore the friction coefficient f is fixed, Fig. 8.16 can be used to find the optimal machine design parameter d. In some case, multi polishing heads with different d may be supplied by the equipment vendor to the IC manufacturers. Fig. 8.16 can help them determine the correct polishing head for a fixed process.

A final question is whether or not the retaining ring will alleviate the edge effects. From the above analysis, it possibly will since the retaining ring moves the zero atmospheric pressure to the edge of the ring. Simulations can be done using the Eqs. 8.15, 8.16 and 8.17 by replacing R_w with R_w plus ring size.

It is noteworthy that the above model based on solid-fluid-solid contact is developed for the polishing head with a gimbaled carrier. For a head with a floating carrier [8.8], the model may not work.

8.3.3 Air Pressure Distribution over the Platen-Belt Interface in the Linear Type Machine

In the linear type of machines, air is distributed from the platen below the belt, Fig. 8.2. The idea is similar to that of distributing slurry from the bottom of the pad. By controlling the flow rate of the airs in different zones below the belt, the pressure distribution across the wafer might be controlled.

Therefore, there are two interfaces in a linear type configuration; one is the wafer-belt interface, and the other, the belt-platen interface. An interesting point is that the pressure model developed based on solid-fluid-solid contact may be applicable to the air pressure distribution prediction. If assuming that the air flow obeys Reynold's equation, the formulations developed in the last section can be used to evaluate the air pressure distribution over the belt-platen interface, with platen size taking the place of wafer size. The air pressure should be asymmetric to balance the moment M generated by the friction force over wafer-belt interface.

8.4 Optimization of WIWNU from the Viewpoint of Consumable Effects

In earlier sections, the optimization of WIWNU is discussed from the perspective of pressure and velocity distribution. An alternative method, based on Eq. 8.4, is to reduce the sensitivity of MRR on the pressure distribution by adjusting Preston's coefficient K_e and MRR_0. The basis of this method can be shown schematically in Fig. 8.17. Lines 1, 2, and 3 represent revised Preston's equation $MRR = K_e PV + MRR_0$ under three different consumable recipes. Suppose that the velocity is uniform, the average down pressure over the wafer surface is P_{avg}, the maximum pressure is P_{max} and the minimum pressure is P_{min} and that this pressure distribution does not change with the consumable recipe. Then, apparently, using Eq. 8.4, the non-uniformity is smaller under recipes 1 and 3 than that under recipe 2.

Changing K_e and MRR_0 in Eq. 8.4 actually requires the change of C_3 and C_2 in the material removal rate Eqs. 3.22 and 5.7-a developed in Chapters 3 and 5. Substitution of them into non-uniformity Eq. 8.3 yields

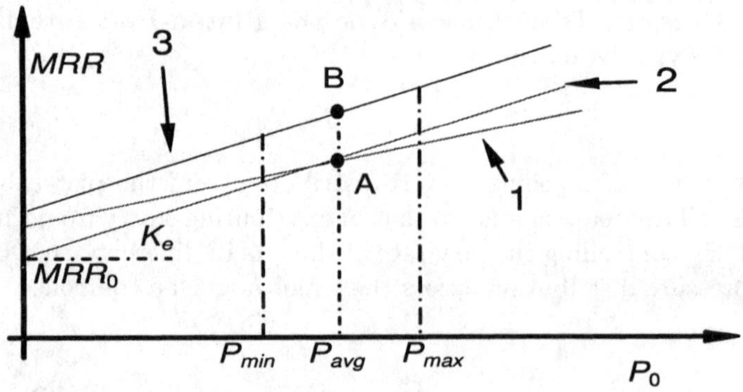

Fig. 8.17. The material removal rate equation for different consumable combinations

$$\mathrm{WIWNU} = \frac{MRR_{max} - MRR_{min}}{MRR_{avg}}$$
$$= \frac{C_3\left[1-\Phi\left(3-C_2P_{max}^{1/3}\right)\right]\sqrt{P_{max}} - C_3\left[1-\Phi\left(3-C_2P_{min}^{1/3}\right)\right]\sqrt{P_{min}}}{C_3\left[1-\Phi\left(3-C_2P_{avg}^{1/3}\right)\right]\sqrt{P_{avg}}}$$
$$= \frac{\left[1-\Phi\left(3-C_2P_{max}^{1/3}\right)\right]\sqrt{P_{max}} - \left[1-\Phi\left(3-C_2P_{min}^{1/3}\right)\right]\sqrt{P_{min}}}{\left[1-\Phi\left(3-C_2P_{avg}^{1/3}\right)\right]\sqrt{P_{avg}}} \qquad (8.18)$$

It can be seen from Eq. 8.18 that only parameters related to C_2 influence the WIWNU. These include: *pad material, pad topography* and *abrasive size distribution.* Ideally, the production rate should not be sacrificed for a WIWNU improvement. Therefore, the average material removal rate should be increased or remain unchanged after the WIWNU improvement [4]. An increase of C_3 can lead to a larger average material removal rate. However, it is noteworthy that a mathematical analysis of Eqs. 8.18 and Eq. 3.22 in Chapter 3 reveals that if MRR_{avg} remains unchanged, the minimum WIWNU that can be realized by changing C_2 (or optimizing consumable parameters) is

$$0.25\left(P_{max} - P_{min}\right)/P_{avg} \times 100\% \qquad (8.19)$$

When increasing MRR_{avg}, the minimum WIWNU that can be realized is even larger.

[4] This means a change of consumable recipe from 3 to 1 in Fig. 8.17 is undesirable even with a WIWNU improvement.

Based on the above discussions, a three-step process can be used to improve WIWNU:

i. Optimize the consumable parameters including pad topography, pad material and abrasive size until the WIWNU requirement is satisfied. Usually, a smaller C_2 yields a smaller WIWNU.
ii. Check the average material removal rate. If it is not smaller than that before the adjustment, the optimization is successful.
iii. If the average removal rate after the adjustment becomes smaller, increase the density of slurry, dilution ratio, slurry abrasive weight concentration, slurry oxidizer concentration, relative velocity or other parameters in C_3, that should not negatively influence the already improved WIWNU, to increase the *MRR*.

Increasing the density of slurry, dilution ratio of slurry, slurry abrasive concentration, and slurry chemical concentration in step 3 above, however, may increase costs. This is generally undesirable from a cost of ownership (CoO) perspective. The pad topography, pad material and abrasive size distribution influence the WIWNU (C_2) and *MRR* simultaneously. Therefore, by selecting proper pad topography and abrasive size distribution parameters, an optimization may be realized without step 3 and this would lower costs. The effects of pad topography, pad materials and abrasive size distribution on both *MRR* and WIWNU also imply that it is critical to keep them constant for a reproducible and controllable CMP process.

The following example shows the sensitivity of the WIWNU on the consumable parameter C_2. It is often observed that the WIWNU decreases when using a harder pad. This WIWNU difference may not be attributed to the pressure distribution since the BEM simulation in section 8.3.1 (Fig. 8.9) shows that it is almost independent of the pad Young's modulus. Instead, it may be attributed to the different values of C_2. From Chapter 5, it is known that C_2 is 0.5 for IC1000 perforated pad, which has a Young's modulus close to 5MPa; see Table 6.1. The pad hardness H_p usually increases with the Young's modulus E_p and we can assume that there is an approximately linear relationship between pad Young's modulus and pad hardness. From the formulation in Chapter 3, the values of

$$C_2 = \frac{0.25 \times (4/3)^{2/3}(x_{avg} + 3\sigma)\,(1/H_p + 1/H_w)\,E_p^{2/3}}{\sigma b_1} \tag{8.20}$$

where b_1 is a roughness parameter, x_{avg} and σ the average size and standard deviation of the abrasive size distribution, and H_w the wafer hardness.

Considering that pad hardness H_p is much smaller than wafer hardness H_w and assuming the roughness parameter is the same for all pads, C_2 will be inversely proportional to $E_p^{1/3}$, i.e., $0.5\times(5/4)^{1/3} = 0.539, 0.5\times(5/3)^{1/3} = 0.593$, $0.5\times(5/2)^{1/3} = 0.679$ and $0.5\times(5/1)^{1/3} = 0.855$ for the pad Young's modulus 4, 3, 2 and 1MPa, respectively. In addition, C_3 is proportional to E_p; see Eq. 3.22. Substituting the pressure distribution data in Fig. 8.9, values of E_p (or C_3) and values of C_2 into the MRR formulation 3.22 yields the normalized MRR distribution as shown in Fig. 8.18. It can be seen that with the increase of pad Young's modulus, the material removal slightly increases. This reflects the competition between the material removal by each single abrasive and the portion of active abrasives [5]. The with-in wafer non-uniformity of pressure distribution [6] and material removal rate distribution after 5mm wafer edge exclusion are shown in Fig. 8.19 [7] schematically. It can be seen that the material removal rate non-uniformity decreases with the pad Young's modulus. while the pressure distribution non-uniformity remains the same for the five pads. From the above discussion, the apparent decrease in the WIWNU, might, therefore, be attributed the smaller sensitivity of material removal on the pressure distribution, or smaller consumable parameter C_2.

Note that the discussion in this section is applicable to the solid-solid contact mode only since the particle-scale model developed is based on solid-solid contact between wafer and pad.

8.5 Feature-Die-Wafer Scale Integration

With the wafer-scale pressure distribution known, the die and feature-scale outputs such as dishing, erosion and step height reduction at any wafer locations can be evaluated by inputting the pressure into the feature-scale model developed in Chapter 6. Figs. 8.20 and 8.21 show the simulation results of the

[5] When the pad hardness is larger, the portion of active abrasives is smaller. However, a larger Young's modulus yields a larger contact pressure and therefore larger material removal by a single abrasive. The final effect is a slight increase of material removal with Young's modulus.

[6] It is calculated by replacing MRR in Eq. 8.1 with pressure.

[7] Note that the across wafer pressure distribution is evaluated based on the linear elasticity assumption. In this case, the distribution does not change with the down pressure. The WIWNU as a function of the down pressure in Fig. 8.19, i.e., WIWNU increases with the down pressure, is obtained based on this assumption. When we assume the pad is a non-linear elastic material, the pressure distribution might be more uniform under higher down pressures. This might lead to a smaller WIWNU for higher down pressures.

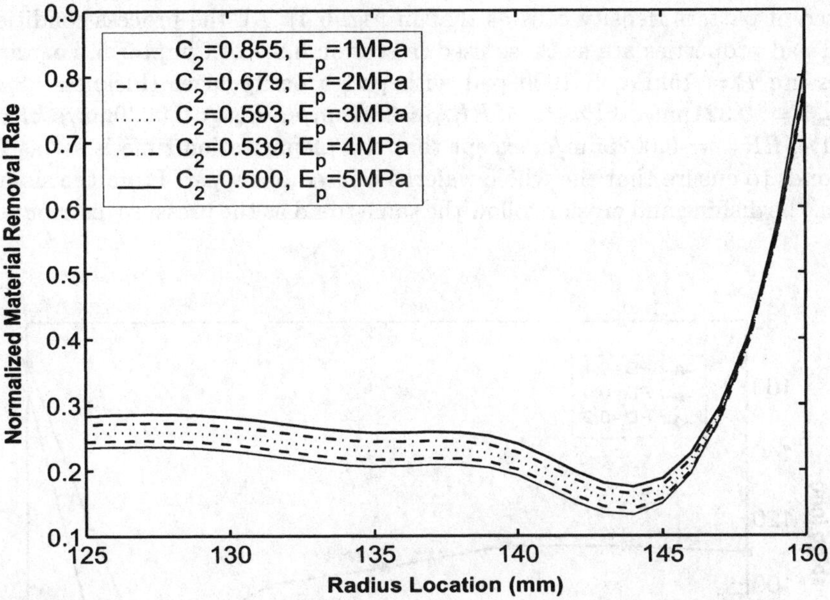

Fig. 8.18. Material removal rate distribution for five pads with different Young's modulus

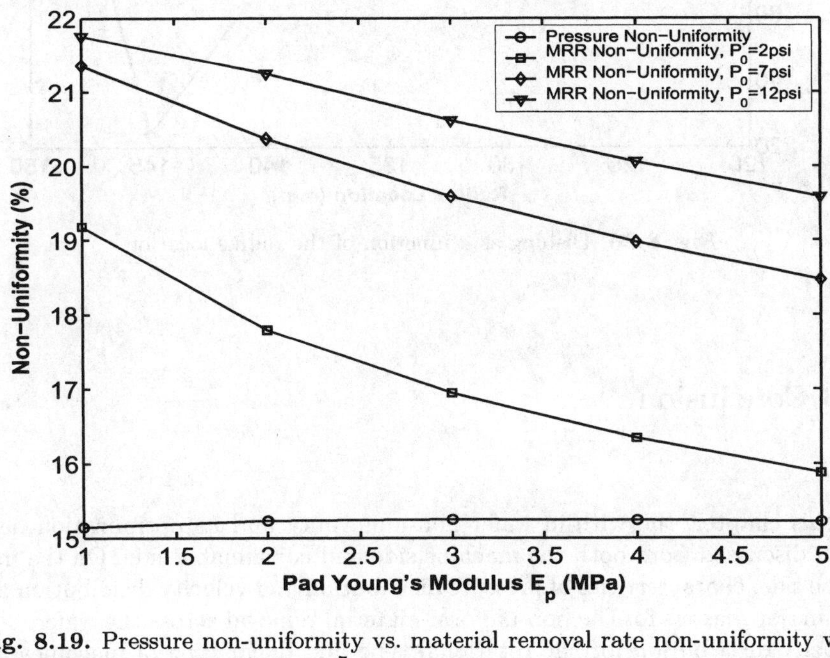

Fig. 8.19. Pressure non-uniformity vs. material removal rate non-uniformity with increase of pad Young's modulus [7]

dishing and erosion at the edge of a 300mm wafer (assuming the die is composed of pattern density cells as that in Fig. 6.4). All the process conditions and pad properties are as those used in section 6.8.7 of Chapter 6, i.e., down pressure P_0= 15kPa, IC1000 pad with pad asperity layer thickness 58μm, K_{ecu} = 0.321nm/s$\cdot kPa^{-1}$, MRR_{cu0}=1.01nm/s, K_{eox}=0.0029nm/s$\cdot kPa^{-1}$ and MRR_{ox0}=-0.0025nm/s, except that the planarization time is set as 220 seconds to ensure that the whole wafer is cleared of copper. From the simulation, the dishing and erosion follow the same trend as the pressure distribution.

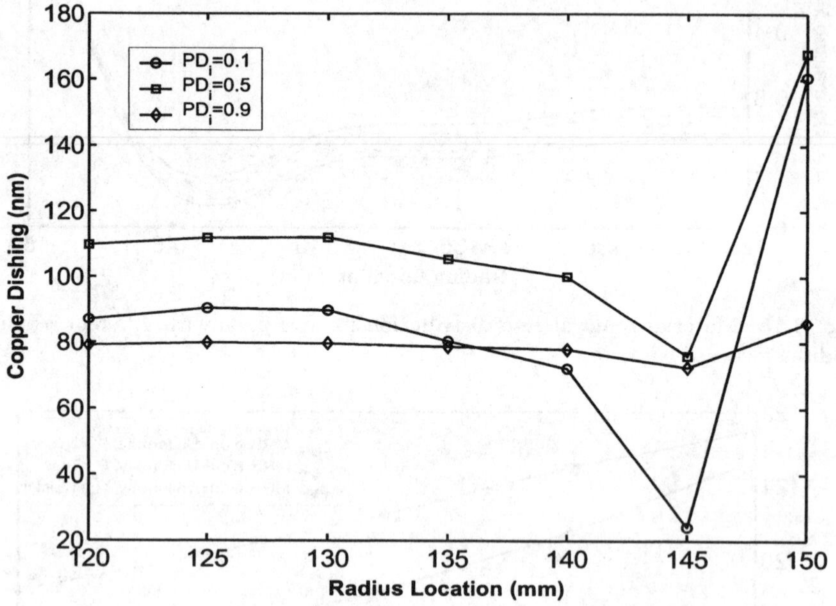

Fig. 8.20. Dishing as a function of the radius location

8.6 Conclusion

In this chapter, the with-in wafer non-uniformity and its optimization have been discussed from both the machine side and consumable side. On the machine side, characteristics of pressure distribution and velocity distribution are the major reasons for the non-uniform material removal across the wafer. The velocity distribution for the rotational type and linear type of machines are analyzed. The rotational part of the velocity is the contributor to the velocity

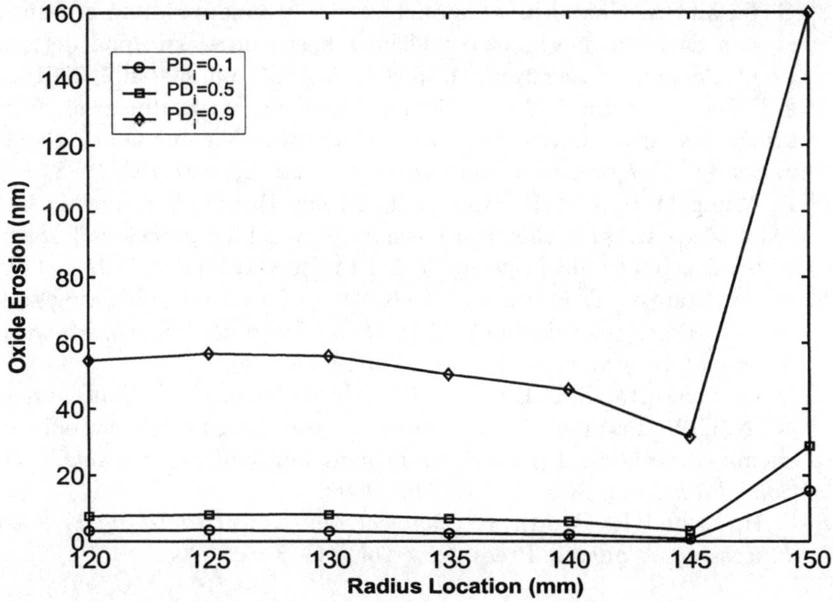

Fig. 8.21. Erosion as a function of the radius location

non-uniformity. The pressure non-uniformity in cases of both solid-solid contact and solid-fluid-solid contact are discussed. Adding a retaining ring around the wafer carrier is an effective way to reduce the pressure non-uniformity at the edge of the wafer in the solid-solid contact mode. In the solid-fluid-solid contact mode, the friction coefficient is identified as the major contributor to the non-uniformity. A process window of pressure and velocity in terms of small non-uniformity can be obtained by combining the model proposed in this chapter and the Stribeck curve of the friction coefficient as a function of polishing parameter. On the consumable side, a systematic method can be applied to optimize the WIWNU in solid-solid contact mode based on the material removal model developed in previous chapters.

8.7 References

[8.1] K. L. Johnson, *Contact Mechanics*, Cambridge University Press, Cambridge, U. K., 1985.

[8.2] Y. Moon, "Mechanical aspects of the material removal mechanism in chemical mechanical polishing (CMP)," *Ph.D. Dissertation, Department of Mechanical Engineering, University of California at Berkeley*, Berkeley, CA, U. S. A., 1999.

[8.3] G. Fu and A. Chandra, "A model for wafer scale variation of removal rate in chemical mechanical polishing based on elastic pad deformation," *Journal of Electronic Materials*, Vol. **30**, pp. 400-407, 2001.

[8.4] J. F. Luo, Y. J. Liu and E. J. Berger, "Analysis of two-dimensional thin structures (from micro- to nano-scales) using the boundary element method," *Computational Mechanics*, Vol. **22**, pp. 404-412, 1998.

[8.5] D. Wang D, J. Lee, K. Holland,T. Bibby, Beaudoin S and T. Cale, "Von Mises stress in chemical-mechanical polishing processes," *Journal of the Electrochemical Society*, Vol. **144**, pp. 1121-1127, 1997.

[8.6] M. M. Touzov, T. Fujita and T. K. Doy, "Novel retaining ring to reduce CMP edge exclusion," *2001 IEEE International Semiconductor Manufacturing Symposium*, pp. 337-340, Oct. 2001.

[8.7] D. G. Thakurta, C. L. Borst, D. W. Schwendeman, R. J. Gutmann and W. N. Gill, "Pad porosity, compressibility and slurry delivery effects in chemical-mechanical planarization: modeling and experiments," *Thin Solid Films*, Vol. **366**, pp. 181-190, 2000.

[8.8] S. H. Li and R. O. Miller, *Chemical Mechanical Polishing in Silicon Processing*, Academic Press, New York, U. S. A., 2000.

9

SUMMARY AND FUTURE WORK

9.1 Summary

Chemical mechanical planarization models at three scales, namely, particle scale, feature and die scales and wafer scale, are presented in this work.

9.1.1 Particle-Scale Modeling

The consumable parameters, from the slurry chemicals, including the oxidizer concentration, slurry pH value and slurry buffering agents, to the nano-scale slurry abrasives, including the abrasive materials, size distribution, and the pad parameters, including the pad materials and topography, play important roles in the material removal in CMP. One of the key contributions of this work is to propose a comprehensive particle-scale model to identify and explain the roles and interactions of consumable parameters including slurry abrasives, slurry chemicals, polishing pad and wafer materials during the material removal process. The critical parameters identified include:

i. **Polishing Pad** Material Properties (including hardness and Young's modulus)
ii. **Polishing Pad** Roughness/Topography
iii. **Abrasive** Size
iv. **Abrasive** Size Distribution
v. **Abrasive** Weight Concentration
vi. **Abrasive** Shape
vii. **Slurry Chemicals** contributing to the passivation rate and the material properties of the passive layer such as the hardness.

The consumable parameters contribute to the material removal through two parameters, one, the number of *active* abrasives, and the other, the material removal by a single *active* abrasive. The interactions among the slurry abrasives, wafer, polishing pad and slurry chemicals can be connected to each other through these two parameters.

The roughness/topography and material properties of the pad are critical in the CMP material removal process. They affect the material removal from the perspectives of both the number and the material removed by a single abrasive. The number of abrasives over the wafer-pad interface is larger when the pad is rougher. The contact area between the wafer and pad is smaller when the pad is rougher and stiffer. A smaller wafer-pad contact area causes a larger contact pressure and therefore larger forces on abrasives, leading to a larger material removal by a single abrasive.

The abrasive size is another important element in the model. It also affects the material removal from the perspective of both the number of abrasives and the material removal by a single abrasive. Under the same abrasive weight concentration, when the abrasive size is larger, the number of abrasives over the wafer-pad interface is smaller, while the material removal by a single *active* abrasive is larger. The final material removal is the result of the competition between these two parameters.

Besides the average size of abrasives, the size distribution of abrasives is critical in the model. The portion of the *active* abrasives depends on this size distribution. A gap g between the wafer and pad is formed in the contact region occupied by abrasives. It is a function of the contact pressure and pad hardness. When the contact pressure is larger or the pad is softer, the gap is smaller, which captures more *active* abrasives. However, the average size of the *active* abrasives drops. Therefore, the effect of the gap and, correspondingly, that of contact pressure and pad hardness on the material removal is twofold. Besides the force applied on the single abrasives, the portion of *active* abrasives and their size is the second point where the contact pressure (pad topography, pad Young's modulus) and pad hardness come into the model.

The slurry chemicals affect the material removal primarily by changing the material properties of the wafer surface and therefore the material removed by a single abrasive. The surface layer is proposed to be a bilayer structure. Two parameters, one the passivation rate, and the other the hardness of the surface layer are used to represent the effects of slurry chemicals. When the passivation rate is higher, the softer portion is larger. Therefore, the material removal rate is larger.

An implication of the model is that the effects of the consumable parameters should be evaluated under a comprehensive framework, considering the competition among different interactions. This model, by separating the material removal into two elements, helps to locate the consumable parameters and their effects correctly. In addition, based on this model, a systematic method can be attempted for the optimization of CMP, specially, the material removal rate (*MRR*) and non-uniformity (NU). This kind of particle-scale model can help guide process development and experimentation for consumable suppliers and IC manufacturers.

9.1.2 Feature- and Die-Scale Modeling

A better understanding of the formation mechanism of the topography related issue such as the pattern density effects, dishing and erosion, is required for the IC designer to optimize the IC design at the beginning of the design stage, and for the process engineer to optimize the process systematically at the fabrication stage. The contribution of this work in this respect is to propose a step height reduction model as a function of time, pattern density and line width based on linear viscoelasticity. The dishing results are more accurate in comparison with those based on linear elasticity. This model can be applied under the framework of the MIT semi-empirical CMP model to evaluate the die-level topography evolution, as discussed in section 6.9.3 of Chapter 6.

9.1.3 Wafer-Scale Modeling

At the wafer scale, this work covers the velocity distribution and pressure distribution for both the rotational type and linear type of machines based on both solid-solid contact and solid-fluid-solid contact. The edge effects are simulated. The wafer-scale model is particularly helpful to the CMP equipment suppliers, to optimize their machine design.

9.2 Future Work

Several interesting topics that can be studied in the near future are as follows:

9.2.1 Further Development and Calibration of the Particle-Scale Model

The particle-scale models at the current stage are based on a number of assumptions. Further calibration is needed in a manufacturing environment. Tabulation of the calibrated parameters, similar to that in oxidation and chemical vapor deposition (CVD) processes, will be helpful for the process development and optimization. The research in this direction will need collaboration with major semiconductor manufacturers and foundries.

9.2.2 'Swarm' Simulation

The 'swarm' is originally a term to describe a biological group with interactive cooperative behaviors among the group members. Nowadays, the 'swarm' simulation has become a term on the modeling of 'swarm'-like behavior of a multi-particle dynamical system where the communications among the members are modeled via interaction forces. Fig. 9.1 demonstrates an example that a desired swarm-like behavior: avoiding obstacles and reaching a target, may be realized by optimizing the interaction of swarm members through a 'swarm' simulation [9.1].

The key of a 'swarm' simulation is to optimize the interactions of the swarm members so that a desired overall goal is realized [9.1]. The slurry abrasives over the wafer-pad interface can be modeled as a multiparticle 'swarm'-like dynamical system. The interaction forces among the slurry particles could be very complicated. They may not only be mechanical forces, but also chemical forces, either attractive or repulsive. In this work, we have modeled the mechanical interactions of slurry abrasive particles in solid-solid contact of the pad and wafer. The effects of chemical forces among particles and wafers with an origin in chemical parameters such as zeta potential and pH values are not considered. An optimization of the chemical interaction forces through a 'swarm' simulation similar to that in [9.1] may obtain the desired overall goal of CMP, say, the maximum number of abrasive particles (like swarms in Fig. 9.1) which impact the wafer surface (like a target in Fig. 9.1) after avoiding the pad asperities (like obstacles in Fig. 9.1), and therefore a maximum material removal rate. These optimized chemical interaction forces then can provide guidance on slurry chemical designs as well as pad topography and abrasive designs.

For more details on the computational design of the swarms, the readers are referred to [9.1]. The research in this direction can help integrate chemical effects into a more detailed particle-scale material removal model, particularly, in the solid-fluid-solid contact mode of wafer, slurry and pad, and offer more insights into chemical effects and mechanical-chemical interactions in CMP.

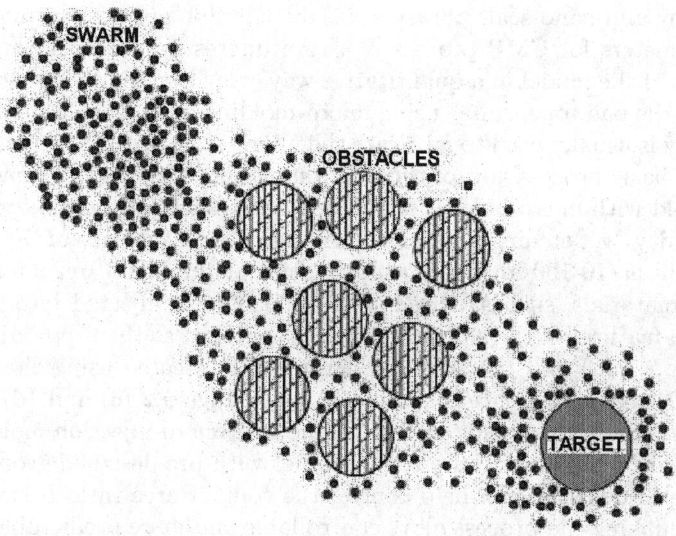

Fig. 9.1. Schematic showing desired swarm-like behavior: avoiding obstacles and reaching target [9.1]

9.2.3 Interconnect Modeling for EDA

The final goal of modeling efforts in CMP is to facilitate the development of electronic design automation (EDA) software tools in terms of design for manufacturability. With the interconnect resistors and capacitances becoming the dominant factors in IC delay for the next-generation ICs, an accurate prediction of topography of chips after copper deposition and CMP process is critical for circuit performance prediction and optimization. Dummy features have been inserted into the lower pattern density area to compensate the pattern density non-uniformity. However, the dummy features degenerate the circuit performance considering, for example, the associated parasite capacitances. The topography evolution model developed may be combined with filling algorithms to obtain the optimal dummy filling, as that discussed in section 6.9.4 of Chapter 6.

9.2.4 Systematic Design and Micro-Fabrication of CMP Consumables

Micro-Injection Molding of Polishing Pad Topography

A stable and predictable CMP process requires full control of the consumable parameters. Based on the particle-scale model, the micro-scale polishing pad topography and nano-scale abrasive size distribution are the two most important parameters for CMP process [9.2]. An interesting research topic, which makes use of the model in a quantitative way, would be the precise design and control of the pad topography using micro-molding technology. Micro-molding technology is usually used to fabricate the micro-scale features of plastic materials. The basic process flow of the micro-molding technology is shown in Fig. 9.1. A mold with micro-holes is fabricated using micro-fabrication technology including dry/wet etching and micro-drilling. The feature size of the holes can be as small as 10-100μm if the aspect ratios of the holes are not too large. Polymer materials, such as those used for pads, are injected into the mold, and micro-features of the silicon mold are replicated as the topography of the pad. Figs. 9.2 (a) and (b) show topographies fabricated using three dimensional (3D) micro-fabrication technology [9.3]. Figs 9.2 (c) and (d) show topographies prototyped using micro-drilling and micro-injection molding with a feature size of 100μm. These topographies with pre-designed geometry and asperity grid densities can help control the contact area ratio between wafer and pad, making the process more controllable and more predictable. Besides the above process using a silicon mold, a micro-replication technology using silicone rubber as mold materials [9.4] is becoming a potential method for pad topography control as well.

Abrasive Particle Size Control using Nano-Fabrication Technology

Besides the pad topography design, another topic might be the precise design and control of the abrasive size using nano-fabrication technology. Nano-milling is one of the potential candidates [9.5] for this. Using nano-milling, the abrasive size and its distribution may be controlled as function of milling temperature and intensity. The 'swarm' simulation may be used to obtain the optimal process conditions of the nano-milling as well.

9.2.5 Green CMP

Environmental concerns associated with manufacturing are becoming increasingly important in the U.S.A., Europe and Asia as evidenced by the increased

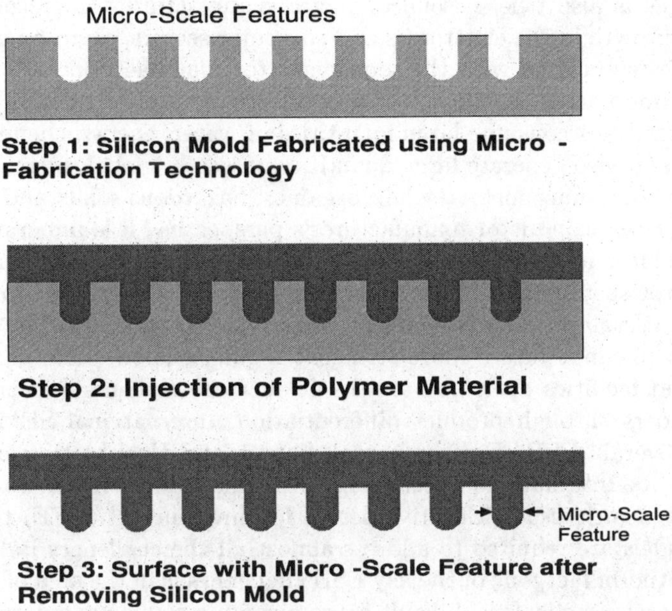

Fig. 9.2. Basic process flow of the micro-molding technology

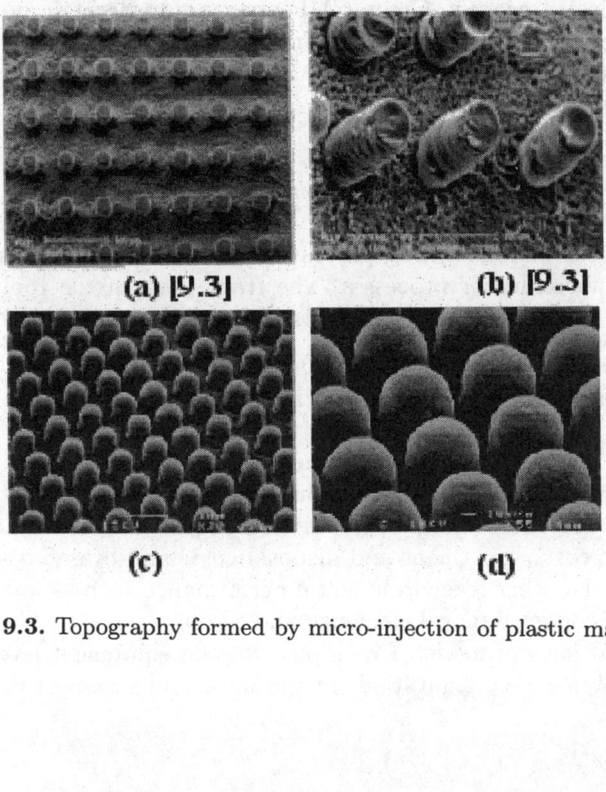

Fig. 9.3. Topography formed by micro-injection of plastic materials

attention being given to this by government and non-government organizations. This is also true in semiconductor manufacturing. In recent years, with rapid growth in the electronics and computer sectors, environmental and health issues associated with the semiconductor manufacturing industry are growing in importance. Semiconductor processes are typically highly resource intensive (facilities consume large quantities of water, energy, chemicals and materials) and also generate large quantities of waste (global warming gases, wastewater containing metals/solids/organics, hazardous solids and sludges, etc). From a semiconductor manufacturer's perspective, it is imperative that industry reduces its environmental footprint i. as part of industry's responsibility towards sustainable industrial development; ii. as a response to growing public and customer awareness of and sensitivity towards environmental issues; iii. to preempt future more stringent regulations on waste generation, disposal and facilities operation and iv. to increase revenues and emerge as market leaders through product differentiation. International SEMATECH identifies several "ESH (Environmental, Safety, and Health) Difficult Challenges", in its International Technology Roadmap [9.6] on Environmental, Safety, and Health issues. Effective design for environment (DFE) tools and process models are required to address almost all the challenges italicized in Table 9.1. And in fact, one of the key ESH challenges is in the area of ESH design and measurement. It is difficult for semiconductor manufacturers to take concrete steps to improve environmental health and safety if they are unable to measure ESH impact at the process level and feed measurements back through the design process. Green CMP attempts to develop tools to bridge this gap in ESH assessment by developing integrated ESH process models for the CMP process.

The Environmental Value Systems (EnV-S) Analysis [9.7] is a design tool developed at UC-Berkeley to evaluate cost, performance and ESH impacts of semiconductor manufacturing process sequences. It has been applied to a number of semiconductor processes. The framework of the tool for copper CMP process is shown in Fig. 9.4 and is composed of three levels of models:

i. Facility level of model. This is the highest level model in the framework. The primary inputs of this model include the facility configuration, interaction with other modules, cost and other infrastructure parameters including recycling and reuse systems and manufacturing performance of CMP equipment including wafer outputs, equipment down time and so on. There are two outputs of the model, one is the cost including energy and water costs, treatment and disposal costs and overall cost of ownership, and the other is environmental performance such as water recycling efficiency, copper removal efficiency and so on.
ii. Equipment level of model. The inputs for the equipment level model include equipment configuration, recipe data and chemical data and the

Table 9.1. Design for environmental, safety and health (DFESH) challenges extracted from the International Technology Roadmap for Semiconductors [9.6]

Challenge	≥65nm/Before 2007	<65nm/After 2007
Chemicals, Materials and Equipment Management	New chemical assessment Chemical data collection Chemical reduction *Environmental Management*	*Chemical use information* Rapid introduction of chemical and materials
Resource Conservation	Reduce energy, water, chemicals and material use *Waste recycle* (increase in resource use as the result of increasing process complexity will require that efficient waste recycling methods be developed)	Reduce water, chemicals and material use (emphasis *on resource sustainability will grow*)
Climate Change Mitigation	*Reduce energy use of process equipment* *Reduce energy use of manufacturing facility* (need to design energy efficient facilities to offset the increasing energy requirements of higher class cleanrooms)	*Reduce energy use* *(the importance of reducing energy use for climate change will increase)*
Design for Environmental Safety and Health (DFESH)	*Evaluate and quantify ESH impact* *(need integrated way to evaluate and quantify ESH impact of process, chemicals, and process equipment, and to make ESH a design parameter in development procedures for new equipment and processes)*	*Evaluate and quantify ESH impact* *(need integrated ESH design in development of new equipment and processes)*

outputs include manufacturing information such as throughput, downtime, consumables use, waste generated, etc. This information is then used at the facility scale to evaluate overall environmental impacts.

iii. CMP process models. Process models that are capable of integrating the key parameters of the process with performance are critical to the understanding required to analyze trade-offs in process development and operation to insure minimal possible environmental impact. These process models, such as the CMP model discussed in this book, is the lowest level of model in Env-S for assessing the whole environment, safety and health parameters of a process. At this level, inputs are related to process parameters, consumables information and wafer information. The outputs of the CMP process model, most importantly the material removal rate, will be inputs to the equipment level models.

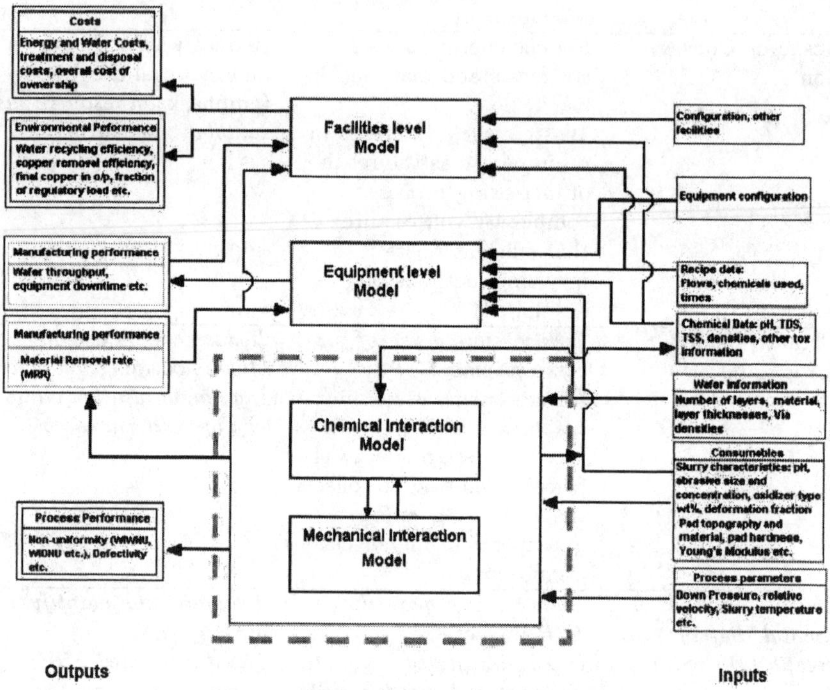

Fig. 9.4. Framework of the green CMP model

In summary, with the IC fabrication technology going to the 65nm node, a more predictable, controllable and reproducible CMP process is needed. Modeling and simulation of CMP at particle-scale, feature- and die-scales and wafer-scale and their integration will be critical to satisfy the requirements of this new era. The modeling efforts in this work provide the basics for the future research and development in these areas.

9.3 References

[9.1] T. I. Zohdi, "Computational design of swarms," *International Journal for Numerical Methods in Engineering*, Vol. **57**, pp. 2205-2219, 2003.

[9.2] J. Luo and D. Dornfeld, "Optimization of CMP from the viewpoint of consumable effects," *Journal of the Electrochemical Society*, Vol. **150**, pp. G807-G815, 2003.

[9.3] X. Zhang, "3-D micro-manufacturing," *Department of Mechanical Engineering, University of California at Los Angeles*, personal communication, 2001.

[9.4] S. Chung, Y. Im, H. Kim, H. Jeong and D. A. Dornfeld, "Evaluation of micro-replication technology using silicone rubber molds and its applications," *International Journal of Machine Tools and Manufacturing*, Vol. **43**, pp. 1337-1345, 2003.

[9.5] B-C. Kim, H. Uono, T. Fuse, T. Ishihara and M. Senna, "Fabrication of nano-sized Si powders with a narrow size distribution by two-step milling," *Journal of Materials Research*, Vol. **18**, pp.1368-1373, 2003.

[9.6] *International Technology Roadmap for Semiconductors, Environmental Safety and Health*, International SEMATECH, Feb. 2003.

[9.7] N. Krishnan, "Design for environment in semiconductor manufacturing," *Ph.D. Dissertation, Department of Mechanical Engineering, University of California at Berkeley*, Berkeley, CA, U. S. A., 2003.

A

BOUNDARY ELEMENT METHOD

A.1 Introduction

The Boundary Element Method (BEM) is a numerical method to solve the boundary value problem by the boundary integral equation (BIE) method. Since the potential problem was first formulated in terms of a direct BIE and solved numerically by Jaswon [A.1] 30 years ago, extensive research effort has been made in BIE formulations of problems in mechanics and their numerical solution schemes. The direct formulation of BIE for the elastostatic problem was first presented by Rizzo [A.2] following the work of Jaswon. Later, the formulation was extended successfully by Cruse and Rizzo [A.3] and Cruse [A.4] to the elastodynamics. The name Boundary Element Method appeared in the late 70's in an attempt to make an analogy with the finite element method (FEM). Today, the BIE/BEM has gained a number of applications in many fields of computational mechanics, such as the fields of wave propagation, heat transfer, diffusion and convection, fluid flow, fracture mechanics, electrical problems, geomechanics, plates and shells, inelastic problems, contact problems, design sensitivity and optimization and inverse problems [A.5-A.14].

As a method based on transferring the governing differential equations into integral equations over the boundary, in linear elastostatics, the BEM leads to an integral constraint equation over the surface of the body, relating boundary tractions (pressures) to boundary displacements. Then the algebraic system of equations is solved for the unknown displacements and tractions (pressures) on the boundary. In this appendix we discuss the general formulation of the CBIE in linear elastostatics briefly. Detailed information about the formulation can be found in references [A.15-A.18].

A.2 Governing Differential Equation of Elastostatics

Employing the usual tensor notation, where a repeated subscript denotes summation over its range and a comma denotes partial differentiation, we can write the governing differential equation of equilibrium for an infinitesimal element in a elastic body as:

$$\sigma_{ij,j} + f_i = 0 \tag{A.1}$$

in which σ_{ij} are the stress components and f_i the body force vector.

Hooke's law for a homogeneous isotropic material states:

$$\sigma_{ij} = \frac{Ev}{(1-2v)(1+v)}\delta_{ij}\varepsilon_{kk} + \frac{E\varepsilon_{ij}}{1+v} \tag{A.2}$$

where $\varepsilon_{ij,}$, E and v are the strain, Young's modulus and Poisson's ratio, respectively and δ_{ij} is the Kronecher delta symbol.

The strains and displacements are related by:

$$\varepsilon_{ij} = \frac{1}{2}(u_{i,j} + u_{j,i}) \tag{A.3}$$

Substituting Eqs. A.2 and A.3 into Eq. A.1, we can express the equilibrim equation for as a function of the displacements components:

$$u_{i,jj} + \frac{1}{1-2v}u_{j,ji} + \frac{2f_i(1-v)}{E} = 0 \tag{A.4}$$

This is called the Navier equation. It is the governing differential equation of elasticity which has to be solved subject to certain boundary conditions. After solving this equation, the unknown tractions (pressures) and displacements on the boundary can be obtained.

A.3 Betti's Reciprocal Theorem and Kelvin's Solution

Betti's general method of integrating the equations of elasticity is one of the classical methods of solving the Navier equation A.4. Consider two equilibrium states of a 2-D elastic body, one with displacements, u_i, due to a body force

f_i, and surface tractions t_i, and the other with displacements, u_i', due to a body force f_i' and surface tractions t_i'. Then the Betti's reciprocal theorem for a two-dimensional domain D with boundary L, as shown in Figure A.1, is given by:

$$\int_L t_i u_i' dL + \int_D f_i u_i' dD = \int_L t_i' u_i dL + \int_D f_i' u_i dD \tag{A.5}$$

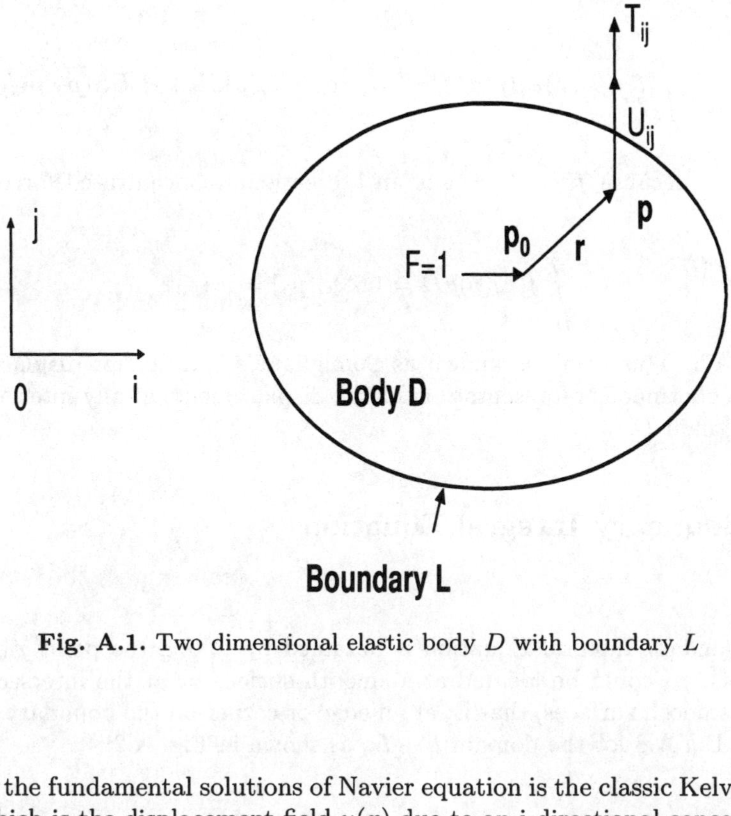

Fig. A.1. Two dimensional elastic body D with boundary L

One of the fundamental solutions of Navier equation is the classic Kelvin solution which is the displacement field $u(x)$ due to an i-directional concentrated unit force F within an elastic body. For plane strain problems [A.1-A.2], the Kelvin solution is:

$$U_{ij} = \frac{(1+v)}{4\pi E(1-v)}\left[(3-4v)\ln\frac{1}{r}\delta_{ij} + \frac{\partial r}{\partial x_i}\frac{\partial r}{\partial x_j}\right] \tag{A.6}$$

$$T_{ij} = -\frac{1}{4\pi(1-v)r}\left\{\begin{array}{l} \frac{\partial r}{\partial n}\left[(1-2v)\delta_{ij} + 2\frac{\partial r}{\partial x_i}\frac{\partial r}{\partial x_j}\right] \\ +(1-2v)(n_i\frac{\partial r}{\partial x_i} - n_j\frac{\partial r}{\partial x_i}) \end{array}\right\} \tag{A.7}$$

where U_{ij} and T_{ij} represent the jth directional displacement and traction at any point p caused by a ith directional unit load F applied at point p_0, r the distance between p_0 and p, and n_i the directional cosines of the normal n, Fig. A.1.

By utilizing Kelvin's concentrated load solution as the prime states in Eq. A.5, an integral equation is obtained, which relates the displacements at an interior point p_0, to the displacements and tractions on the surface:

$$u_i(p_0) = -\int_L T_{ij}(p_0,p)u_j(p)dL + \int_L U_{ij}(p_0,p)t_j(p)dL + \int_D U_{ij}(p_0,p)f_j(p)dD \tag{A.8}$$

Note that because $f' = F = 1$ is an i-directional concentrated force at p_0, there is

$$\int_D f_i' u_i(p)dD = 1 \times u_i(p_0) = u_i(p_0) \tag{A.9}$$

in Eq. A.5. This result is known as Somigliana's identity for displacements and is a continuous representation for the displacements at any interior point p_0 in domain D.

A.4 Boundary Integral Equation

The Boundary Integral Equation is obtained by taking the point p_0 to the boundary. p_0 could be located at a smooth surface or at the intersection of several smooth surfaces, that is, at an edge or corner on the boundary. Let us employ Eq. A.8 for the domain $L + L_\eta$ as shown in Fig. A.2:

$$u_i(p_0) + \int_{L+L_\eta} T_{ij}(p_0,p)u_j(p)dL = \int_{L+L_\eta} U_{ij}(p_0,p)t_j(p)dL + \int_{D+D_\eta} U_{ij}(p_0,p)f_j(p)dD \tag{A.10}$$

where D is the domain with the boundary L, D_η is the exclusion area of a circle with radius η centered at p_0 and boundary L_η. Let $\eta \to 0$, the following boundary equation is derived:

$$C_{ij}(p_0)u_j(p_0) = \int_L [U_{ij}(p_0,p)t_j(p) - T_{ij}(p_0,p)u_j(p)]\, dL + \int_D U_{ij}(p_0,p)f_j(p)dD \tag{A.11}$$

where $C_{ij}(p_0)$ is a 2×2 matrix depending on the smoothness of the curve L at the point p_0. Note that $C_{ij}(p_0)$=0 when $i \neq j$.

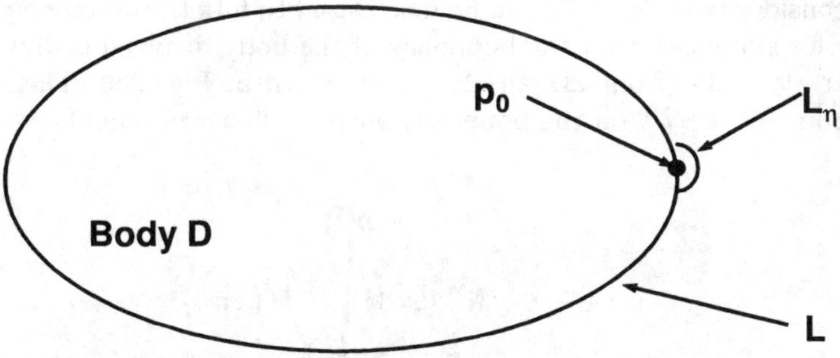

Fig. A.2. Elastic body D with p_0 on the boundary L

The above equation is the boundary constraint equation relating surface tractions to surface displacements, and includes the effect of body forces. If the body force at any place is zero, Eq. A.11 can be written as:

$$C_{ij}(p_0)u_j(p_o) = \int_L [U_{ij}(p_0,p)t_j(p) - T_{ij}(p_0,p)u_j(p)]\, dL \tag{A.12}$$

A.5 Numerical Formulation of Boundary Integral Equation

In general, the analytical solution to the boundary integral equation A.12 for bodies of complicated shape is difficult to obtain and a numerical method which reduces the integral equation to linear algebraic equations has to be used. In order to solve the integral equation numerically, the boundary will be discretized into a series of elements (Here isoparametric elements are used. For detailed information on isoparametric elements, please review books on FEM or BEM.) over which displacements and tractions are written in terms of their values at a series of nodal points. Writing the discretized form of A.12

for every nodal point, a system of linear algebraic equations is obtained. Once the boundary conditions are applied the system can be solved to obtain all the unknown values and consequently an approximate solution to the boundary value is obtained.

Let us consider how Eq. A.12 can be discretized to find the system equations. Assume for simplicity that the boundary of the body involved is divided into N linear elements (from $\Delta L_1$to ΔL_N) as shown in Fig. A.3. There are N nodes from N^1 to N^N on the boundary when N linear elements are used.

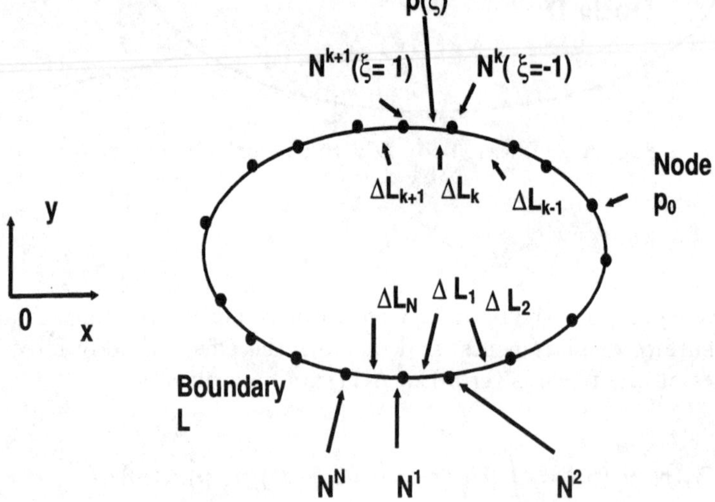

Fig. A.3. An elastic body with disretized boundary L

Then, Eq. A.12 may be written approximately as:

$$C_{ij}(p_0)u_j(p_0) = \sum_{k=1}^{N} \int_{\Delta L_k} U_{ij}(p_0, p)t_j(p)dL - \sum_{k=1}^{N} \int_{\Delta L_k} T_{ij}(p_0, p)u_j(p)dL \tag{A.13}$$

There are now two types of integrals to be carried out over the elements ΔL_k, i.e.,

$$\int_{\Delta L_k} U_{ij}(p_0,p)t_j(p)dL(p) = \int_{-1}^{1} U_{ij}(p_0,p)t_j(\xi)J(\xi)d\xi$$

$$\int_{\Delta L_k} T_{ij}(p_0,p)u_j(p)dL(p) = \int_{-1}^{1} T_{ij}(p_0,p)u_j(\xi)J(\xi)d\xi \tag{A.14}$$

where ξ is the natural coordinate corresponding to point p and $J(\xi)$ is the Jacobian function [A.20]. We assume that the linear element ΔL_k has two nodes N^k and N^{k+1} respectively. Then the value of u_j and t_j at any point p on the element can be defined in terms of their nodal values and two linear interpolation functions (also called as shape functions) Φ_1 and Φ_2, which are given in terms of the natural coordinate ξ, i.e.

$$\begin{aligned} t_j(\xi) &= \Phi_1(\xi)t_j^k + \Phi_2(\xi)t_j^{k+1} \\ u_j(\xi) &= \Phi_1(\xi)u_j^{k+1} + \Phi_2(\xi)u_j^{k+1} \end{aligned} \tag{A.15}$$

where t_j^k, u_j^k, t_j^{k+1} and u_j^{k+1} are the nodal tractions and displacements in the jth direction respectively.

Substitution of Eq. A.15 into Eq. A.14 yields:

$$\begin{aligned} \int_{\Delta L_k} U_{ij}(p_0,p)t_j(p)dL = &\left[\int_{-1}^{1} U_{ij}(p_0,p)\Phi_1(\xi)J(\xi)d\xi\right] t_j^k \\ &+ \left[\int_{-1}^{1} U_{ij}(p_0,p)\Phi_1(\xi)J(\xi)d\xi\right] t_j^{k+1} \end{aligned} \tag{A.16}$$

$$\begin{aligned} \int_{\Delta L_k} T_{ij}(p_0,p)u_j(p)dL = &\left[\int_{-1}^{1} T_{ij}(p_0,p)\Phi_1(\xi)J(\xi)d\xi\right] u_j^k \\ &+ \left[\int_{-1}^{1} T_{ij}(p_0,p)\Phi_1(\xi)J(\xi)d\xi\right] u_j^{k+1} \end{aligned} \tag{A.17}$$

Applying the above procedure to every element, Eq. A.13 can be written as:

$$C_{ij}(p_O)u_j(p_O)$$
$$= \left[G_{ij}^1, G_{ij}^2, \cdots G_{ij}^k, \cdots G_{ij}^{N-1}, G_{ij}^N\right] \left[t_j^1, t_{j,}^2, \cdots t_j^k, \cdots t_j^{N-1}, t_j^N\right]^T -$$
$$\left[H_{ij}^{1'}, H_{ij}^{2'}, \cdots H_{ij}^{k'} \ldots H_{ij}^{N-1'}, H_{ij}^{N'}\right] \left[u_j^1, u_{j,}^2, \cdots u_j^k, \cdots u_j^{N-1}, u_j^N\right]^T \quad \text{(A.18)}$$

where t_j^k and u_j^k is the jth directional traction and displacement on node N^k, and G_{ij}^k and H_{ij}^k are values evaluated directly or indirectly through Eq. A.14 or Eqs. A.16 and Eq. A.17. These integrals can be calculated using numerical integration formula such as ordinary Gauss-quadrature rules when p_O is not on element ΔL_k:

$$\int_{\Delta L_k} U_{ij}(p_O, p)t_j(p)dL = \int_{-1}^{1} U_{ij}(p_O, p)t_j(\xi)J(\xi)d\xi$$
$$= \sum_{m=1}^{M} U_{ij}(p_O, p_m)t_j(p_m)J(\xi_m)$$
$$\int_{\Delta L_k} T_{ij}(p_O, p)u_j(p)dL = \int_{-1}^{1} T_{ij}(p_O, p)u_j(\xi)J(\xi)d\xi$$
$$= \sum_{m=1}^{M} T_{ij}(p_O, p_m)u_j(p_m)J(\xi_m) \quad \text{(A.19)}$$

where ξ is the natural coordinate corresponding to integration point p_m, $J(\xi)$ the Jacobian function and M the number of Gaussian integration points.

In the following two situations the ordinary Gauss-quadrature applied to Eq. A.19 cannot evaluate the integral accurately and efficiently.

i. The p_O is on the element. In this case, the integral will become a singular integral (Note the singular terms $1/r$ and $\ln(1/r)$ in Eqs. A.6 and A.7) and special methods have to be used. Fortunately mature methods such as the rigid body translation method and special logarithm Gaussian quadrature [A.21] can be employed to deal with them respectively.
ii. The p_O is not on the element but close to the element. In this case, the integrals will become nearly singular and the ordinary Gauss-quadrature is no longer practical since a huge number of integration points is needed. The evaluation of the nearly singular integral is the most important work in BEM modeling of thin structures. Most of the earlier methods are either inefficient or cannot provide accurate results when the ratio of the p_O-element distance to the element size is smaller than 10^{-2}. They are not

very useful in dealing with thin structures, such as thin films or coatings, with thickness to length ratio in the micro- or nano-scale. In [A.19], a semi-analytical method which can evaluate such integrals accurately and efficiently is discussed in detail.

After handling the above integrals properly and assuming that p_0 is the kth node on the boundary L, then Eq. A.18 can be written as:

$$\begin{aligned} C_{ij}(k)u_j^k = [G_{ij}^1, G_{ij}^2, G_{ij}^3, ...G_{ij}^k, ... \ G_{ij}^{N-1}, G_{ij}^N][t_j^1, \ t_{j,}^2 t_j^3, ... \ t_j^{N-1}, t_j^N]^T \\ -[H_{ij}^{1'}, H_{ij}^{2'}, H_{ij}^{3'}, ...H_{ij}^{k'} ... \ H_{ij}^{N-1'}, H_{ij}^{N'}][u_j^1, u_{j,}^2, u_j^3, ... \ u_j^{N-1}, u_j^N]^T \end{aligned} \tag{A.20}$$

where $C_{ij}(k) = C_{ij}(p_0)$. Assembling Eq. A.20 generates the following equation:

$$\begin{aligned} [G_{ij}^1, G_{ij}^2, G_{ij}^3, ...G_{ij}^k, ... \ G_{ij}^{N-1}, G_{ij}^N] \ [t_j^1, \ t_{j,}^2 \ t_j^3, ... \ t_j^{N-1}, t_j^N]^T \\ = [H_{ij}^1, H_{ij}^2, H_{ij}^3, ...H_{ij}^k ... \ H_{ij}^{N-1}, H_{ij}^N][u_j^1, \ u_{j,}^2 \ u_j^3, ... \ u_j^{N-1}, u_j^N]^T \end{aligned} \tag{A.21}$$

where sub-matrix $H_{ij}^k = H_{ij}^{k'} - C_{ij}(k)$. Note that G_{ij} and H_{ij} are 2×2 matrix in two-dimensional elasticity.

A.6 System Equations

By allowing every node (from $N^1$to N^N; see Fig. A.3) on the boundary L to be p_0, N equations like A.21 can be obtained. Assembling them together in matrix form gives the global system equations $[H][U]= [G][T]$, i.e.

$$
\begin{bmatrix}
H^1_{11n=1} & H^1_{12n=1} \cdots & H^k_{11n=1} & H^k_{12n=1} & \cdots & H^N_{11n=1} & H^N_{12n=1} \\
H^1_{21n=1} & H^1_{22n=1} \cdots & H^k_{21n=1} & H^k_{22n=1} & \cdots & H^N_{21n=1} & H^N_{22n=1} \\
\cdots\cdots\cdot & & & & & & \\
\cdots\cdots\cdot & & & & & & \\
H^1_{11n=k} & H^1_{12n=k} \cdots & H^k_{11n=k} & H^k_{12n=k} & \cdots & H^N_{11n=k} & H^N_{12n=k} \\
H^1_{21n=k} & H^1_{22n=k} \cdots & H^k_{21n=k} & H^k_{22n=k} & \cdots & H^N_{21n=k} & H^N_{22n=k} \\
\cdots\cdots\cdot & & & & & & \\
\cdots\cdots\cdot & & & & & & \\
H^1_{11n=N} & H^1_{12n=N} \cdots & H^k_{11n=N} & H^k_{12n=N} & \cdots & H^N_{11n=N} & H^N_{12n=N} \\
H^1_{21n=N} & H^1_{22n=N} \cdots & H^k_{21n=N} & H^k_{22n=N} & \cdots & H^N_{21n=N} & H^N_{22n=N}
\end{bmatrix}
\begin{bmatrix} u^1_1 \\ u^1_2 \\ \cdot\cdot \\ \cdot\cdot \\ u^k_1 \\ u^k_2 \\ \cdot\cdot \\ \cdot\cdot \\ u^N_1 \\ u^N_2 \end{bmatrix}
=
$$

$$
\begin{bmatrix}
G^1_{11n=1} & G^1_{12n=1} \cdots & G^k_{11n=1} & G^k_{12n=1} & \cdots & G^N_{11n=1} & G^N_{12n=1} \\
G^1_{21n=1} & G^1_{22n=1} \cdots & G^k_{21n=1} & G^k_{22n=1} & \cdots & G^N_{21n=1} & G^N_{22n=1} \\
\cdots\cdots\cdot & & & & & & \\
\cdots\cdots\cdot & & & & & & \\
G^1_{11n=k} & G^1_{12n=k} \cdots & G^k_{11n=k} & G^k_{12n=k} & \cdots & G^N_{11n=k} & G^N_{12n=k} \\
G^1_{21n=k} & G^1_{22n=k} \cdots & G^k_{21n=k} & G^k_{22n=k} & \cdots & G^N_{21n=k} & G^N_{22n=k} \\
\cdots\cdots\cdot & & & & & & \\
\cdots\cdots\cdot & & & & & & \\
G^1_{11n=N} & G^1_{12n=N} \cdots & G^k_{11n=N} & G^k_{12n=N} & \cdots & G^N_{11n=N} & G^N_{12n=N} \\
G^1_{21n=N} & G^1_{22n=N} \cdots & G^k_{21n=N} & G^k_{22n=N} & \cdots & G^N_{21n=N} & G^N_{22n=N}
\end{bmatrix}
\begin{bmatrix} t^1_1 \\ t^1_2 \\ \cdot\cdot \\ \cdot\cdot \\ t^k_1 \\ t^k_2 \\ \cdot\cdot \\ \cdot\cdot \\ t^N_1 \\ t^N_2 \end{bmatrix}
\tag{A.22}
$$

The vectors U and T in above equation represent all the values of nodal displacements and tractions before applying boundary conditions. These conditions can be introduced by rearranging the columns in H and G, passing all unknowns to a vector X on the left hand side. This gives the final system of equations, i.e.

$$[A]X = [B] \tag{A.23}$$

All boundary values will be fully determined by solving the above system. Furthermore, by applying the boundary values into Eq. A.8, the displacement value in any domain point can be obtained.

A.7 BEM vs. FEM

For computational modeling of elastic bodies/structures, besides BEM, another conventional method is the finite element method (FEM) [A.20]. In comparison with FEM, BEM has the following advantages:

i. For the analysis of thin structures, FEM is usually based on plate and shell theories. However, most plate and shell theories are based on various assumptions about the geometry, loading and deformation of the structure, and therefore the accuracy and reliability of the FEM for thin structures is in doubt. This is especially true for the stress analysis of thin structures since plate and shell models can not predict the normal stresses (contact stresses) accurately. Without the above assumptions, BEM is in general more accurate in stress analysis of thin structures such as the silicon wafer in CMP [A.19].
ii. In BEM only the boundary of the body/structure needs to be meshed while in FEM the whole body needs to be meshed. Therefore, the memory requirement of BEM is lower and the computation time is less.
iii. In BEM, remeshing is minimal when the geometry of the body/structure is changed. For instance, when the thickness of the wafer is changed, there is no need to remesh the boundary of the wafer but only to change the size of the boundary elements on the left and right sides of the wafer. In FEM, however, the whole body/domain needs to be remeshed. Therefore, BEM is more convenient than FEM on problems of geometry optimizations.

A.8 References

[A.1] M. Jaswon,"Integral equation methods in potential theory, I." *Proc. Roy. Soc. Ser. A.*, Vol. **275**, pp. 23-32, 1963.

[A.2] F. J. Rizzo, "An integral equation approach to boundary value problems of classical elastostatics," *Quarterly of Applied Mathematics*, Vol. **25**, pp. 83-95, 1967.

[A.3] T. A. Cruse and F. J. Rizzo, "A direct formulation and numerical solution of the general transient elastodynamic problem I," *Journal of Math. Analysis and Applications*, Vol. **22**, p. 244, 1968.

[A.4] T. A. Cruse, "A direct formulation and numerical Solution of the general transient elastodynamic problem II," *Journal of Math. Analysis and Applications*, Vol. **22**, pp. 341, 1968.

[A.5] R. D. Ciskowski and C. A. Brebbia, *Boundary Element Methods in Acoustics*, Elsevier Applied Science, New York, U. S. A., 1991.

[A.6] L. C. Wrobel and C. A. Brebbia, *Boundary Element Methods in Heat Transfer*, Elsevier Applied Science, New York, U. S. A., 1992.

[A.7] P. K. Banerjee and L. Morino, *Boundary Element Methods in Nonlinear Fluid Dynamics*, Elsevier Applied Science, New York, U. S. A., 1990.

[A.8] T. A. Cruse, *Boundary Element Analysis in Computational Fracture Mechanics*, Kluwer Academic Publishers, Boston, U. S. A., 1988.

[A.9] S-K. Hsu, "Boundary element modeling of electric and displacement fields in two dimensional piezoelectric material," *M.S. Thesis, Dept.*

of Mechanical and Aerospace Engineering, *Case Western Reserve University*, Cleveland, OH, U. S. A., 1994.

[A.10] W. S. Venturini, *Boundary Element Method in Geomechanics*, Springer-Verlag, New York, U. S. A., 1983.

[A.11] D. E. Beskos, *Boundary Element Analysis of Plates and Shells*, Springer-Verlag, New York, U. S. A., 1991.

[A.12] G. Karami, *A Boundary Element Method for Two Dimensional Contact Problems*, Springer-Verlag, New York, U. S. A., 1989.

[A.13] J. C. F. Telles, *The Boundary Element Method for Inelastic Problems*, Springer-Verlag, New York, U. S. A., 1983.

[A.14] D. B. Ingham and L. C. Wrobel, *Boundary Integral Formulations for Inverse Analysis*, Computational Mechanics Publication, Boston, U. S. A., 1997.

[A.15] P. K. Banerjee, *The Boundary Element Methods in Engineering*, McGraw-Hill, New York, U. S. A., 1994.

[A.16] C. A. Brebbia and S. Walker, *Boundary Element Techniques in Engineering*, Newnes-Butterworths, London, U. K., 1980.

[A.17] C. A. Brebbia and J. Dominguez, *Boundary Elements: an Introductory Course*, McGraw-Hill, New York, U. S. A., 1989.

[A.18] J. H. Kane, *Boundary Element Analysis in Engineering Continuum Mechanics*, Prentice-Hall, New Jersey, U. S. A., 1994.

[A.19] J. F. Luo, Y. J. Liu and E. J. Berger, "Analysis of two dimensional thin structures (from micro- to nano-scales) using the boundary element method," *Computational Mechanics*, Vol. **22**, pp. 404-412, 1998.

[A.20] R. D. Cook, D. S. Malkus and M. E. Plesha, *Concepts and Applications of Finite Element Analysis*, John Wiley & Sons, New York, U. S. A., 1989.

[A.21] D. V. Griffiths and I. M. Smith, *Numerical Methods for Engineers: A Programming Approach*, CRC Press, Boca Raton, Florida, U. S. A., 1991.

Index

Printing: Strauss GmbH, Mörlenbach
Binding: Schäffer, Grünstadt